T0360432

Exactly Solvable Models in Many-Body Theory

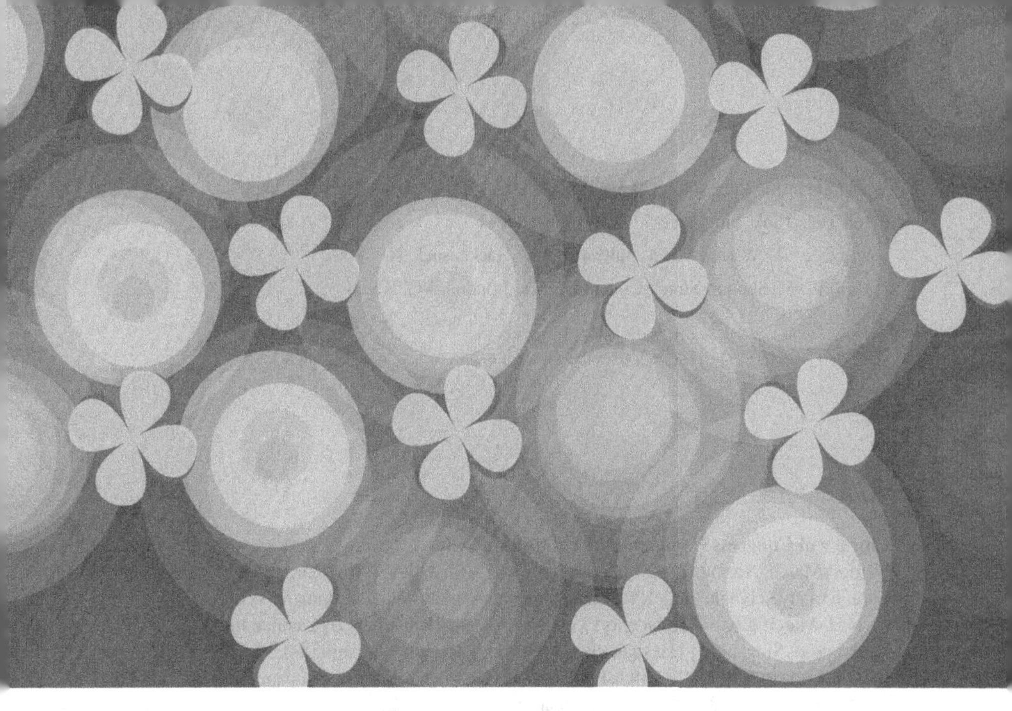

Exactly Solvable Models in Many-Body Theory

N H March
Oxford University, Oxford, UK
University of Antwerp, Antwerp, Belgium

G G N Angilella
Università di Catania, Catania, Italy
Scuola Superiore di Catania, Catania, Italy
CNISM, CNR-IMM, INFN, Catania, Italy

World Scientific

EW JERSEY • LONDON • SINGAPORE • BEIJING • SHANGHAI • HONG KONG • TAIPEI • CHENNAI • TOKYO

Published by

World Scientific Publishing Co. Pte. Ltd.
5 Toh Tuck Link, Singapore 596224
USA office: 27 Warren Street, Suite 401-402, Hackensack, NJ 07601
UK office: 57 Shelton Street, Covent Garden, London WC2H 9HE

Library of Congress Cataloging-in-Publication Data
Names: March, Norman H. (Norman Henry), 1927– author. | Angilella, G. G. N., author.
Title: Exactly solvable models for cluster and many-body condensed matter systems / N.H. March (Oxford University), G.G.N. Angilella (University of Catania, Italy).
Description: Singapore ; Hackensack, NJ : World Scientific Publishing Co. Pte. Ltd., [2016]
Identifiers: LCCN 2016013314| ISBN 9789813140141 (hardcover ; alk. paper) | ISBN 9813140143 (hardcover ; alk. paper)
Subjects: LCSH: Condensed matter--Mathematical models. | Many-body problem--Mathematical models. | Microclusters--Mathematical models.
Classification: LCC QC173.454 .M365 2016 | DDC 530.4/1--dc23
LC record available at https://lccn.loc.gov/2016013314

British Library Cataloguing-in-Publication Data
A catalogue record for this book is available from the British Library.

Printed in Singapore

To Renato Pucci,

colleague esteemed and friend invaluable

Preface

About half a decade ago, we edited a book on 'Many-body theory' (March and Angilella, 2009) which, however, included only papers from particular research groups, and mostly from that of one of us (NHM).

Subsequently, we have felt that a many-body book on 'Exactly solvable models' might provide a useful addition to presently available literature. In this context, there is already a valuable book with such a title by D. C. Mattis (1993). However, this is dominantly concerned with Ising-like Hamiltonians.

This book reviews several theoretical, mostly exactly solvable, models for selected systems in condensed states of matter, including the solid, liquid, and disordered states, and for systems of few or many bodies, both with boson, fermion, or anyon statistics. Some attention is devoted to models for quantum liquids, including superconductors and superfluids. Open problems in relativistic fields and quantum gravity are also briefly reviewed.

This book ranges almost comprehensively, but concisely, across several fields of theoretical physics of matter at various degrees of correlation and at different energy scales, with relevance to molecular, solid-state, and liquid-state physics, as well as to phase transitions, particularly for quantum liquids. Mostly exactly solvable models are presented, with attention also to their numerical approximation and, of course, to their relevance for experiments.

After a general introduction to density matrices in Chapter 1, exactly or nearly exactly solvable models are presented for several few-particle systems. These are arranged according to the statistics of these assemblies, going from fermions and their relevance for the inhomogeneous electron liquid in atoms, molecules, and small clusters (Chapter 2), to bosons, with specific reference to Efimov trimers in nuclear and condensed matter assemblies (Chapter 3), to anyon statistics, and its possible relevance for low-dimensional systems (Chapter 4).

Selected problems in several many-body systems in condensed matter are then reviewed, prominent examples being superconductivity and superfluidity (Chapter 5), isolated impurities in a solid (Chapter 6), and some structural and electronic properties of liquids (especially liquid metals, Chapter 7). Disorder and its implications on transport in solids is treated in Chapter 8.

The last three Chapters address more general topics, in some sense transverse with respect to the specific problems being treated in the preceding Chapters, *viz.* statistical field theory, including selected examples of exactly solvable models, especially in low dimensions, and a more general discussion of critical exponents related especially to dimensionality (Chapter 9), relativistic field theory and relativistic field equations (Chapter 10), and a brief summary of the open problems concerning the construction of a fully quantum theory of gravity (Chapter 11).

Several Appendices at the end of the monograph deal with special aspects of the topics listed above, or with specific models which we have felt should be better addressed to separately.

No doubt, interested readers who are already expert in selected areas of the present book will note omissions, for all of which then the authors must take responsibility, and will attempt to make additions if and when the opportunity arises. We shall much appreciate any related comments from interested readers.

May 2016

N. H. March,
Oxford University, Oxford, UK
University of Antwerp, Antwerp, Belgium

G. G. N. Angilella,
Dipartimento di Fisica e Astronomia,
and Scuola Superiore di Catania,
Università di Catania, Catania, Italy
CNISM, CNR-IMM, INFN, Catania, Italy

Acknowledgments

The book embraces topics which have been of mutual interest for the authors over some two decades of close collaboration.

This has been possible through the generous support and hospitality of several institutions, including the Department of Physics and Astronomy, University of Catania, Italy, the Department of Physics, University of Antwerp, Belgium, the International Centre for Theoretical Physics (ICTP), Trieste, Italy, the Scuola Normale Superiore, Pisa, Italy, the Istituto Nazionale di Fisica Nucleare (INFN), Sez. di Catania, Italy, the Laboratori Nazionali del Sud (INFN), Catania, Italy, the Scuola Superiore di Catania, Italy, which the authors, jointly, would like to gratefully acknowledge.

NHM would also like to acknowledge the Department of Theoretical, Atomic, and Optical Physics at the University of Valladolid, the Nano-Bio Spectroscopy Group of the University of the Basque Country and the Donostia International Physics Centre (DIPC) in Donostia/San Sebastian, Spain, the Department of Theoretical Physics of the University of Debrecen, Hungary, the Department of Theoretical Physics of the University of Szeged, Hungary, the German Cancer Research Centrum (DKFZ) in Heidelberg, Germany, the School of Physics and Astronomy of the University of Cardiff, UK, the Department of Marine Sciences of the Texas A&M University at Galveston, TX, USA, for support and warm hospitality.

The authors would also like to acknowledge J. A. Alonso, J. V. Alvarez, C. Amovilli, N. Andrei, N. Andrenacci, A. Balaban, G. Balestrino, F. Bartha, A. Bianconi, A. Cabo, R. Citro, F. Claro, G. Compagnini, G. Forte, A. Grassi, A. Holas, I. A. Howard, D. J. Klein, M. Knapp-Mohammady, V. E. Kravtsov, A. La Magna, D. Lamoen, F. E. Leys, G. M. Lombardo, C. C. Matthai, K. Morawetz, Á. Nagy, M. Parrinello, G. Pastore, G. Senatore,

F. M. D. Pellegrino, G. Piccitto, R. Pucci, A. Rubio, F. Siringo, R. Squire, A. Sudbø, M. P. Tosi, C. Van Alsenoy, J. Van Dick, V. E. Van Doren, A. A. Varlamov, Zhidong Zhang.

The image on the cover is inspired by the d-wave orbitals of ultracold fermionic atoms in a two-dimensional optical lattice. Such a system has been predicted to behave as a topological semimetal (Sun *et al.*, 2011). Thanks are due to Danilo 'Tif' Giuffrida for a graphical interpretation thereof.

Contents

Chapter 1

Low-order density matrices

1.1 Low-order spinless density matrix theory

We start by considering a generic assembly of N electrons, characterized by the Hamiltonian (in atomic units)

$$H = T + U + V, \tag{1.1a}$$

$$T = -\frac{1}{2}\sum_{i=1}^{N} \nabla_i^2, \tag{1.1b}$$

$$U = \sum_{i<j} u(\mathbf{r}_i, \mathbf{r}_j), \quad u(\mathbf{r}_j, \mathbf{r}_i) = u(\mathbf{r}_i, \mathbf{r}_j), \tag{1.1c}$$

$$V = \sum_{i=1}^{N} v(\mathbf{r}_i). \tag{1.1d}$$

Here, as usual, T is the total kinetic energy operator, U denotes the electron-electron (e-e) repulsion energy operator, and V denotes the electron-nucleus (e-n) attraction energy operator, respectively. In realistic situations, one obviously has

$$u(\mathbf{r}, \mathbf{r}') = \frac{1}{|\mathbf{r} - \mathbf{r}'|}, \tag{1.2a}$$

$$v(\mathbf{r}) = -\sum_{J=1}^{M} \frac{Z_J}{|\mathbf{r} - \mathbf{R}_J|}, \tag{1.2b}$$

(for a molecule, cluster, or solid characterized by M nuclei of charge Z_J, placed at fixed positions $\mathbf{R}_J$, $J = 1, \ldots M$), although the explicit functional form of u and v may be kept as indicated, in order to also embrace model systems.

The Nth-order density matrix (DM) associated with a normalized eigenfunction $\Psi(\mathbf{x}_1, \ldots \mathbf{x}_N)$ of the Schrödinger equation $H\Psi = E\Psi$ is then defined as

$$\gamma_N(\mathbf{x}_1, \ldots \mathbf{x}_N; \mathbf{x}'_1, \ldots \mathbf{x}'_N) = \Psi(\mathbf{x}_1, \ldots \mathbf{x}_N)\Psi^*(\mathbf{x}'_1, \ldots \mathbf{x}'_N), \tag{1.3}$$

where $\mathbf{x}_i = (\mathbf{r}_i, s_i)$ is a shorthand notation for coordinates and spin variables of the ith electron. More generally, the pth-order reduced DM (pDM, with $p < N$) is defined by integrating out $(N - p)$ coordinates as (see ter Haar, 1961; Holas and March, 1995, and references therein)

$$\gamma_p(\mathbf{x}_1, \ldots \mathbf{x}_p; \mathbf{x}'_1, \ldots \mathbf{x}'_p) = \binom{N}{p} \int d\mathbf{x}_{p+1} \cdots d\mathbf{x}_N$$
$$\times \gamma_N(\mathbf{x}_1, \ldots \mathbf{x}_p, \mathbf{x}_{p+1} \ldots \mathbf{x}_N; \mathbf{x}'_1, \ldots \mathbf{x}'_p, \mathbf{x}_{p+1} \ldots \mathbf{x}_N), \tag{1.4}$$

where, consistently with the above notation, $\int d\mathbf{x}_i \equiv \int d\mathbf{r}_i \sum_i$. In cases where spin is not of interest, one considers a spin-averaged or spinless DM defined as

$$\rho_p(\mathbf{r}_1, \ldots \mathbf{r}_p; \mathbf{r}'_1, \ldots \mathbf{r}'_p) = \sum_{s_1, \ldots s_p} \gamma_p(\mathbf{x}_1, \ldots \mathbf{x}_p; \mathbf{x}'_1, \ldots \mathbf{x}'_p)\big|_{\text{all } s'_i = s_i}, \tag{1.5}$$

while the diagonal elements of the spinless DM are denoted by

$$n_p(\mathbf{r}_1, \ldots \mathbf{r}_p) = \rho_p(\mathbf{r}_1, \ldots \mathbf{r}_p; \mathbf{r}_1, \ldots \mathbf{r}_p), \tag{1.6}$$

with $n_p \geq 0$. The basic quantity of interest in Density Functional Theory (DFT) is thus $n(\mathbf{r}) = n_1(\mathbf{r}) = \rho_1(\mathbf{r}; \mathbf{r})$, where the subscript '1' can be omitted, for simplicity.

Expressing the Hamiltonian in second quantization as

$$H = -\frac{1}{2} \int d\mathbf{x}\, \hat{\psi}^\dagger(\mathbf{x}) \nabla^2 \hat{\psi}(\mathbf{x}) + \frac{1}{2} \int d\mathbf{x} d\mathbf{x}' \, \frac{\hat{\psi}^\dagger(\mathbf{x})\hat{\psi}^\dagger(\mathbf{x}')\hat{\psi}(\mathbf{x}')\hat{\psi}(\mathbf{x})}{|\mathbf{x} - \mathbf{x}'|}, \tag{1.7}$$

where

$$\hat{\psi}^\dagger(\mathbf{x}) = \sum_k \varphi_k^*(\mathbf{x}) \hat{c}_k^\dagger, \tag{1.8a}$$

$$\hat{\psi}(\mathbf{x}) = \sum_k \varphi_k(\mathbf{x}) \hat{c}_k \tag{1.8b}$$

are creation and annihilation quantum field operators at $\mathbf{x}$, respectively, with $\hat{c}_k^\dagger$ and $\hat{c}_k$ fermion creation and annihilation operators, $\{\hat{c}_k, \hat{c}_{k'}\} = \{\hat{c}_k^\dagger, \hat{c}_{k'}^\dagger\} = 0$, $\{\hat{c}_k, \hat{c}_{k'}^\dagger\} = \delta_{kk'}$, in the single-particle spin-orbitals $\varphi_k(\mathbf{x})$, the reduced pDM can be viewed as the quantum average of the operator

$$\hat{\gamma}_p(\mathbf{x}_1, \ldots \mathbf{x}_p; \mathbf{x}'_1, \ldots \mathbf{x}'_p) = \frac{1}{p!} \hat{\psi}^\dagger(\mathbf{x}'_p) \cdots \hat{\psi}^\dagger(\mathbf{x}'_1) \hat{\psi}(\mathbf{x}_1) \hat{\psi}(\mathbf{x}_p). \tag{1.9}$$

The same Hamiltonian can then be expressed in terms of 1DM and 2DM as

$$H = -\frac{1}{2}\int d\mathbf{x}d\mathbf{x}'\,\delta(\mathbf{x}-\mathbf{x}')\nabla^2\hat{\gamma}_1(\mathbf{x};\mathbf{x}') + \int d\mathbf{x}d\mathbf{x}'\,\frac{\hat{\gamma}_2(\mathbf{x},\mathbf{x}';\mathbf{x},\mathbf{x}')}{|\mathbf{x}-\mathbf{x}'|}. \quad (1.10)$$

Similarly, the total (*e.g.* ground-state) energy of the system can be expressed as

$$E = -\frac{1}{2}\int d\mathbf{x}d\mathbf{x}'\,\delta(\mathbf{x}-\mathbf{x}')\nabla^2\gamma_1(\mathbf{x};\mathbf{x}') + \int d\mathbf{x}d\mathbf{x}'\,\frac{\gamma_2(\mathbf{x},\mathbf{x}';\mathbf{x},\mathbf{x}')}{|\mathbf{x}-\mathbf{x}'|}. \quad (1.11)$$

This immediately generalizes also in presence of an external, one-body potential.

Moreover, since the Hamiltonian contains only (at most) 2-body interactions, and given the fact that a $(p-1)$DM can be related to a pDM as

$$\begin{aligned}\gamma_{p-1}(&\mathbf{x}_1,\ldots\mathbf{x}_{p-1};\mathbf{x}'_1,\ldots\mathbf{x}'_{p-1}) \\ &= \frac{p}{N-p+1}\int d\mathbf{x}_p\,\gamma_p(\mathbf{x}_1,\ldots\mathbf{x}_{p-1},\mathbf{x}_p;\mathbf{x}'_1,\ldots\mathbf{x}'_{p-1},\mathbf{x}_p), \quad (1.12)\end{aligned}$$

one can then express the total energy E as a functional of γ_2 alone, *i.e.* of its 2-body correlation functions. Minimizing such a functional would then result in a variational characterization of the ground-state energy of the system, provided the reduced density matrices over which the variation is being taken are properly conditioned (see Section 1.1.2). This is basically the original motivation which led Coulson (1960) to favor a quantum many-body theory wholly based on density matrices, thereby somehow anticipating density functional theory (see also the discussion in Sec. 5.1.3).

1.1.1 *Natural orbitals and occupation numbers*

The single-particle spin-orbitals $\varphi_k(\mathbf{x})$ in Eqs. (1.8) are called *natural orbitals.* They are supposed to form a complete, orthonormal basis set in the single-particle Hilbert space, and in terms of them it is possible to diagonalize the 1DM operator as

$$\hat{\gamma}_1(\mathbf{x};\mathbf{x}') = \sum_k \varphi^*_k(\mathbf{x}')\varphi_k(\mathbf{x})c^\dagger_k c_k. \quad (1.13)$$

Due to the appearance of the number operator $c^\dagger_k c_k$, the eigenvalues n_k of the 1DM operator in such a basis are called *occupation numbers.* It is then possible to express the 1DM in terms of natural orbitals and occupation numbers as

$$\gamma_1(\mathbf{x};\mathbf{x}') = \sum_k n_k \varphi_k(\mathbf{x})\varphi_k^*(\mathbf{x}'). \tag{1.14}$$

This can be viewed as the kernel of an integral operator in spin-coordinate space, whose associated eigenvalue equation reads

$$\int d\mathbf{x}' \gamma_1(\mathbf{x};\mathbf{x}')\varphi_k(\mathbf{x}') = n_k \varphi_k(\mathbf{x}). \tag{1.15}$$

The natural orbitals form a complete basis set, which one may also assume to be orthonormal:

$$\int d\mathbf{x}\, \varphi_k^*(\mathbf{x})\varphi_{k'}(\mathbf{x}) = \delta_{kk'}, \tag{1.16a}$$

$$\sum_k \varphi_k(\mathbf{x})\varphi_k^*(\mathbf{x}') = \delta(\mathbf{x}-\mathbf{x}'). \tag{1.16b}$$

1.1.2 *N-representability*

In the minimization of the total energy $E = E[\gamma_2]$, Eq. (1.11), regarded as a functional of the 2DM γ_2, it has to be ensured that all density matrices can be represented by an ensemble of properly antisymmetrized N-particle wave functions. This problem is known as the N-representability problem.

A set of necessary and sufficient conditions for the N-representability of 1DMs has been actually found by Coleman (1963) (see also Löwdin, 1955, for an alternative proof of their necessity). However, the N-representability problem is still an open one for 2DMs, although some recent progress has been achieved (Mazziotti, 1998; Mazziotti and Erdahl, 2001). (See also Kummer, 1967, for a mathematical formulation of the problem.)

When expressed in terms of natural orbitals and occupation numbers, as in Eq. (1.14), a necessary and sufficient condition for a 1DM to be N-representable is (Coleman, 1963)

$$0 \leq n_k \leq 1, \tag{1.17a}$$

$$\operatorname{tr}\hat{\gamma}_1 \equiv \sum_k n_k = N. \tag{1.17b}$$

Obviously, Eq. (1.17a) reflects Pauli exclusion principle, while Eq. (1.17b) is a consequence of the fixed total number of particles. In particular, a system of non-interacting particles has $n_k = 1$ for exactly N distinct values of k, and $n_k = 0$ otherwise. This is consistent with the fact that the ground state of a system of N non-interacting particles (fermions) is a globally antisymmetric Slater determinant,

$$\Phi_K(\mathbf{x}_1, \ldots \mathbf{x}_N) = \frac{1}{\sqrt{N!}} \det(\varphi_{k_1}, \ldots \varphi_{k_N}), \tag{1.18}$$

whose entries are natural orbitals, and where $K = \{k_i\}$, with $k_i < k_{i+1}$. Such Slater determinants can still be employed to express the state of N interacting particles, but then an infinite series thereof is required, say

$$\Psi(\mathbf{x}_1, \ldots \mathbf{x}_N) = \sum_K c_K \Phi_K(\mathbf{x}_1, \ldots \mathbf{x}_N), \tag{1.19}$$

with $\sum_K |c_K|^2 = 1$, and each occupation number has then to satisfy strictly the conditions above, Eqs. (1.17).

Using the expansion Eq. (1.19) for the N-particles wave-function in terms of Slater determinants, one can express the 1DM as

$$\begin{aligned}\gamma_1(\mathbf{x};\mathbf{x}') &= N \sum_{KL} c_K^* c_L \int d\mathbf{x}_2 \cdots d\mathbf{x}_N \, \Phi_K^*(\mathbf{x}', \mathbf{x}_2, \ldots \mathbf{x}_N) \Phi_L(\mathbf{x}, \mathbf{x}_2, \ldots \mathbf{x}_N) \\ &= \sum_{KL} c_K^* c_L \sum_{k \in K, \ell \in L} \varphi_k^*(\mathbf{x}') D_{k\ell}^{(KL)} \varphi_\ell(\mathbf{x}),\end{aligned} \tag{1.20}$$

where $D_{k\ell}^{(KL)}$ is a minor of first order of

$$D^{(KL)} = \int d\mathbf{x}_1 \cdots d\mathbf{x}_N \, \Phi_K^*(\mathbf{x}_1, \ldots \mathbf{x}_N) \Phi_L(\mathbf{x}_1, \ldots \mathbf{x}_N). \tag{1.21}$$

One finds therefore that the 1DM γ_1 is indeed N-representable,

$$\gamma_1(\mathbf{x};\mathbf{x}') = \sum_k n_k \varphi_k^*(\mathbf{x}') \varphi_k(\mathbf{x}), \tag{1.22}$$

where the occupation numbers n_k are given by

$$n_k = \sum_{k \ni K, k \ni L} c_K^* D_{kk}^{(KL)} c_L = \sum_{k \ni K} |c_K|^2, \tag{1.23}$$

where use has been made of the orthonormality of the natural orbitals, and therefore that $D_{kk}^{(KL)} = \delta_{KL}$. In particular, condition (1.17a) is fulfilled, since the sum runs only over those coefficients c_K such that the respective determinant contains the orbital k. Similarly, condition (1.17b) is also fulfilled, since

$$\sum_k n_k = \sum_k \sum_{k \ni K} |c_K|^2 = \sum_K N |c_K|^2 = N, \tag{1.24}$$

because of normalization. We refer the reader to Löwdin (1955) for a proof of the sufficiency of the conditions, Eqs. (1.17), for the N-representability of a 1DM. This proof relies on the convexity of the set of ensemble N-representable 1DM, where an ensemble N-representable 1DM is a sum of pure N-representable 1DMs, $\gamma_1 = \sum_j c_j \gamma_1^{(j)}$, where $\gamma_1^{(j)}(\mathbf{x};\mathbf{x}') = N \int d\mathbf{x}_2 \cdots d\mathbf{x}_N \, \Psi_j^*(\mathbf{x}', \mathbf{x}_2, \ldots \mathbf{x}_N)\Psi_j(\mathbf{x}, \mathbf{x}_2, \ldots \mathbf{x}_N)$, and Ψ_j denotes a pure state.

The problem of N-representability and its computational implications for quantum chemistry has been analyzed by Liu *et al.* (2007).

1.2 Gilbert theorem

The first part of Eq. (1.11) shows that the kinetic energy part of the total (*e.g.* ground-state) energy is uniquely determined by the 1DM $\gamma_1(\mathbf{x};\mathbf{x}')$. The theorem by Hohenberg and Kohn (1964), which is at the basis of density functional theory (DFT), guarantees a stronger result, with a seemingly smaller input, namely that all information about that system (including its *total* energy, both in the ground state, and in the excited states) is determined by the system's single-particle density $n(\mathbf{x})$. However, besides being N-representable, such density $n(\mathbf{x})$ must also be V-representable, *i.e.* the functional $E = E[n]$ has to be varied over all density distributions n containing an integral number of electrons, which can *actually* be realized by *some* external potential. Clearly, V-representability is a more stringent condition than N-representability alone (see Levy, 1979, for a discussion).

A theorem by Gilbert (1975) shows then that it is possible to establish a one-to-one correspondence between the 1DM and the *ground-state* wavefunction of the system. Therefore, any observable of the system in its ground-state can be written as a functional of the 1DM. While Gilbert theorem applies only to the ground state, it is only required that the functional $E = E[\gamma_1]$ be varied among N-representable 1DMs. Moreover, Gilbert theorem generalizes Hohenberg and Kohn theorem to the case of nonlocal external potentials, thereby including also model potentials, such as pseudopotentials.

Indeed, let us consider the Schrödinger equation associated with the Hamiltonian in Eqs. (1.1),

$$H\Psi \equiv (T + U + V)\Psi = E\Psi, \tag{1.25}$$

where the external potential V, Eq. (1.1d), is replaced by a more general, non-local potential acting on the wave-function Ψ as

$$V\Psi \equiv \int d\mathbf{x}_1' \cdots d\mathbf{x}_N' \sum_j v(\mathbf{x}_j, \mathbf{x}_j')\Psi(\mathbf{x}_1', \ldots \mathbf{x}_N'). \tag{1.26}$$

The case of local potentials can be embedded in the expression above by replacing $\sum_j v(\mathbf{x}_j, \mathbf{x}'_j) \mapsto \sum_j \prod_{k\neq j} \delta(\mathbf{x}_k - \mathbf{x}'_k) v(\mathbf{x}_j, \mathbf{x}'_j)$.

The direct part of Gilbert theorem is immediately achieved, as solving Eq. (1.25) for a given external potential V yields a ground-state wavefunction Ψ, which in turn defines a unique 1DM γ_1, through Eq. (1.4), with $p = 1$:

$$V \to \Psi \to \gamma_1.$$

Inverting the above chain of implications is however not entirely possible. For example, even in the trivial case of N non-interacting fermions, whose ground-state wavefunction is a Slater determinant, and whose 1DM can be expressed in terms of natural orbitals by Eq. (1.14) with all occupation numbers $n_k = 1$, if one adds to the non-interacting Hamiltonian a potential term $v(\mathbf{x}, \mathbf{x}') = \langle \mathbf{x}|f(\hat{\gamma}_1)|\mathbf{x}'\rangle$, where $f(\hat{\gamma}_1)$ is an arbitrary function of the 1DM operator, one may then show that the ground-state wavefunction is still a Slater determinant (and γ_1 is of course unaltered). Therefore, at variance with a V-representable density in Hohenberg and Kohn theorem, it is not possible to derive a one-to-one correspondence between an N-representable 1DM and the Hamiltonian. However, it is still possible to infer a one-to-one correspondence between γ_1 and the ground-state Ψ:

$$V \not\leftarrow \Psi \leftarrow \gamma_1.$$

The proof proceeds by *reduction ad absurdum,* as in Hohenberg and Kohn theorem. Suppose that two (physically) distinct wavefunctions $\Psi \not\equiv \Psi'$ yield the same 1DM, $\gamma_1 \equiv \gamma'_1$. Such wavefunctions must be therefore the ground-state eigenfunctions of two Hamiltonians, H and H' say, differing by the external potential, $V \not\equiv V'$. Denoting their respective eigenenergies by $E = \langle\Psi|H|\Psi\rangle$ and $E' = \langle\Psi'|H'|\Psi'\rangle$, by the Ritz variational principle one has

$$E = \langle\Psi|H|\Psi\rangle < \langle\Psi'|H|\Psi'\rangle$$
$$< E' + \int d\mathbf{x}d\mathbf{x}'\, \gamma'_1(\mathbf{x}';\mathbf{x})[v(\mathbf{x},\mathbf{x}') - v'(\mathbf{x},\mathbf{x}')]. \quad (1.27)$$

Exactly the same reasoning can be inverted, starting from E' now, and writing

$$E' = \langle\Psi'|H'|\Psi'\rangle < \langle\Psi|H'|\Psi\rangle$$
$$< E + \int d\mathbf{x}d\mathbf{x}'\, \gamma_1(\mathbf{x}';\mathbf{x})[v'(\mathbf{x},\mathbf{x}') - v(\mathbf{x},\mathbf{x}')]. \quad (1.28)$$

Adding the last two equations together, and making use of the initial assumption that $\gamma_1 \equiv \gamma_1'$, one obtains that

$$E + E' < E' + E, \tag{1.29}$$

which is clearly absurd.

One may then show that the ground-state energy is variationally characterized as the minimum of the functional $E[\gamma_1]$,

$$E \leq \min_{\operatorname{tr} \hat{\gamma}_1 = 1} E[\gamma_1], \tag{1.30}$$

over all N-representable 1DM (Gilbert, 1975).

1.2.1 *Exchange and correlation energy in terms of first order reduced density matrices*

While both the kinetic energy, as already emphasized, the Hartree energy, and the external potential energy can be expressed explicitly as a functional of the 1DM $\gamma_1(\mathbf{x}';\mathbf{x})$, this is not in general possible for the exchange and correlation energy, $E_{\rm xc}[\gamma_1]$. One explicit approximation is of course the Hartree-Fock exchange and correlation energy, whose expression in terms of the 1DM reads

$$E_{\rm xc}^{\rm HF}[\gamma_1] = -\frac{1}{2} \int d\mathbf{x} d\mathbf{x}' \, \frac{|\gamma_1(\mathbf{x}';\mathbf{x})|^2}{|\mathbf{x}-\mathbf{x}'|}. \tag{1.31}$$

In terms of natural orbitals and occupation numbers, it reads

$$E_{\rm xc}^{\rm HF}[\gamma_1] = -\frac{1}{2} \sum_{kk'} n_k n_{k'} \int d\mathbf{x} d\mathbf{x}' \, \frac{\varphi_k(\mathbf{x})\varphi_k^*(\mathbf{x}')\varphi_{k'}(\mathbf{x}')\varphi_{k'}^*(\mathbf{x})}{|\mathbf{x}-\mathbf{x}'|}. \tag{1.32}$$

Another quite popular explicit approximation has been proposed by Müller (1984) (see also Buijse and Baerends, 2002). A generalization thereof reads

$$E_{\rm xc}^{\rm M}[\gamma_1] = -\frac{1}{2} \int d\mathbf{x} d\mathbf{x}' \, \frac{(\gamma_1^*(\mathbf{x}';\mathbf{x}))^p \, (\gamma_1(\mathbf{x}';\mathbf{x}))^{1-p}}{|\mathbf{x}-\mathbf{x}'|}, \tag{1.33}$$

for some $0 < p < 1$, where in the original formulation $p = \frac{1}{2}$. Its expression in terms of natural orbitals and occupation numbers is analogous to Eq. (1.32), but with $n_k n_{k'} \mapsto (n_k n_{k'})^{1/2}$, as a consequence of the fact that the operator 1DM has been replaced by its square root. The proposal of Goedecker and Umrigar (1998) consists instead in the substitution $n_k n_{k'} \mapsto (n_k n_{k'})^{1/2}(1-\delta_{kk'}) + n_k^2 \delta_{kk'}$. Frank *et al.* (2007) emphasize that such Müller's functional possesses several advantages, among which that of being a convex functional of its argument, *viz.* $E_{\rm xc}^{\rm M}[\lambda\gamma_1 + (1-\lambda)\gamma_1'] \leq \lambda E_{\rm xc}^{\rm M}[\gamma_1] + (1-\lambda)E_{\rm xc}^{\rm M}[\gamma_1']$, for $0 \leq \lambda \leq 1$.

Other useful approximations favor different powers of the 1DM, such as that of Sharma *et al.* (2008), which has been shown to follow more closely the dependence of the total energy on the number of particles, especially for finite systems. Yet other approximations propose exchange and correlation functionals, where the weight of the Hartree-Fock term in Eq. (1.32) in terms of occupation numbers is replaced by suitable combinations of n_k and $n_{k'}$, to the effect of better modeling weak versus strong occupancy, or bond correlation (Gritsenko *et al.*, 2005). Modified versions including cumulant expansions have been proposed by Piris (2006); Piris *et al.* (2011), by Rohr *et al.* (2008, 2010), and by Marques and Lathiotakis (2008).

1.3 Differential virial theorem

1.3.1 *Definition of kinetic energy density*

In terms of the DM ρ most quantities of interest can be defined, such as the kinetic-energy tensor

$$t_{\alpha\beta}(\mathbf{r};[\rho]) = \frac{1}{4}\left(\frac{\partial^2}{\partial r'_\alpha \partial r''_\beta} + \frac{\partial^2}{\partial r'_\beta \partial r''_\alpha}\right)\rho(\mathbf{r}';\mathbf{r}'')\Bigg|_{\mathbf{r}'=\mathbf{r}''=\mathbf{r}}. \tag{1.34}$$

This is a real, symmetric tensor, whose trace is the non-negative kinetic-energy density (a scalar)

$$t(\mathbf{r}) = \sum_\alpha t_{\alpha,\alpha}(\mathbf{r}) \geq 0. \tag{1.35}$$

The total kinetic energy in the state Ψ can thus be expressed as

$$T_\psi \equiv \langle\Psi|T|\Psi\rangle = \int d^3\mathbf{r}\, t(\mathbf{r}). \tag{1.36}$$

1.3.2 *Derivation of differential virial theorem*

Following Holas and March (1995), we now derive a generalization of the virial theorem for an electron assembly, in the presence of interaction. Our starting point is the Schrödinger equation associated with the Hamiltonian, Eq. (1.1):

$$H\Psi = (T + U + V)\Psi = E\Psi. \tag{1.37}$$

Since H is a real operator, this equation must be separately obeyed by the real and imaginary parts of the eigenstate Ψ (either ground- or excited-state), written as $\Psi = \Psi_{\rm I} + i\Psi_{\rm II}$. Making use of the fact that T is a differential operator, while $U+V$ is an algebraic (multiplicative) operator, one may thus write

$$U + V - E = -\frac{T\Psi_{\rm I}}{\Psi_{\rm I}}, \tag{1.38a}$$

$$U + V - E = -\frac{T\Psi_{\rm II}}{\Psi_{\rm II}}. \tag{1.38b}$$

Differentiating both sides of Eq. (1.38a) with respect to $r_{1\alpha}$, one obtains (Holas and March, 1995)

$$\partial_{1\alpha}v(\mathbf{r}_1) + \sum_{j=2} \partial_{1\alpha}u(\mathbf{r}_1, \mathbf{r}_j)$$
$$= \frac{1}{2}\frac{1}{\Psi_{\rm I}}\sum_{i=1}^{N}\sum_{\beta}\partial_{1\alpha}\partial_{i\beta}^2\Psi_{\rm I} - \frac{1}{2}\frac{\partial_{1\alpha}\Psi_{\rm I}}{\Psi_{\rm I}^2}\sum_{i=1}^{N}\sum_{\beta}\partial_{i\beta}^2\Psi_{\rm I}. \tag{1.39}$$

An analogous equation can be obtained by operating similarly on Eq. (1.38b). Multiplying these equations by $\Psi_{\rm I}^2$ and $\Psi_{\rm II}^2$, respectively, and adding these two equations, one arrives at

$$|\Psi|^2\partial_{1\alpha}\left[v(\mathbf{r}_1) + \sum_{j=2}^{N} u(\mathbf{r}_1, \mathbf{r}_j)\right]$$
$$= \sum_{i=1}^{N}\sum_{\beta}\left[\frac{1}{4}\partial_{1\alpha}\partial_{i\beta}^2|\Psi|^2 - \partial_{i\beta}\left(\partial_{1\alpha}\Psi_{\rm I}\partial_{i\beta}\Psi_{\rm I} + \partial_{1\alpha}\Psi_{\rm II}\partial_{i\beta}\Psi_{\rm II}\right)\right]. \tag{1.40}$$

Next, one sums over the spin variable s_1, eliminates $\mathbf{x}_2, \ldots \mathbf{x}_N$ by averaging (integrating and summing, respectively) over the corresponding spatial and spin coordinates, and includes an overall factor of N to obtain

$$n(\mathbf{r}_1)\partial_{1\alpha}v(\mathbf{r}_1) + \sum_{j=2}^{N} N\sum_{s_1}\int d\mathbf{x}_2\cdots d\mathbf{x}_N\,|\Psi|^2\partial_{1\alpha}u(\mathbf{r}_1, \mathbf{r}_j) = \frac{1}{4}\nabla_1^2\partial_{1\alpha}n(\mathbf{r}_1)$$
$$-2\sum_{\beta}N\sum_{s_1}\int d\mathbf{x}_2\cdots d\mathbf{x}_N\,\frac{1}{2}\partial_{1\beta}\left(\partial_{1\alpha}\Psi_{\rm I}\partial_{1\beta}\Psi_{\rm I} + \partial_{1\alpha}\Psi_{\rm II}\partial_{1\beta}\Psi_{\rm II}\right)$$
$$+\sum_{j=2}^{N}\sum_{\beta}N\sum_{s_1}\int d\mathbf{x}_2\cdots d\mathbf{x}_N$$
$$\times\,\partial_{j\beta}\left[\frac{1}{4}\partial_{1\alpha}\partial_{j\beta}|\Psi|^2 - \left(\partial_{1\alpha}\Psi_{\rm I}\partial_{j\beta}\Psi_{\rm I} + \partial_{1\alpha}\Psi_{\rm II}\partial_{j\beta}\Psi_{\rm II}\right)\right]. \tag{1.41}$$

The integrand in the last term contains the divergence with respect of $\mathbf{r}_j$: $\sum_\beta \partial_{j\beta}[\ldots]$, of a function of Ψ and its derivatives. Since Ψ is a solution of the Schrödinger equation, Ψ and its derivatives vanish as $|\mathbf{r}_j| \to \infty$. The last integral over $d\mathbf{r}_j$ therefore equals zero, when converted into a surface integral, the surface of integration being the surface at infinity. The expression in the middle term can be recognized as the kinetic-energy density tensor $t_{\alpha\beta}(\mathbf{r}_1)$, defined in Eq. (1.34), while the generic contribution to the first term can be rewritten as

$$N(N-1)\sum_{s_1}\int d\mathbf{x}_2\cdots d\mathbf{x}_N\,|\Psi|^2\partial_{1\alpha}u(\mathbf{r}_1,\mathbf{r}_2)$$
$$= 2\int d\mathbf{r}_2\, n_2(\mathbf{r}_1,\mathbf{r}_2)\partial_{1\alpha}u(\mathbf{r}_1,\mathbf{r}_2). \quad (1.42)$$

Finally, Eq. (1.41) reduces to

$$n(\mathbf{r})\partial_\alpha v(\mathbf{r}) + 2\int d\mathbf{r}' n_2(\mathbf{r},\mathbf{r}')\partial_\alpha u(\mathbf{r},\mathbf{r}') = \frac{1}{4}\nabla^2\partial_\alpha n(\mathbf{r}) - 2\sum_\beta \partial_\beta t_{\alpha\beta}(\mathbf{r}). \quad (1.43)$$

This equation represents a differential form of the (integral) virial theorem, here including also e-e and n-n interactions which can (but not necessarily) be of Coulombic nature.

In order to obtain the global (integral) version of the virial theorem, one may operate by $\int d^3\mathbf{r}\sum_\alpha r_\alpha$ on both sides of Eq. (1.43) to obtain

$$2T = \int d^3\mathbf{r}\, n(\mathbf{r})\mathbf{r}\cdot\nabla v(\mathbf{r}) + 2\int d^3\mathbf{r}d^3\mathbf{r}'\, n_2(\mathbf{r},\mathbf{r}')\mathbf{r}\cdot\nabla u(\mathbf{r},\mathbf{r}'), \quad (1.44)$$

which can be reduced further in the case of a Coulombic e-e repulsion as

$$2T + E_{ee} = \int d^3\mathbf{r}\, n(\mathbf{r})\mathbf{r}\cdot\nabla v(\mathbf{r}). \quad (1.45)$$

The latter was known to Clausius in classical mechanics, who wrote

$$2\langle T\rangle = -\langle \mathbf{r}\cdot\mathbf{F}\rangle, \quad (1.46)$$

where T denotes kinetic energy, $\mathbf{F}$ the vector force, and the averages are with respect to time. The quantity $\mathbf{r}\cdot\mathbf{F}$ appearing in Eq. (1.46) was termed by Clausius the *virial* of the force $\mathbf{F}$.

Starting from Eq. (1.43), Holas *et al.* (2005) derived a sum rule for the exchange-correlation (xc) force in DFT,

$$\mathbf{F}_{\mathrm{xc}} = -\nabla v_{\mathrm{xc}}(\mathbf{r}), \quad (1.47)$$

where $v_{\mathrm{xc}}(\mathbf{r})$ is the exchange-correlation potential, in the form

$$\langle n\mathbf{F}_{\mathrm{xc}}\rangle = 0, \quad (1.48)$$

which could then be used to test the consistency of existing approximate density functionals used in applications.

Chapter 2

Solvable models for small clusters of fermions

2.1 Ground-state energy as a functional of the electron density

In principle, an alternative route exists to the determination of the ground-state energy of atoms, molecules, and clusters to that of solving the many-body Schrödinger equation. The basic idea goes back to Thomas-Fermi statistical theory (Thomas, 1926; Fermi, 1927, 1928). In particular, these authors used semiclassical theory to write the electronic ground-state energy in the form

$$E[n(\mathbf{r})] = c_k \int n^{\frac{5}{3}}(\mathbf{r})\, d\mathbf{r} + \int n(\mathbf{r}) V_{\text{ext}}(\mathbf{r})\, d\mathbf{r} + \frac{1}{2} e^2 \int \frac{n(\mathbf{r}) n(\mathbf{r}')}{|\mathbf{r} - \mathbf{r}'|}\, d\mathbf{r}'. \quad (2.1)$$

Here, for the example of a single atom having nuclear charge Ze, the external potential entering Eq. (2.1) is given by

$$V_{\text{ext}}(\mathbf{r}) = -\frac{Ze^2}{r}. \quad (2.2)$$

If we invoke the variational principle

$$\delta(E - \mu N) = 0, \quad (2.3)$$

with N the total number of electrons, where the variation δ is performed with respect to $n(\mathbf{r})$, then the equation for the Lagrange multiplier μ is readily obtained from Eq. (2.1) as

$$\mu = \frac{5}{3} c_k n^{\frac{2}{3}}(\mathbf{r}) + V_{\text{ext}}(\mathbf{r}) + e^2 \int \frac{n(\mathbf{r}')}{|\mathbf{r} - \mathbf{r}'|}\, d\mathbf{r}'. \quad (2.4)$$

In Eq. (2.4), μ has the physical significance of the electron chemical potential, which is constant throughout the entire inhomogeneous electron cloud of density $n(\mathbf{r})$ in the finite atomic assemblies under consideration.

The kinetic constant c_k was taken from the known theory of the uniform Fermi electron gas of constant density, and is given by

$$c_k = \frac{3h^2}{10m}\left(\frac{3}{8\pi}\right)^{\frac{2}{3}}. \tag{2.5}$$

As an immediate example, Thomas and Fermi (TF, below) in their pioneering work considered neutral atoms, the energy of which, say $E_{\mathrm{TF}}(Z)$, was determined in the subsequent study of Milne (1927) as

$$E_{\mathrm{TF}}(Z) = -0.7687Z^{\frac{7}{3}} \tag{2.6}$$

in atomic units, e^2/a_0, with $a_0 = \hbar^2/me^2$ the Bohr radius. This resulted from combining Eq. (2.4) with the Poisson equation of electrostatics, leading to an ordinary nonlinear differential equation for the spherical self-consistent potential $V(\mathbf{r})$, related to $n(\mathbf{r})$ by

$$\nabla^2 V(\mathbf{r}) = 4\pi e^2 n(\mathbf{r}). \tag{2.7}$$

(For an historical perspective on TF theory, see March (1983b); for the mathematical foundations of TF theory, see Lieb and Simon (1977).)

Subsequently, Hohenberg and Kohn (1964) completed the TF work by showing that the ground-state energy of the atoms, molecules, and clusters was a unique (albeit unknown) functional of the ground-state electron density $n(\mathbf{r})$ (see, *e.g.* Parr and Yang, 1989, for a review on DFT). Since this functional $E[n(\mathbf{r})]$ is at the heart of modern density functional theory (DFT), we turn immediately below to give an exact model example for the case of a two-electron (only) spin-compensated atom.

2.2 Moshinsky model for a two-electron atom

The simplest model of a two-electron spin-compensated atom goes back at least to Moshinsky (1952). He assumed harmonic forces for both 'confinement to an origin' with external potential $V_{\mathrm{ext}} = \frac{1}{2}kr^2$, and also for interparticle interactions, denoted more generally below $u(r)$, with $r = |\mathbf{r}_2 - \mathbf{r}_1|$ the interparticle distance. From Moshinsky's ground-state wave function, March (2002) derived the first-order spinless density matrix (1DM) $\gamma(\mathbf{r}_1, \mathbf{r}_2)$ as a functional of its diagonal density $n(r)$, enabling the total correlated kinetic energy to be thereby obtained as a functional of the density $n(r)$. But the first important solvable model, proposed by Holas *et al.* (2003, HHM below), generalized the above Moshinsky model by retaining harmonic confinement, but taking a general interparticle interaction $u(r)$. Then, the above class of models is defined by the Hamiltonian

$$H = -\frac{\hbar^2}{2m}(\nabla_1^2 + \nabla_2^2) + \frac{1}{2}k(r_1^2 + r_2^2) + u(r). \tag{2.8}$$

Introducing the center of mass (CM) coordinate $\mathbf{R} = \frac{1}{2}(\mathbf{r}_1 + \mathbf{r}_2)$, and observing that

$$r_1^2 + r_2^2 = 2R^2 + \frac{1}{2}r^2, \tag{2.9a}$$

$$\nabla_1^2 + \nabla_2^2 = \frac{1}{2}\nabla_{\mathbf{R}}^2 + \nabla_{\mathbf{r}}^2, \tag{2.9b}$$

the HHM Hamiltonian, Eq. (2.8) can be separated into a sum of two contributions as

$$H = H_{\mathrm{CM}}(\mathbf{R}) + H_{\mathrm{rm}}(\mathbf{r}), \tag{2.10}$$

one for the center of mass motion,

$$H_{\mathrm{CM}}(\mathbf{R}) = -\frac{\hbar^2}{2M}\nabla_{\mathbf{R}}^2 + \frac{1}{2}M\omega^2 R^2, \tag{2.11}$$

and one for the relative motion (rm),

$$H_{\mathrm{rm}}(\mathbf{r}) = -\frac{\hbar^2}{2m_{\mathrm{rm}}}\nabla_{\mathbf{r}}^2 + \frac{1}{2}m_{\mathrm{rm}}\omega^2 r^2 + u(r). \tag{2.12}$$

In Eqs. (2.11), (2.12), $M = 2m$ is the total mass of the electrons, $m_{\mathrm{rm}} = \frac{1}{2}m$ is the reduced mass of the relative motion, and $\omega^2 = k/m$ is the harmonic frequency. The ground-state wave function can be factorized correspondingly as

$$\Psi(\mathbf{r}_1\sigma_1, \mathbf{r}_2\sigma_2) = \Psi_{\mathrm{CM}}(\mathbf{R})\psi_{\mathrm{rm}}(\mathbf{r})\chi_S(\sigma_1, \sigma_2), \tag{2.13}$$

where one chooses the spin factor χ_S to represent the singlet state as

$$\chi_S(\sigma_1, \sigma_2) = \frac{1}{\sqrt{2}}[\alpha(\sigma_1)\beta(\sigma_2) - \alpha(\sigma_2)\beta(\sigma_1)], \tag{2.14}$$

with α, β single-particle spinors. The remaining factors in Eq. (2.13) then satisfy the decoupled eigenvalue equations

$$H_{\mathrm{CM}}(\mathbf{R})\Psi_{\mathrm{CM}}(\mathbf{R}) = E_{\mathrm{CM}}\Psi_{\mathrm{CM}}(\mathbf{R}), \tag{2.15a}$$

$$H_{\mathrm{rm}}(\mathbf{r})\psi_{\mathrm{rm}}(\mathbf{r}) = E\psi_{\mathrm{rm}}(\mathbf{r}). \tag{2.15b}$$

The properly normalized, spherically symmetric ground-state solution of the CM problem, Eq. (2.15a), is readily shown to have a Gaussian shape, and can be immediately given as

$$\Psi_{\mathrm{CM}}(\mathbf{R}) \equiv \Psi_{\mathrm{CM}}(R) = a_{\mathrm{h}}^{-\frac{3}{2}}\pi^{-\frac{3}{4}}\exp\left(-\frac{1}{2}\frac{R^2}{a_{\mathrm{h}}^2}\right), \tag{2.16a}$$

$$E_{\mathrm{CM}} = \frac{3}{2}\hbar\omega, \tag{2.16b}$$

where a_{h} is a characteristic length associated with harmonic confinement, given by

$$a_{\mathrm{h}} = \left(\frac{\hbar}{M\omega}\right)^{\frac{1}{2}}. \tag{2.17}$$

2.2.1 Schrödinger equation for the relative motion

The rm problem, Eq. (2.15b), can be reduced to a single-particle central field analogue, associated with the radial Schrödinger equation

$$\left(-\frac{\hbar^2}{2m_{\rm rm}}\frac{d^2}{dr^2} + v_{\rm eff}(r)\right)\psi(r) = E\psi(r), \tag{2.18}$$

where

$$v_{\rm eff}(r) = \frac{1}{2}m_{\rm rm}\omega^2 r^2 + u(r), \tag{2.19a}$$

$$\psi(r) = (4\pi)^{1/2} r\psi_{\rm rm}(r), \tag{2.19b}$$

where it has been observed that $\psi_{\rm rm}(\mathbf{r}) \equiv \psi_{\rm rm}(|\mathbf{r}|) = \psi_{\rm rm}(r)$, and the normalization condition for $\psi(r)$ reduces to $\int_0^\infty dr\, \psi^2(r) = 1$.

2.2.2 First-order spinless density matrix

We are now in a position to calculate the first-order spinless density matrix (DM) $\rho_1(\mathbf{r};\mathbf{r}')$, defined in Sec. 1.1, associated with the ground-state solution of the Moshinsky model, Eq. (2.13). Following Holas *et al.* (2003), one first evaluates its diagonal $\rho_1(\mathbf{r};\mathbf{r})$, *viz.* the single-particle number density

$$n(\mathbf{r}) = 2\int d^3\mathbf{r}'\Psi^2_{\rm CM}\left(\frac{1}{2}|\mathbf{r}+\mathbf{r}'|\right)\psi^2_{\rm rm}(|\mathbf{r}-\mathbf{r}'|). \tag{2.20}$$

Making use of Eq. (2.16a) for $\Psi_{\rm CM}(R)$, the angular integrations can be performed to obtain $n(r) \equiv n(\mathbf{r})$ as (Holas *et al.*, 2003)

$$n(r) = \frac{8}{\sqrt{\pi}}e^{-r^2/a_{\rm h}^2}\int_0^\infty dy\, y^2 e^{-y^2/4}\psi^2_{\rm rm}(a_{\rm h}y)\frac{\sinh(ry/a_{\rm h})}{ry/a_{\rm h}}. \tag{2.21}$$

Numerical evaluation of Eq. (2.21) for the Hookean atom confirms Gaussian confinement for this particular model atom, as expected (cf. Fig. 1 of Holas *et al.*, 2003).

Coming back to the off-diagonal generalization of $n(\mathbf{r})$, *viz.* the first-order spinless density matrix $\rho_1(\mathbf{r};\mathbf{r}')$, following the general definition set out in Sec. 1.1, Holas *et al.* (2003) find

$$\rho_1(\mathbf{r};\mathbf{r}') = 2\int d^3\mathbf{r}''\Psi_{\rm CM}\left(\frac{1}{2}|\mathbf{r}+\mathbf{r}''|\right)\psi_{\rm rm}(|\mathbf{r}-\mathbf{r}''|) \times \Psi_{\rm CM}\left(\frac{1}{2}|\mathbf{r}'+\mathbf{r}''|\right)\psi_{\rm rm}(|\mathbf{r}'-\mathbf{r}''|). \tag{2.22}$$

Introducing the relative and mean coordinates $\mathbf{b} = \mathbf{r}-\mathbf{r}'$ and $\mathbf{c} = \frac{1}{2}(\mathbf{r}+\mathbf{r}')$, respectively, and changing the integration variable to $\mathbf{x} = \mathbf{r}''-\mathbf{c}$, Holas *et al.*

(2003) are able to perform one angular integration implicit in Eq. (2.22), and present the first-order spinless density matrix as

$$\rho_1(\mathbf{r};\mathbf{r}') = \frac{4\pi}{(a_\mathrm{h}\sqrt{\pi})^3} e^{-c^2/a_\mathrm{h}^2} e^{-b^2/4a_\mathrm{h}^2}$$
$$\times \int_0^\infty dx\, x^2 e^{-x^2/4a_\mathrm{h}^2} \int_{-1}^1 dt\, e^{-ct_0xt/a_\mathrm{h}^2} I_0 \left(\frac{c\sqrt{1-t_0^2}}{a_\mathrm{h}} \frac{x\sqrt{1-t^2}}{a_\mathrm{h}} \right)$$
$$\times \psi_\mathrm{rm}\left(\sqrt{\frac{1}{4}b^2 + bxt + x^2} \right) \psi_\mathrm{rm}\left(\sqrt{\frac{1}{4}b^2 - bxt + x^2} \right), \quad (2.23)$$

where $I_0(z)$ is the modified Bessel function of first kind of order zero (Gradshteyn and Ryzhik, 1994). A part from the two integrations that must still be performed, Eq. (2.23) of Holas *et al.* (2003) has the merit of showing explicitly that the first-order spinless density matrix depends on the argument vectors $\mathbf{r}$, $\mathbf{r}'$ via the invariants b, c, and $t_0 = \mathbf{b}\cdot\mathbf{c}/bc$, and functionally on $n(\mathbf{r})$ [which is in turn functionally related to $\psi_\mathrm{rm}(\mathbf{r})$, via Eq. (2.21)], as expected from the general derivation in Sec. 1.1.

2.2.3 *Kinetic energy density*

The knowledge of the first-order spinless density matrix allows to evaluate the kinetic energy density within this class of models, whose definition, introduced in Sec. 1.3.1, can be rewritten as

$$t(\mathbf{r}) \equiv t(\mathbf{r};[n]) = \frac{\hbar^2}{2m} \nabla_\mathbf{r} \cdot \nabla_{\mathbf{r}'} \rho_1(\mathbf{r};\mathbf{r}')|_{\mathbf{r}'=\mathbf{r}} . \quad (2.24)$$

In terms of the kinetic energy density, the total kinetic energy simply amounts to $\int d^3\mathbf{r} t(\mathbf{r})$. In the case of this class of models (including the Moshinsky and the Hookean atom), characterized by a separable Schrödinger equation for the CM and the rm motions, Eqs. (2.15), the total kinetic energy can also be separated as the sum of two contributions as

$$T = T_\mathrm{CM} + T_\mathrm{rm} \equiv \frac{\hbar^2}{2M} \int d^3\mathbf{R} \left(\frac{d}{dR} \Psi_\mathrm{CM}(R) \right)^2 + \frac{\hbar^2}{2m_\mathrm{rm}} \int d^3\mathbf{r} \left(\frac{d}{dr} \psi_\mathrm{rm}(r) \right)^2, \quad (2.25)$$

where $T_\mathrm{CM} = \frac{1}{2} E_\mathrm{CM} = \frac{3}{4}\hbar\omega$. In particular, for the Moshinsky atom, Holas *et al.* (2003) quote the following analytical result for the kinetic energy density

$$t(r) = \frac{1}{2} n(r) \left[\frac{3}{2} \frac{(\alpha-1)^2}{\alpha} - \frac{2\alpha - 1}{\alpha} \ln \frac{n(r)}{n(0)} \right], \quad (2.26)$$

where $n(r)$ is implicitly given by Eq. (2.21), and $\alpha^{-1} = 2 - \pi[n(0)/2]^{2/3} = \frac{1}{2}(1 + \sqrt{1 + 2K})$, for an harmonic interparticle potential defined by $u(r) = \frac{1}{2}Kr^2$.

2.2.4 *Virial equation*

By applying an earlier method by Holas and March (1999) based on scaling, Holas *et al.* (2003) derive the virial equation (cf. Sec. 1.1) for this class of model atoms as

$$2T = 2V - U_1', \tag{2.27}$$

where

$$U_1' = \int d^3\mathbf{r} \left(-r\frac{du(r)}{dr}\right) \psi_{\mathrm{rm}}^2(r). \tag{2.28}$$

More in detail, for the separate CM motion, one evidently finds

$$T_{\mathrm{CM}} = V_{\mathrm{CM}} = \frac{1}{2}E_{\mathrm{CM}}, \tag{2.29}$$

while for the rm motion Holas *et al.* (2003) find therefore

$$2T_{\mathrm{rm}} = 2V_{\mathrm{rm}} - U_1', \tag{2.30}$$

with total rm energy

$$E = T_{\mathrm{rm}} + V_{\mathrm{rm}} + U, \tag{2.31}$$

where

$$U = \int d^3\mathbf{r}\, u(r)\psi_{\mathrm{rm}}^2(r). \tag{2.32}$$

2.3 An exactly solvable model for a two-electron spin-compensated atom

One of the few cases in which DFT can be exactly solved is a harmonically confined He-like artificial atom, given by Crandall *et al.* (1984, CWB, below). In that model atom, the Coulombic repulsion $V(r_{12}) = 1/r_{12}$ between the electrons is replaced by a repulsive interaction of the form

$$V(r_{12}) = \lambda\frac{1}{r_{12}^2}, \tag{2.33}$$

where $r_{12} = |\mathbf{r}_1 - \mathbf{r}_2|$ is the modulus of the relative position of the two electrons and $\lambda > 0$ is a parameter, measuring the strength of the inter-electron repulsion. In order to allow for bound states, an additional external confining potential is added, in the form of the harmonic potential

$$V_{\text{ext}}(\mathbf{r}) \equiv V_{\text{h}}(\mathbf{r}) = \frac{1}{2} m\omega^2 r^2. \tag{2.34}$$

In terms of the external potential, the energy functional can be written as

$$E[n(\mathbf{r})] = F[n(\mathbf{r})] + \int n(\mathbf{r}) V_{\text{ext}}(\mathbf{r})\, d\mathbf{r}, \tag{2.35}$$

where the implicit functional

$$F[n(\mathbf{r})] = T[n(\mathbf{r})] + V[n(\mathbf{r})] \tag{2.36}$$

includes the kinetic and interparticle interaction energies, whereas the variational principle yields for the chemical potential

$$\frac{\delta F[n]}{\delta n(\mathbf{r})} + V_{\text{ext}}(\mathbf{r}) = \mu. \tag{2.37}$$

Following Gál and March (2009), we first observe that there exists a generalization of the virial theorem (Slater, 1972; Janak, 1974) (see also Chapter 1), which enables one to express F as

$$F[n(\mathbf{r})] = \frac{1}{2} \int n(\mathbf{r}) \mathbf{r} \cdot \nabla V_{\text{ext}}(\mathbf{r})\, d\mathbf{r}. \tag{2.38}$$

This is due to the fact that the sum of the kinetic and interparticle interaction functionals, Eq. (2.36), is a homogeneous function of degree 2 with respect to coordinates. In particular, one recovers that F reduces to the kinetic energy density functional $T[n(\mathbf{r})]$ in the limit of noninteracting electrons.

In the specific case of harmonic confinement, Eq. (2.34), Eq. (2.38) yields

$$F_{\text{h}}[n(\mathbf{r})] = \int n(\mathbf{r}) V_{\text{h}}(\mathbf{r})\, d\mathbf{r}. \tag{2.39}$$

Gál and March (2009) next exploit a result by Capuzzi *et al.* (2005a), which characterizes the spherically symmetric electron ground-state density $n(r)$ of the CWB model via the linear ordinary differential equation

$$\frac{\hbar}{4m\omega} r n''(r) + \left(\frac{\hbar}{2m\omega} + \frac{3}{2} r^2 \right) n'(r) + r \left(\frac{3}{2} - \alpha + \frac{2m\omega}{\hbar} r^2 \right) n(r) = 0, \tag{2.40}$$

where

$$\alpha = \frac{1}{2} \left(-1 + \sqrt{1 + \frac{4m}{\hbar^2} \lambda} \right) \tag{2.41}$$

is an increasing function of the interparticle repulsion strength λ, and can be therefore used as an alternative measure of such interparticle repulsion. The ground-state energy (Crandall *et al.*, 1984)

$$E_{\mathrm{CWB}} = (\alpha + 3)\hbar\omega \tag{2.42}$$

can then be expressed in terms of the solution of Eq. (2.40) as (Capuzzi *et al.*, 2005a; Gál and March, 2009)

$$E_{\mathrm{CWB}} = \frac{4\int n(r) V_{\mathrm{h}}(r)\, d\mathbf{r}}{\int n(r)\, d\mathbf{r}}, \tag{2.43}$$

which confirms Eq. (2.39), when $\int n(r)\, d\mathbf{r} = 2$.

From Eq. (2.40), Gál and March (2009) were able to derive an explicit expression for $F[n(\mathbf{r})]$ for the CWB model, which does not contain the external (harmonic) potential explicitly. This reads (Gál and March, 2009)

$$\begin{aligned} F_{\mathrm{CWB}}[n(\mathbf{r})] = \frac{\hbar^2}{16m} \int \left(\frac{n'^2(r)}{n(r)} + \frac{r}{2} \frac{n'(r) n''(r)}{n(r)} \right) d\mathbf{r} \\ + \frac{3}{16}\hbar\omega \int \left(r^2 \frac{n'^2(r)}{n(r)} - 9n(r) \right) d\mathbf{r}, \end{aligned} \tag{2.44}$$

where the frequency ω of the harmonic confining potential can be further be eliminated via Eqs. (2.42) and (2.43) as

$$\hbar\omega = \frac{4}{\omega + 3} \frac{F_{\mathrm{CWB}}[n(\mathbf{r})]}{\int n(r)\, d\mathbf{r}}. \tag{2.45}$$

Gál and March (2009) find therefore the explicit form of the internal energy density functional for the CWB model as

$$F_{\mathrm{CWB}}[n(\mathbf{r})] = \frac{\hbar^2}{16m} \frac{\mathcal{N}[n(\mathbf{r})]}{\mathcal{D}[n(\mathbf{r})]}, \tag{2.46a}$$

$$\mathcal{N}[n(\mathbf{r})] = \int \left(\frac{n'^2(r)}{n(r)} + \frac{r}{2} \frac{n'(r) n''(r)}{n(r)} \right) d\mathbf{r}, \tag{2.46b}$$

$$\begin{aligned} \mathcal{D}[n(\mathbf{r})] = 1 + \frac{27}{4(\alpha + 3)} \\ - \frac{3}{4(\alpha + 3)} \frac{1}{\int n(r)\, d\mathbf{r}} \int r^2 \frac{n'^2(r)}{n(r)}\, d\mathbf{r}. \end{aligned} \tag{2.46c}$$

More generally, in view of non-spherically symmetric confining potentials, Eq. (2.46a) is still applicable, but now with

$$\mathcal{N}[n(\mathbf{r})] = \frac{1}{2} \int \left(\mathbf{r} \cdot \nabla n(\mathbf{r}) \frac{\nabla^2 n(\mathbf{r})}{n(\mathbf{r})} \right) d\mathbf{r}, \tag{2.47a}$$

$$\begin{aligned} \mathcal{D}[n(\mathbf{r})] = 1 + \frac{27}{4(\alpha + 3)} \\ - \frac{3}{4(\alpha + 3)} \frac{1}{\int n(\mathbf{r})\, d\mathbf{r}} \int \frac{(\mathbf{r} \cdot \nabla n(\mathbf{r}))^2}{n(\mathbf{r})}\, d\mathbf{r}. \end{aligned} \tag{2.47b}$$

Equation (2.46a) provides an explicit density functional, valid for any strength of the harmonic confinement, and any strength of the interparticle repulsion. Gál and March (2009) finally observe that such a functional is homogeneous of degree one in the electron density, $n(r)$. However, in the limits $\lambda = 0$ and $\alpha = 0$, it does not reduce to the von Weizsäcker kinetic energy functional (von Weizsäcker, 1935)

$$T_{\mathrm{vW}}[n(\mathbf{r})] = \frac{\hbar^2}{8m} \int \frac{|\nabla n(\mathbf{r})|^2}{n(\mathbf{r})} \, d\mathbf{r}, \tag{2.48}$$

although $T_{\mathrm{vW}}[n(\mathbf{r})]$ is the exact degree 1 homogeneous noninteracting kinetic-energy density functional for one-level systems.

2.4 Low-order density gradient contributions to the kinetic energy

The contributions from the density gradient to the internal energy density functional $F_{\mathrm{CWB}}[(n(\mathbf{r})]$, Eq. (2.46a), within the CWB exactly solvable model have been studied more in detail by Geldof *et al.* (2013). The two-particle wavefunction $\Psi(\mathbf{r}_1, \mathbf{r}_2)$ corresponding to the ground-state solution of the CWB model can be factorized into two terms, related to the motion of the centre of mass (CM) and to the relative motion, respectively, as

$$\Psi(\mathbf{r}_1, \mathbf{r}_2) = \Psi_{\mathrm{CM}}(R)\psi(r_{12}), \tag{2.49}$$

where $\mathbf{r}_{12} = \mathbf{r}_1 - \mathbf{r}_2$ is the relative coordinate, as already defined, and $\mathbf{R} = \frac{1}{2}(\mathbf{r}_1 + \mathbf{r}_2)$ is the CM coordinate. One finds a normalized Gaussian for $\Psi_{\mathrm{CM}}(R)$, as expected for the ground-state of any two-particle system in an harmonic confining potential [cf. Eq. (2.16a)], whereas for the relative motion part of the CWB model Capuzzi *et al.* (2005a) explicitly find

$$\psi(r) = \left(\frac{2}{\Gamma(\alpha + \frac{3}{2})} \right)^{\frac{1}{2}} \left(\frac{m\omega}{2\hbar} \right)^{\frac{1}{2}\alpha + \frac{3}{4}} r^{1+\alpha} \exp\left(-\frac{m\omega}{4\hbar} r^2 \right). \tag{2.50}$$

The resulting density, solution of Eq. (2.40), can then be reduced to quadratures as (Capuzzi *et al.*, 2005a,b)

$$n(r) = \frac{1}{2\pi^{\frac{3}{2}} 4^{\alpha} \Gamma(\alpha + \frac{3}{2})} \frac{e^{-r^2/a_{\mathrm{h}}^2}}{a_{\mathrm{h}}^2 r} \int_0^{\infty} z^{2\alpha+1} e^{-\frac{1}{2}z^2} \sinh(rz/a_{\mathrm{h}}) \, dz, \tag{2.51}$$

where $a_{\mathrm{h}} = (\hbar/2m\omega)^{\frac{1}{2}}$ is the harmonic confinement length, already defined in Eq. (2.17). In particular, one finds explicitly

$$n(r = 0) = \frac{1}{\pi^{\frac{3}{2}} 2^{\alpha + \frac{1}{2}} a_{\mathrm{h}}}. \tag{2.52}$$

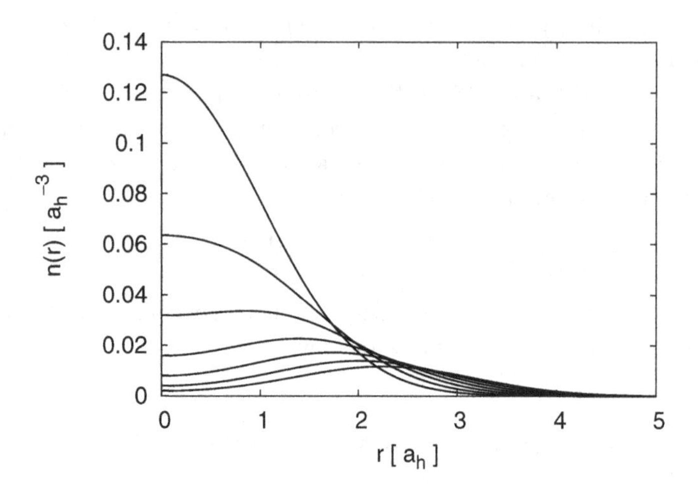

Fig. 2.1 Electron density profile $n(r)$, Eq. (2.51), of the harmonically confined CWB model as a function of $r/a_{\rm h}$, for interparticle repulsion parameter $\alpha = 0 - 6$ (increasing from top to bottom).

Figure 2.1 shows the density profile of the harmonically confined CWB model as a function of $r/a_{\rm h}$, for different values of the interparticle repulsion parameter α. With increasing interparticle repulsion, one recovers a shift of the maximum density towards larger values of the interparticle separation r, and an exponential decrease of its value at $r = 0$, in accordance with Eq. (2.52).

In order to extract the effect of interparticle correlation energy, following Geldof *et al.* (2013), for a spherically symmetric electron density $n(\mathbf{r}) \equiv n(r)$, we define

$$\tau(r) = \frac{n'^2(r)}{n(r)}. \tag{2.53}$$

One can then eliminate the second term in the integrand in Eq. (2.46b) by differentiating $n(r)\tau(r) = n'^2(r)$ to get

$$\frac{2n'(r)n''(r)}{n(r)} = \tau'(r) + \frac{n'(r)}{n(r)}\tau(r). \tag{2.54}$$

This enables us to rewrite essentially the numerator for $F[n(r)]$, Eq. (2.46b), as

$$\mathcal{N}[n(r)] = \int \tau(r)\, d\mathbf{r} + \frac{1}{4}\int r\tau'(r)\, d\mathbf{r} + \frac{1}{4}\int r\frac{\tau(r)n'(r)}{n(r)}\, d\mathbf{r}. \tag{2.55}$$

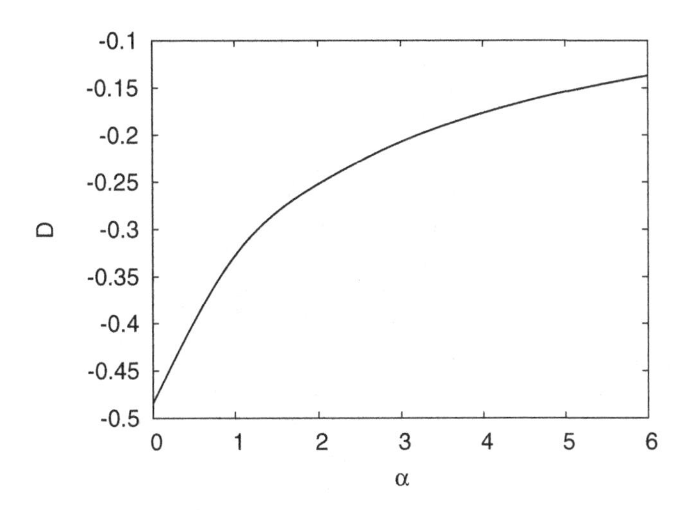

Fig. 2.2 Denominator $\mathcal{D}[n(r)]$, Eq. (2.46c), corresponding to solutions $n(r)$ of the harmonically confined CWB model given by Eq. (2.51) and shown in Fig. 2.1, as a function of the interaction parameter α.

Using spherical symmetry, and integrating the second term on the right-hand side of Eq. (2.55) by parts, one finds eventually

$$\mathcal{N}[n(r)] = \frac{1}{4}\int \left(\tau(r) + r\tau(r)\frac{n'(r)}{n(r)}\right) d\mathbf{r}. \tag{2.56}$$

One may then write the total internal energy functional as

$$F[n(r)] = \frac{\hbar^2}{16m}\int (f_{\rm vW}(r) + \Delta f(r))\, d\mathbf{r} \equiv F_{\rm vW}[n(r)] + \Delta F[n(r)], \tag{2.57}$$

with

$$f_{\rm vW}(r) = \frac{1}{\mathcal{D}[n(r)]}\frac{1}{4}\tau(r) \tag{2.58}$$

the von Weizsäcker kinetic energy density, whereas the remaining term

$$\Delta f(r) = \frac{1}{\mathcal{D}[n(r)]}\frac{1}{4}r\tau(r)\frac{n'(r)}{n(r)} \tag{2.59}$$

is the residual correlation energy, due to interparticle repulsion, in the CWB model.

Figure 2.2 shows $\mathcal{D}[n(r)]$ for the density profiles Eq. (2.51) of the harmonically confined CWB model shown in Fig. 2.1, as a function of the interaction parameter α. Figure 2.3 then shows the von Weiszäcker kinetic energy density $f_{\rm vW}(r)$ and residual correlation energy density $\Delta f(r)$,

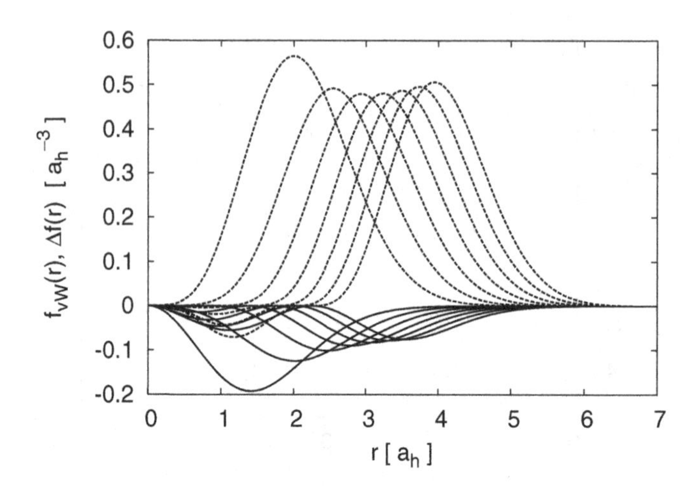

Fig. 2.3 Von Weizsäcker energy density $f_{\text{vW}}(r)$, Eq. (2.58) (solid lines), and correlated energy density $\Delta f(r)$, Eq. (2.59) (dashed lines), building up the total energy functional $F[n(r)]$ in Eq. (2.57) for the harmonically confined CWB model. Interparticle distance r is measured in units of the harmonic confinement length a_{h}. Values of the interaction parameter $\alpha = 0 - 6$ are as in Fig. 2.1.

defined by Eqs. (2.58) and (2.59), respectively. The resulting total energy functional $F[n(r)]$ in Eq. (2.57) is then shown in Fig. 2.4, along with its contributions, as function of the interaction parameter α. One finds that, while ΔF is the main (positive) contribution to F, the (negative) contribution from the von Weiszäcker term (F_{vW}) also gets enhanced, with increasing interaction parameter α.

2.5 Exactly solvable ground-state energies for some series of light atomic ions with non-integral nuclear charges to within statistical error of QMC

Early work of Foldy (1951) was important in emphasizing an intimate correlation between the ground-state energy of neutral atoms of nuclear charge Ze, *viz.* $E(Z)$, and the electrostatic potential created by the electrons, denoted as V_0 below, evaluated at the assumed point nucleus. Before Foldy's study, such a correlation could be extracted analytically from the Thomas-Fermi (TF) semi-classical theory, beginning with the ground-state energy $E(Z)$ of neutral atoms having nuclear charge Ze derived by Milne (1927).

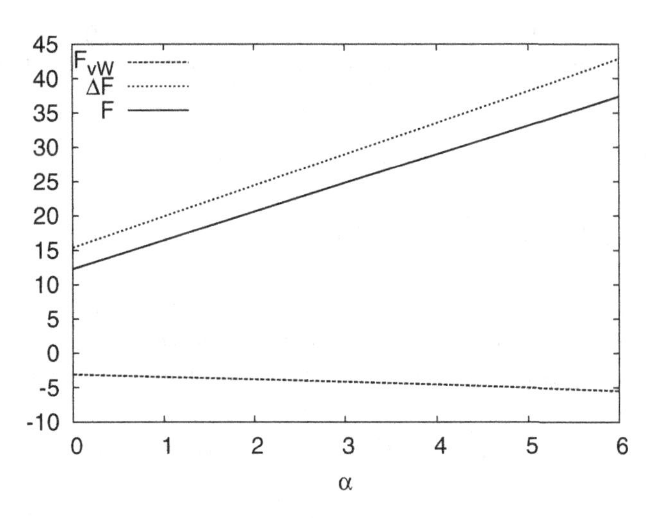

Fig. 2.4 Total (F, solid line), von Weizsäcker (F_{vW}, long-dashed line), and correlation energy functional (ΔF, short-dashed line) in Eq. (2.57), in units of $\hbar^2/16m$, as a function of the interaction parameter α.

His result was explicitly

$$E(Z) = -0.7687\, Z^{7/3} \frac{e^2}{a_0}, \tag{2.60}$$

where $a_0 = \hbar^2/me^2$ denotes Bohr's radius. The electron-nuclear potential energy U_{en} in the TF model is also known (see *e.g.* March and Deb, 1987) to be given by

$$U_{en}(Z) = \frac{7}{3} E(Z). \tag{2.61}$$

But on physical grounds, $U_{en}(Z)$ is simply the interaction energy of the charge Ze sitting in a potential V_0, *i.e.*

$$U_{en}(Z) = ZeV_0. \tag{2.62}$$

Hence, using Eq. (2.62) in Eq. (2.61), the desired correlation is expressed by

$$V_0(Z) = \frac{7}{3} \frac{E(Z)}{Ze} \simeq -1.79\, Z^{4/3} \frac{e}{a_0}, \tag{2.63}$$

which is the precise semi-classical (non-relativistic) prediction for neutral atoms. Unfortunately, the self-consistent TF ground-state density $n_{\mathrm{TF}}(r)$ has two truly major difficulties: (i) it is infinite at the assumed point nucleus, being singular as $r^{-3/2}$, and (ii) it decays as the power law r^{-6} rather than exponentially.

Politzer and Parr (1974) proposed a generalization of Eq. (2.63) to read [see also Politzer's chapter in March and Deb (1987) and Politzer *et al.* (2005)]

$$E(Z, N) = \int_0^Z V_0 dZ'. \tag{2.64}$$

However, it is known that as Z' is varied through an isoelectronic series of atomic ions with N electrons, there is a critical (non-integral) atomic number, say $Z_c(N)$, at which, on further infinitesimal reduction, one electron ionizes, sometimes referred to, somewhat loosely, as a 'phase transition'. Hence, following Amovilli and March (2015), we shall rewrite Eq. (2.64) as

$$E(Z, N) = E(Z_c, N) + \int_{Z_c}^Z V_0 dZ'. \tag{2.65}$$

Obviously, V_0 is to be determined from the ground-state electron density $n(\mathbf{r}, Z', N)$ through the isoelectronic series of atomic ions with N electrons.

It is relevant at this point to emphasize that, at least in principle, $E(Z, N)$ is determined solely by the ground-state density $n(\mathbf{r}, Z, N)$, through the Hohenberg-Kohn theorem (Hohenberg and Kohn, 1964). Unfortunately, the functional $E[n(\mathbf{r})]$ remains unknown, and we bypass this fact here by invoking Quantum Monte Carlo (QMC) calculations for the appropriate range of Z' to very accurately evaluate numerically the integral appearing in Eq. (2.65) (Amovilli and March, 2015).

With this as background, we turn immediately to present the QMC results of Amovilli and March (2015) to insert into Eq. (2.65).

2.5.1 *Discussion of QMC results for He- to B-like isoelectronic series of atomic ions*

Tables 2.1–2.4 record the results of Amovilli and March (2015) for the ground-state energy E and for V_0 from QMC calculations. Both variational (VMC) and diffusion (DMC) have been utilized, for $N = 2$ to 5 (He- to B-like isoelectronic series of atomic ions). V_0 values for the neutral atoms from VMC and DMC are in pretty good agreement. In particular, for the B neutral atom, V_0 values are $-11.41(1)$ and $11.39(1)$ Hartree for VMC and DMC, respectively. The electron-nuclear potential energies U_{en} corresponding to these values follow from Eq. (2.62) for the neutral B atom as -57.07 and -56.95 Hartree. One finds $U_{en}/E = 2.316$ and 2.310, these ratio being somewhat smaller than the TF prediction of $7/3$ in Eq. (2.61).

Table 2.1 Quantum Monte Carlo results for the electrostatic potential at the nucleus for the He-like series of atomic ions for some thirty values (mostly non-integral) of nuclear charge Ze. All quantities are in atomic units. After Amovilli and March (2015).

Z	E (VMC)	V_0 (VMC)	E (DMC)	ZV_0 (DMC)
0.88	−0.38517(1)	−0.8925(10)	−0.38548(1)	−0.7867(20)
0.89	−0.39398(1)	−0.9056(10)	−0.39452(1)	−0.7991(20)
0.90	−0.40306(1)	−0.9146(10)	−0.40347(1)	−0.8213(20)
0.905	−0.40784(1)	−0.9178(10)	−0.40800(1)	−0.8270(20)
0.907	−0.41029(1)	−1.1545(20)	−0.41042(2)	−1.0189(20)
0.91	−0.41378(1)	−1.1563(20)	−0.41380(2)	−1.0349(20)
0.911	−0.41495(1)	−1.1579(20)	−0.41494(2)	−1.0427(20)
0.9113	−0.41528(1)	−1.1589(10)	−0.41524(2)	−1.0533(20)
0.9115	−0.41559(1)	−1.1553(10)	−0.41550(2)	−1.0452(20)
0.912	−0.41598(1)	−1.1609(10)	−0.41611(2)	−1.0614(20)
0.915	−0.41957(1)	−1.1667(10)	−0.41963(2)	−1.0756(20)
0.92	−0.42547(1)	−1.1853(10)	−0.42550(2)	−1.0845(20)
0.93	−0.43743(1)	−1.2097(10)	−0.43744(2)	−1.1244(20)
0.96	−0.47479(1)	−1.2827(10)	−0.47478(2)	−1.2307(20)
1.00	−0.52773(1)	−1.3602(10)	−0.52771(2)	−1.3655(20)
1.10	−0.67475(1)	−1.5713(10)	−0.67476(2)	−1.7379(20)
1.30	−1.02980(1)	−1.9737(10)	−1.02989(2)	−2.5669(30)
1.50	−1.46519(1)	−2.3765(20)	−1.46534(2)	−3.5699(40)
1.70	−1.98061(1)	−2.7767(20)	−1.98074(2)	−4.7168(50)
1.95	−2.73736(1)	−3.2749(30)	−2.73742(2)	−6.4033(60)
1.98	−2.83657(1)	−3.3334(30)	−2.83657(2)	−6.6097(70)
2.00	−2.90364(1)	−3.3776(30)	−2.90373(4)	−6.7484(70)
2.01	−2.93757(1)	−3.4024(30)	−2.93757(3)	−6.8330(70)
2.03	−3.00590(1)	−3.4328(30)	−3.00588(3)	−6.9968(70)
2.05	−3.07503(1)	−3.4745(30)	−3.07509(3)	−7.1379(70)
2.10	−3.25133(1)	−3.5759(30)	−3.25133(4)	−7.5318(80)
2.30	−4.00664(2)	−3.9701(30)	−4.00674(3)	−9.1471(80)
2.65	−5.52082(2)	−4.6765(30)	−5.52087(3)	−12.3791(90)
3.00	−7.27989(1)	−5.3803(40)	−7.27990(3)	−16.1282(10)
3.50	−10.21775(1)	−6.3776(50)	−10.21780(4)	−22.3467(17)
4.00	−13.65555(2)	−7.3775(60)	−13.65562(5)	−29.5164(20)
5.00	−22.03087(3)	−9.3803(70)	−22.03134(10)	−46.8642(27)

In the customary language of DFT, the ground-state energy E and the universal density functional F, on the minimum, are related by

$$E = F + U_{en}. \tag{2.66}$$

Hence for VMC, for neutral B, $F = 32.43$ Hartree, to be compared to $F = 32.30$ Hartree for DMC. We anticipate that the DMC value for F will be somewhat more accurate than that from the variational approach. If T denotes the total kinetic energy including correlation kinetic energy, then

Table 2.2 QMC results for Li-like isoelectronic series of atomic ions for about 20 values of fractional nuclear charge Ze. All quantities are in atomic units. After Amovilli and March (2015).

Z	E (VMC)	V_0 (VMC)	E (DMC)	ZV_0 (DMC)
1.95	2.73407(1)	−3.3204(20)	−2.73597(1)	−6.4446(20)
2.01	2.93414(4)	−3.4207(20)	−2.93757(1)	−6.8702(20)
2.05	3.07421(3)	−3.5127(20)	−3.07612(1)	−7.2125(20)
2.10	3.25455(2)	−3.6534(20)	−3.25508(1)	−7.6669(20)
2.30	4.03046(4)	−4.1283(30)	−4.03236(6)	−9.5009(30)
2.50	4.90202(3)	−4.5836(30)	−4.90269(3)	−11.4711(20)
2.70	5.86398(3)	−5.0385(30)	−5.86469(2)	−13.6113(20)
2.81	6.43108(4)	−5.2894(40)	−6.43253(2)	−14.8618(20)
2.84	6.59192(2)	−5.3555(40)	−6.59216(2)	−15.2186(20)
2.845	6.61868(2)	−5.3647(40)	−6.61900(2)	−15.2732(20)
2.85	6.64559(2)	−5.3799(40)	−6.64584(2)	−15.3327(20)
2.855	6.67250(2)	−5.3917(40)	6.67276(2)	−15.3879(20)
2.86	6.69947(2)	−5.3918(40)	−6.69974(2)	−15.4451(20)
3.00	7.47779(2)	−5.7200(40)	−7.47810(2)	−17.1661(30)
3.50	10.61908(2)	−6.8491(50)	−10.61971(3)	−23.9474(40)
4.00	14.32433(4)	−7.9774(50)	−14.32527(10)	−31.8723(15)
5.00	23.42402(6)	−10.2136(70)	−23.42581(16)	−51.1051(20)

the virial theorem would give as a useful approximation $T = -E$, in each case.

In Table 2.5 we show, for comparison, the separate magnitudes of the two terms on the right-hand side of Eq. (2.65) together with the total energy $E(Z, Z)$ for neutral atoms. It is important here to make contact with the fairly recent work of Politzer *et al.* (2005). These authors, however, did not integrate $V_0(Z', N)$ from the critical value $Z_c(N)$ to Z, but from a chosen value of Z', *viz.* $N-1$. It was the existence of Z_c, at which V_0 is continuous, but non-analytic, which prompted Amovilli and March (2015) to separate into the form of Eq. (2.65). Politzer *et al.* (2005) analyzed a larger interval of nuclear charges. For the four neutral atoms studied by Amovilli and March (2015), their results show an energy which is about −0.2 Hartree lower than the exact one. In the case of Table 2.5, the discrepancy is much lower, ranging from 0.0002 Hartree for He, to 0.01 Hartree for B. It should be noted that Politzer *et al.* (2005) used the B3PW91 functional to compute V_0, while Amovilli and March (2015) used QMC.

It is worthy of note then that one could view Eq. (2.65) as a formally exact summation of the $1/Z$ expansion (Layzer, 1959) to all orders for the difference energy $E(Z, N) - E(Z_c, N)$. Unfortunately, so far no known exact analytic formula exists for $Z_c(N)$, though Cordero *et al.* (2013) give

Table 2.3 QMC results for Be-like isoelectronic series of atomic ions for about 15 values of fractional nuclear charge Ze. All quantities are in atomic units. After Amovilli and March (2015).

Z	E (VMC)	V_0 (VMC)	E (DMC)	ZV_0 (DMC)
2.81	−6.42351(3)	−5.3520(40)	−6.42963(7)	−14.9583(50)
2.84	−6.58742(2)	−5.4496(40)	−6.59033(6)	−15.5090(50)
2.845	−6.61478(3)	−5.4832(60)	−6.61785(6)	−15.5917(50)
2.85	−6.64236(4)	−5.5108(40)	−6.64480(5)	−15.7154(50)
2.855	−6.67001(3)	−5.5267(40)	−6.67227(5)	−15.7704(50)
2.86	−6.69905(3)	−5.5644(40)	−6.70082(6)	−15.8568(50)
2.90	−6.91779(5)	−5.6429(40)	−6.92404(8)	−16.3710(50)
3.00	−7.49685(4)	−5.9034(40)	−7.50053(7)	−17.6939(60)
3.50	−10.76575(5)	−7.1637(50)	−10.16827(8)	−25.0570(70)
3.95	−14.24689(6)	−8.3000(60)	−14.25006(11)	−32.7569(90)
4.00	−14.66461(7)	−8.4101(60)	−14.66791(11)	−33.6570(90)
4.05	−15.08962(7)	−8.5547(60)	−15.09286(11)	−34.5678(90)
4.10	−15.52029(7)	−8.6760(60)	−15.52382(12)	−35.5470(100)
4.30	−17.30648(8)	−9.1839(60)	−17.30963(12)	−39.4211(100)
4.50	−19.19290(8)	−9.6728(60)	−19.19638(12)	−43.5149(110)
4.70	−21.17892(9)	−10.1850(70)	−21.18294(13)	−47.8008(120)
5.00	−24.34664(7)	−10.9277(70)	−24.35036(13)	−54.6127(140)

Table 2.4 QMC results for B-like isoelectronic series of atomic ions for seven values of fractional nuclear charge Ze. All quantities are in atomic units. After Amovilli and March (2015).

Z	E (VMC)	V_0 (VMC)	E (DMC)	ZV_0 (DMC)
3.90	−13.83248(6)	−8.1922(50)	−13.83736(9)	−31.8800(70)
3.95	−14.24454(6)	−8.3012(50)	−14.24887(8)	−32.7797(70)
4.05	−15.08983(8)	−8.5864(60)	−15.09302(8)	−34.7415(100)
4.10	−15.52244(6)	−8.6979(60)	−15.52767(17)	−36.1269(100)
4.30	−17.34745(7)	−9.4313(60)	−17.35668(15)	−40.5444(100)
4.70	−21.34813(8)	−10.566(70)	−21.35827(18)	−49.5728(120)
5.00	−24.64189(8)	−11.414(70)	−24.65164(18)	−56.9498(130)

an empirical fit over a restricted range of N. Also the inequality

$$N - 1 \geq Z_c(N) \geq N - 2 \tag{2.67}$$

is well established.

In summary, the main achievement of the work of Amovilli and March (2015), reviewed in this Section, is to exhibit extensive QMC calculations for four isoelectronic series of light elements. These are finally utilized to determine the ground-state energies $E(Z, N)$ of such ions from Eq. (2.65).

Table 2.5 Displays magnitues of the two terms on the right-hand side of Eq. (2.65). All quantities are in atomic units. After Amovilli and March (2015).

$Z = N$	Z_c	$E_c(Z_c, N)$	$\int V_0$	$E(Z, Z)$	E (DMC)	ratio
2	0.911	−0.41494(2)	−2.4890	−2.9039	−2.90372	0.1667
3	2.000	−2.90372(4)	−4.5801	−7.4838	−7.47810	0.6340
4	2.856	−6.68112(5)	−7.9819	−14.6630	−14.66791	0.8370
5	4.000	−14.6679(10)	−9.9727	−24.6406	−24.65164	1.4708

Chapter 3

Small clusters of bosons

3.1 Efimov trimers

The *Efimov effect* (Efimov, 1970, 1971) relates to the existence of bound states of three bosons in dimensionality $d = 3$, even when the interaction is sufficiently weak to exclude bound states of two such particles. This in particular implies the counterintuitive fact that, if one particle is removed, the other two fall apart.[1] Therefore, one frequently depicts such a 3-body body state as a 'Borromean ring', *i.e.* a knot formed by three interlaced rings, such that it gets broken, if any single such ring is removed.[2] The condition for the existence of such 'Efimov trimers' is that the scattering length a exceeds the range r_0 of the interparticle potential, $|a| \gg r_0$, where the scattering length a is related to the first term of the expansion of the

[1]The Efimov effect addresses the (in)famous 'three-body problem' in nonrelativistic quantum mechanics. It is well-known that such a problem does not in general admit an analytical solution in classical mechanics, and it is all the more remarkable that it possesses an exact solution in quantum mechanics, where only few problems are known to be exactly solvable, and usually with no more than one particle. As Mattuck (1976) poses it, "How many bodies are required before we have a problem? ... In 18th century Newtonian mechanics, the 3-body problem was insoluble. With the birth of general relativity around 1910 and quantum electrodynamics in 1930, the 2- and 1-body problem was insoluble. And within modern quantum field theory, the problem of zero bodies (vacuum) is insoluble. So, if we are out after exact solutions, no bodies at all is already too many!" Also, as one of us (NHM) summarized the question during a seminar at the University of Catania in 2014, "It is 2 or 10^{23}: nothing in between!"

[2]The Borromean rings owe their name to the House of Borromeo, an aristocratic family mostly based in Northern Italy and renowned from the 14th century onwards, who used the knot in its coat of arms. This is still visible in several insignia, along with the family's motto 'humilitas', in the family's palace and gardens at Isola Bella in the Lake Maggiore, Italy, and in the Collegio Borromeo in Pavia, Italy.

following function of the s-wave phase shift δ_k (Schwinger, 1947)

$$k \cot \delta_k = -\frac{1}{a} + \frac{1}{2} r_s k^2 + \dots, \tag{3.1}$$

where $r_s \approx r_0$ is the effective range of the interaction (see Braaten and Hammer, 2006, and Sec. 6.2). In that case, there appears of sequence of 3-body bound states whose energies are distributed almost geometrically in the range between $\hbar^2/ma^2$ and $\hbar^2/mr_0^2$, where m is the mass of the particles. As $|a|$ is increased, new bound states appear in the spectrum at critical values of a that differ by multiplicative factors of e^{π/s_0}, where s_0 depends on the statistics and the mass ratios of the particles. In the case of identical bosons, s_0 is the solution to the transcendental equation

$$s_0 \cosh \frac{\pi s_0}{2} = \frac{8}{\sqrt{3}} \sinh \frac{\pi s_0}{6}, \tag{3.2}$$

resulting in $s_0 \approx 1.00624$, so that $e^{\pi/s_0} \approx 22.7$. As $|a|/r_0 \to \infty$, the number N of 3-body bound states increases asymptotically as

$$N \approx \frac{s_0}{\pi} \log \frac{|a|}{r_0}. \tag{3.3}$$

Since the (modulus of the) scattering length $|a|$ is larger than any other length in the problem, several properties of the system are universal, in the sense that they should not depend on any specific detail of the 2-body potential. The simplest such property is that in the resonant limit in which there are infinitely many arbitrarily-shallow 3-body bound states, the ratio of the binding energies of the successive bound states tends to a universal number as the threshold is approached:

$$E_T^{(n+1)}/E_T^{(n)} = e^{-2\pi/s_0} \approx 1/515.03, \quad n \gg 1. \tag{3.4}$$

Moreover, one expects the Efimov effect to be operating in various systems and at widely different energy scales, such as within nuclear matter, in atomic and molecular systems, and in the solid state. This scale invariance has suggested renormalization group techniques may be relevant to address the Efimov problem (Horinouchi and Ueda, 2015).

The universal low-energy behaviour of atoms with large scattering length has many features in common with critical phenomena (Chapter 9). The scattering length a plays a role analogous to the correlation length ξ. The region of large scattering length is analogous to the critical region, and the resonant limit where $|a| \to \infty$ is analogous to the critical point. In the critical region, $|a|$ is the most important length scale for low-energy observables.

Efimov's effect had been foreshadowed in an earlier paper by Thomas (1935), who had considered the related problem of the stability of the triton, *i.e.* the nucleus of a ^{3}H atom (or tritium), which is composed of three nucleons (one proton and two neutrons: therefore, three bosons). Thomas (1935) studied the bound state corresponding to a 2-body potential with depth V_0 and range r_0, such that $V_0 \to \infty$ and $r_0 \to 0$, while keeping the binding energy E fixed. Using a simple variational argument, Thomas (1935) showed that the energy E_T of the lowest 3-body bound state diverges to ∞ is this limit. Thus the spectrum of 3-body bound states is unbounded from below. This again implies that there can exist 3-body bound states, even though 2-body bound states are not possible. The importance of the result by Thomas (1935) is however limited by the fact that the binding energy and other properties of this deepest 3-body bound state may depend on the details of the interaction potential.

3.1.1 *Hyperspherical coordinates*

While the 2-body problem can be very often reduced to an effective 1-body problem, by use of centre-of-mass and relative coordinates, and of the effective mass, in the case of the 3-body problem it is advantageous to employ the so-called hyperspherical coordinates (Nielsen *et al.*, 2001; Braaten and Hammer, 2006). These are defined in terms of the Jacobi coordinates, which for a system of three particles of equal mass read $(i, j, k = 1, 2, 3)$

$$\mathbf{r}_{ij} = \mathbf{r}_i - \mathbf{r}_j, \tag{3.5a}$$

$$\mathbf{r}_{ij,k} = \mathbf{r}_k - \frac{1}{2}(\mathbf{r}_i + \mathbf{r}_j). \tag{3.5b}$$

Then the hyperradius R is defined as the root-mean-square distance among the three particles, and can be expressed in terms of the Jacobi coordinates above through

$$R^2 = \frac{1}{3}(r_{12}^2 + r_{23}^2 + r_{31}^2) = \frac{1}{2}r_{ij}^2 + \frac{2}{3}r_{ij,k}^2. \tag{3.6}$$

Roughly speaking, the hyperradius R is 'small' is all three particles are close together, and it is 'large' if any single particle is far from any other. The Delves hyperangle $\alpha_k \in [0, \frac{\pi}{2}]$ is defined as (Delves, 1960)

$$\alpha_k = \operatorname{atan}\left(\frac{\sqrt{3}r_{ij}}{2r_{ij,k}}\right), \tag{3.7}$$

where (i, j, k) here is a permutation of $(1, 2, 3)$. One finds $\alpha_k \sim 0$ when particle k is far from particles i and j, and $\alpha_k \sim \frac{\pi}{2}$ where particle k is near the centre of mass of particles i and j. Then the moduli of the Jacobi coordinates can be expressed as

$$r_{ij} = \sqrt{2} R \sin\alpha_k, \tag{3.8a}$$

$$r_{ij,k} = \sqrt{\frac{3}{2}} R \cos\alpha_k. \tag{3.8b}$$

The other two hyperangles can be (symmetrically) obtained from

$$\sin^2\alpha_i = \frac{1}{4}\sin^2\alpha_k + \frac{3}{4}\cos^2\alpha_k + \frac{\sqrt{3}}{2}\sin\alpha_k\cos\alpha_k\,\hat{\mathbf{r}}_{ij}\cdot\hat{\mathbf{r}}_{ij,k}, \tag{3.9}$$

where again (i,j,k) is a permutation of $(1,2,3)$, and $\left|\frac{\pi}{3} - \alpha_k\right| < \alpha_i < \frac{\pi}{2} - \left|\frac{\pi}{6} - \alpha_k\right|$. One may verify the identity

$$\sin^2\alpha_1 + \sin^2\alpha_2 + \sin^2\alpha_3 = \frac{3}{2}. \tag{3.10}$$

The volume element in the Jacobi coordinates reads

$$d^3\mathbf{r}_{ij}d^3\mathbf{r}_{ij,k} = \frac{3\sqrt{3}}{4}R^5\sin^2(2\alpha_k)\,dR\,d\alpha_k\,d\Omega_{ij}\,d\Omega_{ij,k}, \tag{3.11}$$

where $d\Omega_{ij}$, $d\Omega_{ij,k}$ are the differential solid angles associated with the unit vectors $\hat{\mathbf{r}}_{ij}$ and $\hat{\mathbf{r}}_{ij,k}$, respectively.

3.1.2 *Schrödinger equation for the 3-body problem: adiabatic hyperspherical approximation*

The Schrödinger equation for a 3-body system can be written most generally as

$$(T+V)\Psi = E\Psi \tag{3.12}$$

where $\Psi = \Psi(\mathbf{r}_1, \mathbf{r}_2, \mathbf{r}_3)$ is the wavefunction, and

$$T = -\frac{\hbar^2}{2m}\sum_{i=1}^{3}\nabla_i^2, \tag{3.13a}$$

$$V = V(\mathbf{r}_1, \mathbf{r}_2, \mathbf{r}_3) \tag{3.13b}$$

are the total kinetic and potential operators, respectively. If the interaction potential V is translationally invariant, then the number of independent coordinates for the wavefunction in the centre-of-mass reference frame reduces to six. A convenient choice is that of the hyperradius R, and five hyperangular variables, $\Omega \equiv (\alpha_k, \hat{\mathbf{r}}_{ij}, \hat{\mathbf{r}}_{ij,k})$. One finds

$$T = T_R + T_{\alpha_k}, \tag{3.14a}$$

$$V = \frac{\Lambda_{ij,k}^2}{2mR^2} + V(R,\Omega), \tag{3.14b}$$

with

$$T_R = -\frac{\hbar^2}{2m}\left(\frac{\partial^2}{\partial R^2} + \frac{5}{R}\frac{\partial}{\partial R}\right) \tag{3.15a}$$

$$= -\frac{\hbar^2}{2m} R^{-5/2}\left(\frac{\partial^2}{\partial R^2} - \frac{15}{4}\frac{1}{R^2}\right) R^{5/2}, \tag{3.15b}$$

$$T_{\alpha_k} = -\frac{\hbar^2}{2m}\frac{1}{R^2}\left(\frac{\partial^2}{\partial \alpha_k^2} + 4\cot(2\alpha_k)\frac{\partial}{\partial \alpha_k}\right) \tag{3.15c}$$

$$= -\frac{\hbar^2}{2m}\frac{1}{R^2}\left(\frac{\partial^2}{\partial \alpha_k^2} + 4\right)\sin(2\alpha_k), \tag{3.15d}$$

$$\Lambda^2_{ij,k} = \frac{\mathbf{L}^2_{ij}}{\sin^2\alpha_k} + \frac{\mathbf{L}^2_{ij,k}}{\cos^2\alpha_k}. \tag{3.15e}$$

The adiabatic hyperspherical representation then consists in expanding the wavefunction in the new variables $\Psi = \Psi(R,\Omega)$ as

$$\Psi(R,\Omega) = R^{-5/2}\sum_n f_r(R)\Phi_n(R,\Omega), \tag{3.16}$$

with

$$\int d\Omega\, \Phi_n^*(R,\Omega)\Phi_m(R,\Omega)$$
$$\equiv \int_0^{\pi/2} d\alpha_k\, \sin^2(2\alpha_k)\int d\Omega_{ij}\int d\Omega_{ij,k}\, \Phi_n^*(R,\Omega)\Phi_m(R,\Omega)$$
$$= \delta_{nm}, \quad \forall R. \tag{3.17}$$

At a given hyperradius R, the functions $\Phi_n(R,\Omega)$ are solutions to the parametric eigenvalue equation

$$\left[T_{\alpha_k} + \frac{\Lambda^2_{ij,k}}{2mR^2} + V(R,\Omega)\right]\Phi_n(R,\Omega) = V_n(R)\Phi_n(R,\Omega), \tag{3.18}$$

where the eigenvalue $V_n(R)$ parametrically depends on R, and can be interpreted as an effective potential. Projecting onto Φ_m and making use of the orthonormality condition, Eq. (3.17), the Schrödinger equation reduces to a coupled set of eigenvalue equations for the hyperradial functions $f_n(R)$

$$\left[-\frac{\hbar^2}{2m}\left(\frac{\partial^2}{\partial R^2} - \frac{15}{4}\frac{1}{R^2}\right) + V_n(R)\right] f_n(R)$$
$$+ \sum_m \left[2U_{nm}(R)\frac{\partial}{\partial R} + W_{nm}(R)\right] f_m(R) = Ef_n(R), \tag{3.19}$$

where $U_{nm}(R)$ and $W_{nm}(R)$ are effective coupling potentials defined by

$$U_{nm}(R) = -\frac{\hbar^2}{2m}\int d\Omega\, \Phi_n^*(R,\Omega)\frac{\partial}{\partial R}\Phi_m(R,\Omega), \tag{3.20a}$$

$$W_{nm}(R) = -\frac{\hbar^2}{2m}\int d\Omega\, \Phi_n^*(R,\Omega)\frac{\partial^2}{\partial R^2}\Phi_m(R,\Omega). \tag{3.20b}$$

The adiabatic hyperspherical approximation (Macek, 1968; Braaten and Hammer, 2006) then consists in neglecting off-diagonal terms, which allows to decouple the equations as

$$\left[-\frac{\hbar^2}{2m}\left(\frac{\partial^2}{\partial R^2} - \frac{15}{4}\frac{1}{R^2}\right) + V_n(R)2U_{nn}(R)\frac{\partial}{\partial R} + W_{nn}(R)\right] f_m(R) = E f_n(R). \tag{3.21}$$

3.1.3 *Low-energy limit: Faddeev equations*

Assuming for the sake of simplicity that the interaction potential V can be decomposed as the sum of three (translationally invariant) 2-body potentials as

$$V(\mathbf{r}_1, \mathbf{r}_2, \mathbf{r}_3) = V(r_{12}) + V(r_{23}) + V(r_{31}), \tag{3.22}$$

one then looks for solutions in the form (Faddeev, 1961; Braaten and Hammer, 2006)

$$\Psi(\mathbf{r}_1, \mathbf{r}_2, \mathbf{r}_3) = \psi^{(1)}(\mathbf{r}_{23}, \mathbf{r}_{23,1}) + \psi^{(2)}(\mathbf{r}_{31}, \mathbf{r}_{31,2}) + \psi^{(3)}(\mathbf{r}_{12}, \mathbf{r}_{12,3}). \tag{3.23}$$

The corresponding equations (Faddeev equations) are then

$$\left(T_R + T_{\alpha_1} + \frac{\Lambda_{23,1}^2}{2mR^2}\right)\psi^{(1)} + V(r_{23})\left(\psi^{(1)} + \psi^{(2)} + \psi^{(3)}\right) = E\psi^{(1)}, \tag{3.24}$$

and those obtained by cyclic permutation of the wavefunctions indexes. In the limit $\psi^{(2)} = \psi^{(3)} = 0$, the remaining equation for $\psi^{(1)}$ reduces to an equation for the 2-body problem for particles 2 and 3. Thus, the Faddeev wavefunction $\psi^{(1)}$ can naturally account for correlations between particles 2 and 3 as $R \to \infty$, when particles 2 and 3 are both at a large distance from particle 1.

Restricting to the case of total angular momentum $L = 0$ and expanding $\psi^{(1)}$ as

$$\psi^{(1)}(\mathbf{r}_{23}, \mathbf{r}_{23,1}) = \sum_{\ell_x, m_x}\sum_{\ell_y, m_y} f^{(1)}_{\ell_x m_x, \ell_y m_y}(R, \alpha_1) Y_{\ell_x m_x}(\hat{\mathbf{r}}_{23}) Y_{\ell_y m_y}(\hat{\mathbf{r}}_{23,1}), \tag{3.25}$$

where ℓ_x, m_x and ℓ_y, m_y are the quantum numbers associated with orbital angular momentum in the 23 and 23,1 systems, respectively, at low energy one may safely retain only the terms with $\ell_x = \ell_y = 0$. The wavefunction solution of the Schrödinger equation then reduces to

$$\Psi(\mathbf{r}_1, \mathbf{r}_2, \mathbf{r}_3) = \psi(R, \alpha_1) + \psi(R, \alpha_2) + \psi(R, \alpha_3), \tag{3.26}$$

and the corresponding Faddeev equations read

$$(T_R + T_{\alpha_1} - E)\psi(R, \alpha_1) + V(\sqrt{2}R \sin\alpha_1)[\psi(R, \alpha_1) + \psi(R, \alpha_2) - \psi(R, \alpha_3)] = 0, \tag{3.27}$$

together with those equations obtained by a cyclic permutation over α_1, α_2, and α_3. These three equations can be further simplified and reduced to a single equation using the fact that the averages of $\psi(R, \alpha_2)$ and $\psi(R, \alpha_3)$ over the angular variables $\hat{\mathbf{r}}_{23}$ and $\hat{\mathbf{r}}_{23,1}$ can be expressed as an integral operator acting on $\psi(R, \alpha_1)$

$$\begin{aligned} \langle \psi(R, \alpha_2) \rangle_{\hat{\mathbf{r}}_{23}, \hat{\mathbf{r}}_{23,1}} &= \langle \psi(R, \alpha_3) \rangle_{\hat{\mathbf{r}}_{23}, \hat{\mathbf{r}}_{23,1}} \\ &= \frac{2}{\sqrt{3}} \int_{|\alpha_1 - \pi/3|}^{\pi/2 - |\alpha_1 - \pi/6|} \frac{\sin(2\alpha')}{\sin(2\alpha_1)} \psi(R, \alpha')\, d\alpha'. \end{aligned} \tag{3.28}$$

This results in the so-called low-energy Faddeev equation, which is an integro-differential equation for $\psi(R, \alpha)$, *viz.*

$$\begin{aligned} &(T_R + T_\alpha - E)\psi(R, \alpha) \\ &= -V(\sqrt{2}R \sin\alpha) \left[\psi(R, \alpha) + \frac{4}{\sqrt{3}} \int_{|\alpha - \pi/3|}^{\pi/2 - |\alpha - \pi/6|} \frac{\sin(2\alpha')}{\sin(2\alpha)} \psi(R, \alpha')\, d\alpha' \right]. \end{aligned} \tag{3.29}$$

This can in turn be solved using the hyperspherical expansion

$$\psi(R, \alpha) = \frac{1}{R^{5/2} \sin(2\alpha)} \sum_n f_n(R) \phi_n(R, \alpha), \tag{3.30}$$

where ϕ_n is a complete set of functions, solving the eigenproblem

$$\begin{aligned} &-\left(\frac{\partial^2}{\partial \alpha^2} + \lambda_n(R) \right) \phi_n(R, \alpha) \\ &= -\frac{2mR^2}{\hbar^2} V(\sqrt{2}R \sin\alpha) \left(\phi_n(R, \alpha) + \frac{4}{\sqrt{3}} \int_{|\alpha - \pi/3|}^{\pi/2 - |\alpha - \pi/6|} \phi_n(R, \alpha')\, d\alpha' \right), \end{aligned} \tag{3.31}$$

with boundary conditions $\phi_n(R,0) = \phi_n(R,\pi/2) = 0$. Each eigenvalue $\lambda_n(R)$ then defines a distinct channel potential for the hyperradial variable as

$$V_n(R) = [\lambda_n(R) - 4]\frac{\hbar^2}{2mR^2}. \tag{3.32}$$

The scalar product between two such eigenfunctions belonging to different eigenvalues define a matrix $G_{nm}(R)$ parametrically depending on the hyperradius R, defined as

$$G_{nm}(R) = \int_0^{\pi/2} d\alpha\, \phi_n^*(R,\alpha)\phi_m(R,\alpha). \tag{3.33}$$

Such a matrix is in general non-trivial (*i.e.* it is not the identity matrix), as $\phi_n(R,\alpha)$ are eigenfunctions of a non-Hermitean operator. By inserting the expansion, Eq. (3.30), into the low-energy Faddeev equation, Eq. (3.29), projecting onto $\phi_n^*(R,\alpha)$ and then multiplying by the inverse of the matrix $G_{nm}(R)$, one obtains a set of coupled eigenvalue equations for the hyperradial wavefunctions $f_n(R)$

$$\left[-\frac{\hbar^2}{2m}\left(\frac{\partial^2}{\partial R^2} - \frac{15}{4}\frac{1}{R^2}\right) + V_n(R)\right] f_n(R) \\ + \sum_m \left(2P_{nm}(R)\frac{\partial}{\partial R} + Q_{nm}(R)\right) f_m(R) = E f_n(R), \tag{3.34}$$

where the coupling potentials $P_{nm}(R)$ and $Q_{nm}(R)$ are defined by

$$P_{nm}(R) = -\frac{\hbar^2}{2m}\sum_k G_{nk}^{-1}(R)\int_0^{\pi/2} d\alpha\, \phi_k^*(R,\alpha)\frac{\partial}{\partial R}\phi_m(R,\alpha), \tag{3.35a}$$

$$Q_{nm}(R) = -\frac{\hbar^2}{2m}\sum_k G_{nk}^{-1}(R)\int_0^{\pi/2} d\alpha\, \phi_k^*(R,\alpha)\frac{\partial^2}{\partial R^2}\phi_m(R,\alpha). \tag{3.35b}$$

Eqs. (3.34) are similar to Eqs. (3.19) for the adiabatic hyperspherical representation of the 3-body Schrödinger equation. The main difference here is that hyperangular variables have been averaged out, due to the simplifying assumption that the subsystems angular momenta may be neglected. Moreover, if the eigenvalues $\lambda_n(R)$ were constant or varying sufficiently slowly as a function of R, $P_{nm}(R)$ and $Q_{nm}(R)$ can be neglected, and the eigenvalue equations, Eqs. (3.34), decouple, thereby reducing to a radial Schrödinger equation in each of the channel defined by the hyperspherical potentials $V_n(R)$:

$$\left[-\frac{\hbar^2}{2m}\left(\frac{\partial^2}{\partial R^2} - \frac{15}{4}\frac{1}{R^2}\right) + V_n(R)\right] f_n(R) = E f_n(R). \tag{3.36}$$

3.1.3.1 *Solutions in special cases*

To gain further insight, in the limiting case of a vanishing (or constant) 2-body potential $V(r)$, the eigenvalues turn out to be independent of R and explicitly read $\lambda_n(R) = 4(n+1)^2$, $n = 0, 1, 2, \ldots$, with corresponding eigenfunctions $\phi_n(R, \alpha) = \sin[2(n+1)\alpha]$, and hyperspherical potentials $V_n(R) = 4n(n+2)\hbar^2/(2mR^2)$.

Another limiting case where some analytical insight can be obtained is that in which the 2-body potential $V(r)$ is short-ranged (Braaten and Hammer, 2006).

In the case of large scattering length $|a|$, Faddeev equation, Eq. (3.29), can be reduced to the exact transcendental equation for $\lambda_n(R)$ (Efimov, 1971; Fedorov and Jensen, 1993; Braaten and Hammer, 2006)

$$\cos\left(\frac{\pi}{2\sqrt{\lambda}}\right) - \frac{8}{\sqrt{3\lambda}} \sin\left(\frac{\pi}{6\sqrt{\lambda}}\right) = \sqrt{\frac{2}{\lambda}} \frac{R}{a} \sin\left(\frac{\pi}{2\sqrt{\lambda}}\right), \tag{3.37}$$

with infinitely many solutions $\lambda = \lambda_n(R)$, corresponding to the hyperangular wavefunctions

$$\phi_n(R, \alpha) = \sin\left[\left(\frac{\pi}{2} - \alpha\right) \sqrt{\lambda_n(R)}\right] \tag{3.38}$$

(see Nielsen *et al.*, 2001; Braaten and Hammer, 2006, for the generalization to the case $\ell_x = 0$, $\ell_y = 1$). Physical channel eigenvalues coming from Eq. (3.37) have to be obtained numerically, but several asymptotic behaviours can be obtained analytically in the limits $R \to 0$ and $R \to \infty$ (Braaten and Hammer, 2006). In particular, the resulting low-R asymptotic behaviour for the corresponding hyperspherical potential is

$$V_n(R) \approx (\lambda_n(0) - 4) \frac{\hbar^2}{2mR^2}, \quad R \ll |a|, \tag{3.39}$$

thereby resulting in an attractive potential for the lowest channel ($n = 0$), and repulsive for all the other channels.

3.1.4 *Efimov states in the resonant limit*

The Efimov effect consists in the existence of infinitely many 3-body bound states in the resonant limit of large scattering length, $a \to \pm\infty$ (Efimov, 1971; Macek, 1986; Braaten and Hammer, 2006). In this limit, the adiabatic hyperspherical approximation is accurate at all finite values of R, and in particular for bound states $f_n(R)$ with an exponentially decreasing behaviour as $R \to \infty$. These can be found for $n = 0$, which is the only attractive channel, with $\lambda_0(R) = -s_0^2$, where $s_0 \approx 1.00624$ has been defined in Eq. (3.2). Then, the Schrödinger wavefunction in the centre-of-mass reference frame reduces to

$$\Psi(\mathbf{r}_1, \mathbf{r}_2, \mathbf{r}_3) = R^{-5/2} f_0(R) \sum_{i=1}^{3} \frac{\sinh[s_0 \left(\frac{\pi}{2} - \alpha_i\right)]}{\sin(2\alpha_i)}, \tag{3.40}$$

with

$$-\frac{\hbar^2}{2m} \left(\frac{\partial^2}{\partial R^2} + \frac{s_0^2 + \frac{1}{4}}{R^2} \right) f_0(R) = E f_0(R). \tag{3.41}$$

The boundary condition can be provided in the form of a matching condition on the logarithmic derivative $R_0 f'(R_0)/f(R_0)$ at an intermediate hyperradius R_0. The exponentially decreasing solution is

$$f_0(R) = \sqrt{R} K_{is_0}(\sqrt{2}\kappa R), \tag{3.42}$$

where κ is a binding wave number related to the trimer binding energy E_T by

$$E_T = \frac{\hbar^2 \kappa^2}{m}, \tag{3.43}$$

$K_{is_0}(z)$ is a Bessel function with purely imaginary index, so that

$$f_0(R) \approx -\left(\frac{\pi R}{s_0 \sinh(\pi s_0)} \right)^{1/2} \sin\left(s_0 \log(\kappa R) + \alpha_0\right), \quad \kappa R \ll 1, \tag{3.44}$$

where

$$\alpha_0 = -\frac{1}{2} s_0 \log 2 - \frac{1}{2} \arg \frac{\Gamma(1 + i s_0)}{\Gamma(1 - i s_0)}. \tag{3.45}$$

The trimer binding wave numbers can then be presented in the scaling form

$$\kappa_n = (e^{-\pi/s_0})^{n-n_0} e^{-\alpha_0/s_0} \Lambda_0, \tag{3.46}$$

where Λ_0 is implicitly defined by

$$s_0 \log(\Lambda_0 R_0) = \mathrm{acot}\left(\cot[s_0 \log(\kappa R_0) + \alpha_0]\right), \tag{3.47}$$

and n_0 is an arbitrary integer arising from the choice of the branch in the cotangent in the equation above. This implies the geometric behaviour of the Efimov spectrum, which can be cast into the form

$$E_T^{(n)} = (e^{-2\pi/s_0})^{n-n_*} \frac{\hbar^2 \kappa_*^2}{m}, \tag{3.48}$$

with $\kappa_* = \kappa_{n=n_*}$, and $s_0 \log \kappa_* = s_0 \log \Lambda_0 - \alpha_0 \mod \pi$. It follows that the binding energies of two successive Efimov states have the ratio $e^{2\pi/s_0} \approx 515.03$, which is one of the most characteristic signature of the Efimov effect, already anticipated in Eq. (3.4) (Fig. 3.1).

In an Efimov state, $f_0^2(R)$ is to be interpreted as a probability distribution for the hyperradius. Using such a probability distribution, one finds for the mean square hyperradius in the nth Efimov state

$$\langle R^2 \rangle_n = \frac{2(1 + s_0^2)}{3\kappa_n^2} = (e^{2\pi/s_0})^{n-n_*} \frac{2(1 + s_0^2)}{3\kappa_*^2}. \tag{3.49}$$

Thus, $\langle R^2 \rangle_n$ increases with increasing n, by a factor $e^{\pi/s_0} \approx 22.7$.

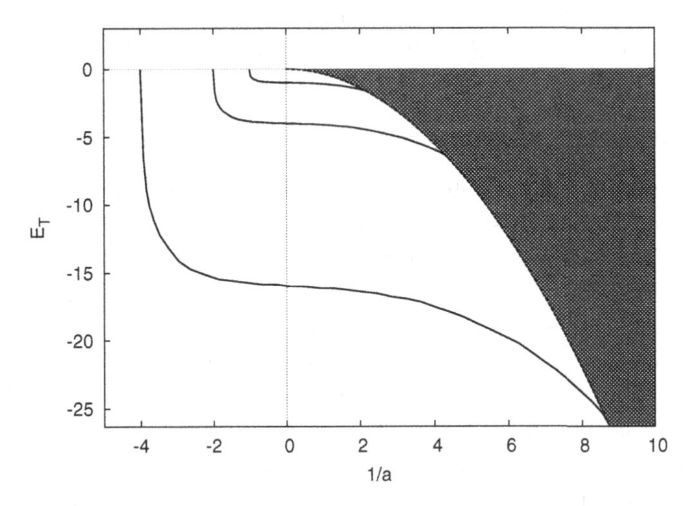

Fig. 3.1 Schematic behaviour of the binding energy E_T for Efimov trimer states as a function of inverse scattering length $1/a$. A nonlinear scale has been applied to both axes, to enhance the clarity of the overall plot. The three-particle continuum corresponds to $E_T > 0$ (unbound three particles), while the hatched region for $a > 0$ and $-\hbar^2/2ma^2 < E_T < 0$ corresponds to the dimer-particle continuum (one bound pair, one unbound particle). The critical scattering length a_* is asymptotically reached when an energy curve for a trimer bound state approaches the dimer-particle continuum. There actually are infinitely many arbitrarily shallow Efimov states, with an accumulation point at zero energy and an asymptotic discrete scaling symmetry with discrete scaling factor $e^{\pi/s_0} \approx 22.7$.

3.2 Efimov resonances and experimental evidence thereof

During the nearly four decades after its theoretical prediction (Efimov, 1971), the observation of the Efimov effect has been proposed in several physical systems, ranging from nuclear matter, to atoms and molecules, halo nuclei (Jensen *et al.*, 2004), trimers of ^{4}He atoms (Lim *et al.*, 1977), and ultracold quantum gases (see Ferlaino and Grimm, 2010; Ferlaino *et al.*, 2011, for a review). The energy scales involved in these systems are also quite different, ranging from MeV in nuclear physics to peV for ultracold atomic systems.

In nearly all cases, the effect is connected with an Efimov resonance, occurring when the scattering length a is close to the critical value, a_* say, corresponding to the threshold energy for the production of a particle-dimer pair (Fig. 3.1). However, in the halo states of atomic nuclei, the occurrence of Efimov states is prevented by the Coulomb interaction (Hammer *et al.*,

2013). The case of trimers of helium atoms and of other atomic species is also controversial (Brühl *et al.*, 2005; Baccarelli *et al.*, 2000).

3.2.1 *Relevance of Efimov trimers for helium clusters*

In particular, it seems of interest here to draw attention to the potential importance of the chemistry of small helium clusters, and in particular of $^4He_2$ and $^4He_3$ for constructing such a (say variational) many-body ground-state wavefunction of liquid ^{4}He. Such a discussion began in the 1980s (March, 1983a; Ghassib and Chester, 1984; March, 1984) and interest in it has been re-opened by quite recent studies of helium dimers and trimers in free space. Thus Grisenti *et al.* (2000) have reported in the vapour, using a diffraction transmission grating technique, the existence of a helium dimer with a 'bond length' of ~ 50 Å, and as a consequence a very tiny binding energy. The He trimer in free space has also been identified experimentally. Of course, no such free-space information can be quite decisive in regard to the dense liquid. However, experience in nuclear physics with the 'α particle' model points to the potential usefulness of building especially ^{4}He trimers into a ground-state many-body wavefunction for liquid He II.

As discussed by March and by Ghassib and Chester (1984), it is possible in such a treatment of the ground state of liquid ^{4}He that Efimov states (Efimov, 1970, 1971) will play a role. It is relevant in this context, though the system is different, to refer to the investigation of Bulgac (2002) on a quantum liquid droplet. He concludes, under some conditions for which the interested reader can refer to the original article, that the effect of three-body correlations can be subsumed into the ground-state energy by using the Efimov states.

Other molecular systems which could be candidates for the observation of Efimov trimers are benzene and hexafluorobenzene (Squire and March, 2014). The possibility of obtaining spectroscopic evidence of trimers in such molecular systems is somehow corroborated by the recently observed photoemission of Cooper pairs from aromatic hydrocarbons (Wehlitz *et al.*, 2012).

3.2.2 *Feshbach resonances in ultracold atoms*

A breakthrough was provided by the suggestion that an Efimov resonance could be obtained by the proximity to a Feshbach resonance (Fano, 1935, 1961; Feshbach, 1958, 1962) in ultracold atoms (Braaten and Hammer,

2001) (see Chin *et al.*, 2010, for a review). There, at variance with nuclear physics systems, coupling to an external magnetic field enables to finely tune the scattering length of the 2-body interaction. A Feshbach resonance requires the existence of bound states in an *open* channel, characterized by a 2-body potential which goes to zero as $r \to \infty$ (say, a van der Waals potential, behaving as $\sim -1/r^6$ as $r \to \infty$). If there exists another channel, say for particles in a different spin state from those in the open channel, characterized by a potential tending to a finite, positive value as $r \to \infty$, then this channel is *closed,* as the only states in it which are energetically accessible are bound states. The weak coupling between the two channels allows transitions for pairs of particles. A large scattering length for particles in the open channel can be obtained by tuning the closed channel potential so as to bring one of its bound states close to the threshold for the open channel. In the case of ultracold alkali atoms (Chin *et al.*, 2010), the open channel consists of a pair of atoms in a specific hyperfine state. The closed channel consists of a pair of atoms in different hyperfine states, with a higher scattering threshold. The weak coupling between the two channels is then provided by the hyperfine interaction. A magnetic field B can then be employed to exploit the different magnetic moments of the two hyperfine states, and reduce the energy gap between the scattering thresholds down to nearly a degeneracy. Near a Feshbach resonance, the scattering length can be approximated by

$$a(B) = a_{bg} \left(1 - \frac{\Delta}{B - B_0} \right), \tag{3.50}$$

where a_{bg} is the scattering length far from the resonance, B_0 defines the Feshbach resonance, and Δ determines the resonance width. Thus, by tuning B close to its value B_0 at the resonance, one can make $|a|$ arbitrarily large, and in particular much larger than the interaction range, in such a way as to realize the conditions under which an Efimov resonance may be observed.

The first observation of the Efimov effect occurred in an optically trapped gas of Cs atoms, at temperatures below 10 nK (Kraemer *et al.*, 2006). There, a triatomic Efimov resonance was observed, as a prominent peak in the 3-body recombination rate. This could be related to a negative value $a_* < 0$ of the scattering length, corresponding to the intersection of the Efimov trimer binding energy curve with the three-particle continuum in Fig. 3.1.

More recent results were associated with an atom-dimer resonance with $a_* > 0$ (merging point between the Efimov trimer curve and the dimer-particle continuum in Fig. 3.1) in ultracold cesium at 40 nK (Knoop *et al.*, 2009). In a later experiment employing an ultracold gas of ^{39}K (Zaccanti *et al.*, 2009), a triatomic resonance for a negative value of the scattering length was again detected, similar to what had been observed with cesium. But two consecutive minima in the 3-body recombination rate were detected, occurring at positive values of the scattering rate in the ratio 25 ± 4. This is agreement, within the experimental error, with the theoretical value of ≈ 22.7 predicted by Efimov, cf. Eq. (3.2), and provides therefore an experimental evidence of Efimov's universal scaling law. Further experiments in ^{7}Li (Gross *et al.*, 2009; Pollack *et al.*, 2009) and in heteronuclear ^{6}Li-^{133}Cs mixtures (Tung *et al.*, 2014) confirmed these results, and suggested that even 4-body resonances might exist, thereby re-kindling the few-body problem (Modugno, 2009; Ferlaino and Grimm, 2010) (but cf. Amado and Greenwood, 1973).

Chapter 4

Anyon statistics with models

The necessity of quantum statistics is motivated by the indistinguishability of identical particles (Reichl, 2009; Fetter and Walecka, 2004). In dimensions $d = 3$ (and possibly higher), the fact that the quantum state $|\psi\rangle$, say, of two identical, non-interacting particles should be invariant with respect to their adiabatic exchange, leads to the representation of $|\psi\rangle$ either as a symmetric, or an antisymmetric superposition of the quantum states of each particle, *i.e.*

$$|\psi\rangle = \frac{1}{\sqrt{2}} \left(|n_1\rangle |n_2\rangle \pm |n_2\rangle |n_1\rangle \right). \tag{4.1}$$

Correspondingly, particles can be classified as either bosons or fermions, respectively, depending on whether they obey the Fermi-Dirac (FD) or the Bose-Einstein (BE) quantum statistics. Eq. (4.1) implies that $P_{1\leftrightarrow 2}|\psi\rangle = \pm|\psi\rangle$, where $P_{1\leftrightarrow 2}$ permutes particles 1 and 2, for bosons (upper sign) and fermions (lower sign), respectively. In particular, if $n_1 = n_2$, then $|\psi\rangle = 0$ for fermions, which is Pauli exclusion principle. The spin-statistics theorem then fixes the spin of fermions and bosons as a semi-odd-integer or integer multiple of $\hbar$, respectively (Fierz, 1939; Pauli, 1940).

Such a restriction, however, can be overcome in $d \leq 2$. In particular, the configuration space of identical particles in $d = 2$ is infinitely connected, whereas it is only doubly connected in $d \geq 3$ dimensions. Therefore, in two dimensions *any* statistics would be possible, in principle, with properties somehow interpolating between those of Fermi-Dirac and of Bose-Einstein statistics. In particular, the fact that $\| P_{1\leftrightarrow 2}|\psi\rangle \| = \| |\psi\rangle \|$ implies in general that $P_{1\leftrightarrow 2}|\psi\rangle = e^{i\varphi}|\psi\rangle$, with $\varphi = \alpha\pi$ a generic (global) phase, not necessarily equal to 0 or π, as is the case for bosons and fermions, respectively. Since this phase is expressed as a fraction α of the phase π for fermions, one also speaks of *fractional* exchange *statistics* (Sec. 4.1). In this context, the number α, ranging from $\alpha = 0$ (bosons) to $\alpha = 1$ (fermions), is also referred to as the statistical parameter. As mentioned above, fractional exchange statistics is usually restricted to two spatial dimensions ($d = 2$) (see Forte, 1992, for a review, and Sec. 4.1 below). However, fractional exchange statistics can be formalized, to some extent, also in $d = 1$ (Ha, 1996; Calabrese and Mintchev, 2007).

Current interest in fractional statistics (Leinaas and Myrheim, 1977; Wilczek, 1982a,b) is motivated by its possible relevance for fractional quantum Hall effect (Laughlin, 1983a,b; Halperin, 1984; Stormer *et al.*, 1999) and high-temperature superconductivity (Laughlin, 1988b,a). In this context, an early application was made by Lea *et al.* (1991, 1992), who discussed the de Haas–van Alphen oscillatory orbital magnetism of a two-dimensional electron gas (2DEG). More recently, it has been proposed that noise experiments in quantum Hall fluids might reveal the existence of elementary excitations obeying fractional statistics (Kim *et al.*, 2005; Camino *et al.*, 2005). Also, topological excitations such as vortex rings (anyonic loops) in the three-dimensional chiral spin liquids may obey fractional non-Abelian statistics (Si and Yu, 2008).

A different concept of fractional statistics, namely *fractional* exclusion *statistics,* is based on the structure of the Hilbert space, rather than the configuration space, of the particle assembly, and is therefore not restricted to $d \leq 2$ (Sec. 4.2). Fractional exclusion statistics has been introduced by Haldane (1991), who considered the ratio

$$g = -\frac{\Delta D}{\Delta N}, \tag{4.2}$$

where ΔD denotes the change in size of the subset of available states in the Hilbert space corresponding to a change ΔN of the number of particles (*i.e.* elementary excitations). Clearly, one has again $g = 0$ for bosons and $g = 1$ for fermions, the latter being a consequence of Pauli exclusion principle. At variance with anyons, particles obeying fractional exclusion statistics are usually dubbed g-ons or exclusons. The distribution function for fractional exclusion statistics has been derived by Wu (1994) (see Sec. 4.2 below). A limiting form of the same distribution function had been derived by March *et al.* (1993) within an approximate chemical collision model (see also March, 1993, 1997). Following Wu's distribution function, the thermodynamics of an ideal gas with fractional exclusion statistics has been studied in some detail in arbitrary dimensions (Joyce *et al.*, 1996; Iguchi, 1997).

The relation between fractional exchange and exclusion statistics is elusive. In particular, it has been emphasized (Nayak and Wilczek, 1994) that in order to derive a consistent statistical mechanics for g-ons, Haldane's generalized exclusion principle, Eq. (4.2), must hold locally in phase space, which indeed applies rigorously to either bosons or fermions. In other words, ΔD in Eq. (4.2) should be related to states of close energy, which is brought about by an effective interaction which is local in momentum space. From this, it has been concluded that anyons are not ideal g-ons, but interacting g-ons (Nayak and Wilczek, 1994). Haldane (1991) provides explicit examples of systems for which fractional exclusion, albeit not exchange, statistics might be applicable. An earlier attempt to relate the parameter g of fractional exclusion statistics and the statistical parameter α of fractional exchange statistics has been made by Murthy and Shankar (1994), on the basis of the virial expansion of the g-ons partition function. More recently, an analytic, monotonic relation $g = g(\alpha)$ has been derived by Speliotopoulos (1997) in the case $g = 1/m$ (for integer m), which is relevant for the fractional quantum Hall effect (see also Canright and Johnson, 1994, for a review).

We end this general introduction to fractional statistics by mentioning some of the existing books on the subject (Wilczek, 1990; Lerda, 1992; Khare, 2005), where the interested reader can of course find more details on this topic and its applications.

4.1 Anyon exchange statistics in $d = 2$ dimensions

Soon after the introduction of fractional statistics (Leinaas and Myrheim, 1977; Wilczek, 1982a,b), it was realized that anyons are related to the braid group, just like bosons and fermions are related to the permutation group (Lerda and Sciuto, 1993; Frau *et al.*, 1994). (See also Frau *et al.*, 1996, for a review.) In other words, in characterizing the quantum state of an assembly of many anyons under the adiabatic exchange of any two of them, it is important to also specify the path along which they are interchanged, *i.e.* the way in which they braid around each other. A typical wave function for N anyons in dimensions $d = 2$ with statistical parameter α would thus be

$$\psi(z_1, z_1^*; \dots z_N, z_N^*) = \prod_{i'<j'} (z_{i'} - z_{j'})^\alpha f(z_1, z_1^*; \dots z_N, z_N^*), \tag{4.3}$$

where $z_{i'} = x_{i'} + iy_{i'}$ denotes the position of particle i' in complex notation, and f is a single-valued function, symmetric under all permutations of its arguments (positions).[1] When particle i' is exchanged with particle j', ψ acquires a global phase $e^{\pm i\alpha\pi}$, depending on whether the exchange is performed clockwise or counterclockwise. It is therefore important to also specify the orientation of the exchange, whenever $\alpha \neq 0$ or $1 \mod 2$. This is reflected in appropriate 'braiding relations' (at variance with commutation or anticommutation relations; see Eqs. (4.35) below) among anyon creation and annihilation operators. The latter can in turn be constructed from the corresponding fermionic (or bosonic) operators by means of a generalized Jordan-Wigner transformation (Jordan and Wigner, 1928).

Following Frau *et al.* (1996), let us denote by $\mathcal{L} = \mathcal{L}(q)$ the Lagrangian of a system of N anyons, depending on their configuration $q = q(t) \equiv (z_1, z_1^*; \ldots z_N, z_N^*)$. Within Feynman's path-integral formulation, the probability amplitude (propagator) that the system evolves from a certain configuration q at time t to the same configuration q at a time t' (a loop in configuration space) is then given by

$$K(q;t,t') = \sum_h \chi(h) \int \mathcal{D}q_h e^{\frac{i}{\hbar}\int_t^{t'} d\tau\, \mathcal{L}}, \tag{4.4}$$

where $\chi(h) = e^{-i\alpha\pi P_h}$ is the complex weight associated to the homotopy class h of the loops considered, and q_h is an element of such class, with parity P_h, *i.e.* the number of counterclockwise exchanges minus the number of clockwise exchanges that occur in h. For bosons, $\alpha = 0$, and all homotopy classes of loops contribute to Eq. (4.4) with the same weight, while for fermions, $\alpha = 1$, and the contribution from a given homotopy class of loops is weighted by ± 1, depending on the parity of the loop.

One explicit way to represent such a complex weight is

$$\begin{aligned}\chi(h) &= \exp\left(-i\alpha \sum_{i'<j'} \left(\Theta_{i'j'}^{(h)}(t') - \Theta_{i'j'}^{(h)}(t)\right)\right) \\ &= \exp\left(-i\alpha \sum_{i'<j'} \int_t^{t'} d\tau\, \dot{\Theta}_{i'j'}^{(h)}(\tau)\right),\end{aligned} \tag{4.5}$$

where $\Theta_{i'j'}^{(h)}(t)$ is the winding angle of particle i' with respect to particle j' measured along the braiding h at time t, and a superscript dot denotes time

[1] One example is of course Laughlin's wavefunction (Laughlin, 1983a), where $f(z_{i'}, z_{i'}^*) \propto \exp(-\sum_{i'} |z_{i'}|^2/4\ell_0^2)$, which was extremely successful in explaining much of the phenomenology of the fractional quantum Hall effect (Stormer *et al.*, 1999).

differentiation. This can be embedded in the Lagrangian used in Eq. (4.4) as an effective, homotopy-dependent term as

$$K(q;t,t') = \sum_h \int \mathcal{D}q_h e^{\frac{i}{\hbar}\int_t^{t'} d\tau \left(\mathcal{L} - \hbar\alpha \sum_{i'<j'} \dot{\Theta}^{(h)}_{i'j'}\right)}. \tag{4.6}$$

The above expression can be interpreted as if $K(q;t,t')$ were decomposed into equally-weighted subamplitudes for bosons described by the effective, homotopy dependent Lagrangian

$$\mathcal{L}_B \equiv \mathcal{L} - \hbar\alpha \sum_{i'<j'} \dot{\Theta}^{(h)}_{i'j'}. \tag{4.7}$$

Equivalently, Eq. (4.6) can also be interpreted as being decomposed into fermion subamplitudes, weighted by ± 1, depending on the loop parity, with fermions being described by the effective, homotopy dependent Lagrangian

$$\mathcal{L}_F \equiv \mathcal{L} - \hbar\alpha' \sum_{i'<j'} \dot{\Theta}^{(h)}_{i'j'}, \tag{4.8}$$

where $\alpha' = \alpha - 1$. This is suggestive that anyon statistics can be described in terms of an 'effective' interaction among ordinary bosons (or fermions), whose dynamics is described by the modified Lagrangians $\mathcal{L}_B$ (or $\mathcal{L}_F$, respectively). These effective particles are in fact called *composite bosons* and *composite fermions*, and the additional interaction among them, having a topological character, is called *statistical interaction*.

4.1.1 *Statistical interaction via Chern-Simons fields*

Following Wilczek (1982a,b) (see also Frau *et al.*, 1996; Altland and Simons, 2007, for a pedagogical introduction), it is possible to describe such statistical interactions via Chern-Simons fields. For the sake of simplicity, one may start with the Lagrangian of N spinless, charged, mutually non-interacting fermions

$$\mathcal{L} = \frac{1}{2} m \sum_{i'=1}^{N} v_{i'}^2 - V(\mathbf{r}_1, \dots \mathbf{r}_N), \tag{4.9}$$

where $\mathbf{v}_{i'} = \dot{\mathbf{r}}_{i'}$ is the velocity of fermion i', and V is an external potential. Classically, Pauli exclusion principle can be mimicked assuming the fermions to be hard-core particles. The action for the particles is then given by

$$S = \int d\mathbf{x}\, \mathcal{L}. \tag{4.10}$$

The particle number density $\rho(\mathbf{x},t)$ can be viewed as the time-like component of a 4-vector j^μ,

$$\rho(\mathbf{x},t) \equiv j^0(\mathbf{x},t) = \sum_{i'=1}^{N} \delta\left(\mathbf{x} - \mathbf{r}_{i'}(t)\right). \tag{4.11}$$

One can correspondingly define the current density as

$$\mathbf{j}(\mathbf{x},t) = \sum_{i'=1}^{N} \mathbf{v}_{i'}(t)\delta\left(\mathbf{x} - \mathbf{r}_{i'}(t)\right). \tag{4.12}$$

The continuity equation then enforces 4-current conservation as

$$\partial_t\rho + \nabla\cdot\mathbf{j} = \partial_\mu j^\mu = 0. \tag{4.13}$$

One then introduces coupling to an Abelian gauge field A_μ via the standard minimal coupling. This amounts to adding to the action a term

$$S_{\text{int}} = -e\int d\mathbf{x}\, j^\mu(x)A_\mu(x), \tag{4.14}$$

where e is the fermion charge, and the free action for the gauge field is given by the Chern-Simons term (Siegel, 1979; Schonfeld, 1981; Deser *et al.*, 1982)

$$S_{\text{CS}} = \frac{\kappa}{2}\int d\mathbf{x}\, \epsilon^{\mu\nu\lambda}A_\mu(x)\partial_\nu A_\lambda(x), \tag{4.15}$$

where κ is a coupling constant. The total action $S_{\text{tot}} = S + S_{\text{int}} + S_{\text{CS}}$ is then invariant under standard Abelian gauge transformations. Moreover, it can be shown that such a total action can be thought of as associated to a modified effective Lagrangian as $\mathcal{L}_F$, Eq. (4.8).

Indeed, by varying S_{tot} with respect to A_0, it is possible to associate, or 'attach', a magnetic field $B = \partial_2 A_1 - \partial_1 A_2$ to the particle density ρ so as to make

$$B = -\frac{e}{\kappa}\rho. \tag{4.16}$$

In the Weyl gauge $A^0 = 0$ and under the additional constraint that $\partial_i A^i = 0$ (transverse spatial components), this is possible with

$$A^i = \sum_{i'=1}^{N} A^i_{i'}(\mathbf{x},t), \tag{4.17}$$

where

$$A^i_{i'}(\mathbf{x},t) \equiv A^i_{i'}\left(\mathbf{r}_1(t),\ldots\mathbf{r}_N(t)\right)\Big|_{\mathbf{r}_{i'}=\mathbf{x}}, \tag{4.18}$$

and

$$A^i_{i'}(\mathbf{r}_1(t), \ldots \mathbf{r}_N(t)) = -\frac{e}{2\pi\kappa} \sum_{j' \neq i'} \epsilon^{ij} \frac{r_{i'j} - r_{j'j}}{|\mathbf{r}_{i'} - \mathbf{r}_{j'}|^2}. \tag{4.19}$$

Therefore, coupling to a Chern-Simons field associates a nonlocal vector potential $\mathbf{A}_{i'}$ to each particle. This determines a magnetic field B that is everywhere zero except at the particle location. Each particle then carries charge e and a magnetic flux $\phi = -e/\kappa$. Thus, whenever a particle turns around another particle, it picks up a phase change via the Aharonov-Bohm effect, because it moves around the flux of the other, and vice versa.

In order to show that this system composed of fermions coupled to Chern-Simons fields does indeed describe anyons with an appropriate statistical parameter, one may note that

$$A^i_{i'}(\mathbf{r}_1, \ldots \mathbf{r}_N) = \frac{e}{2\pi\kappa} \frac{\partial}{\partial r_{i'i}} \sum_{j' \neq i'} \Theta_{i'j'}, \tag{4.20}$$

where the winding angle of particle i' with respect to particle j' is explicitly given by

$$\Theta_{i'j'} = \text{atan}\left(\frac{r_{i'2} - r_{j'2}}{r_{i'1} - r_{j'1}}\right). \tag{4.21}$$

Minimal coupling then implies that the Lagrangian for the interacting system becomes

$$\mathcal{L}' = \mathcal{L} + e \sum_{i'=1}^{N} \mathbf{v}_{i'} \cdot \mathbf{A}_{i'}(\mathbf{r}_1, \ldots \mathbf{r}_N) = \mathcal{L} - \frac{e^2}{2\pi\kappa} \sum_{i'<j'} (v^i_{i'} - v^i_{j'}) \frac{\partial}{\partial r^i_{i'}} \Theta_{i'j'}. \tag{4.22}$$

This coincides with Eq. (4.8), on account of the fact that $(v^i_{i'} - v^i_{j'}) \frac{\partial}{\partial r^i_{i'}} \Theta_{i'j'} = \frac{d}{dt} \Theta_{i'j'}$, with

$$\alpha' = \frac{e^2}{2\pi\kappa}. \tag{4.23}$$

The same result can be derived within second quantization, using the formalism of quantum field theory. Besides confirming the interpretation given above, such a derivation provides also explicit 'braiding' rules for the anyon field operators, which replace the usual commutation and anticommutation rules for boson and fermion field operators, respectively. Following again Frau *et al.* (1996), let $\psi(x) \equiv \psi(\mathbf{x}, t)$ denote the annihilation field operator for a fermion with mass m and charge e. The action for such a field, minimally coupled to a Chern-Simons field, then reads

$$S = \int d\mathbf{x} \left(i\psi^\dagger D_0 \psi + \frac{1}{2m} \psi^\dagger (D_1^2 + D_2^2) \psi + \frac{\kappa}{2} \epsilon^{\mu\nu\lambda} A_\mu \partial_\nu A_\lambda \right), \tag{4.24}$$

where $D_\mu = \partial_\mu + ieA_\mu$ is the (gauge) covariant derivative. Varying S with respect to A_μ, one obtains the analog of the Maxwell equations for the gauge field, *viz.*

$$\epsilon^{\mu\nu\lambda}\partial_\nu A_\lambda = \frac{e}{\kappa} j^\mu, \tag{4.25}$$

where

$$j^0 = \psi^\dagger \psi \equiv \rho, \tag{4.26a}$$

$$j^i = \frac{i}{2m}\left(\psi^\dagger D^i \psi - (D^i\psi)^\dagger \psi\right), \tag{4.26b}$$

which again satisfy the continuity equation, $\partial_\mu j^\mu = 0$. The field potential can be obtained by solving the pseudo-Maxwell equations, Eqs. (4.25), for given source currents j^μ. The zeroth component of Eqs. (4.25) yields

$$\partial_1 A_2 - \partial_2 A_1 = \frac{e}{\kappa}\rho. \tag{4.27}$$

The spatial components can be solved in the appropriate spatially transverse gauge, $\partial_i A^i = 0$, and their solution can be presented in terms of a suitable Green function as

$$A^i(x) = \epsilon^{ij}\frac{\partial}{\partial x^j}\left(\frac{e}{\kappa}\int d\mathbf{x}'\, G(\mathbf{x}-\mathbf{x}')\rho(\mathbf{x}')\right), \tag{4.28}$$

where $\Delta G(\mathbf{x}-\mathbf{x}') = \delta(\mathbf{x}-\mathbf{x}')$, or, in $d=2$ dimensions,

$$G(\mathbf{x}-\mathbf{x}') = \frac{1}{2\pi}\log|\mathbf{x}-\mathbf{x}'|. \tag{4.29}$$

Therefore, one may express the spatial components of the Chern-Simons vector potential as

$$A^i(x) = \frac{e}{2\pi\kappa}\int d\mathbf{x}'\, \frac{\partial}{\partial x_i}\Theta(\mathbf{x}-\mathbf{x}')\rho(\mathbf{x}'), \tag{4.30}$$

where again

$$\Theta(\mathbf{x}-\mathbf{x}') = \operatorname{atan}\left(\frac{x_2 - x_2'}{x_1 - x_1'}\right). \tag{4.31}$$

It can be shown that the line integral of $A^i(x)$ around a closed loop around a given particle yields the flux of the magnetic field attached to that particle.

Upon quantization, standard anticommutation relations for the fermionic fields ψ, $\psi^\dagger$ require that

$$\{\psi(\mathbf{x},t), \psi^\dagger(\mathbf{x}',t)\} = \delta(\mathbf{x}-\mathbf{x}'), \tag{4.32a}$$

$$\{\psi(\mathbf{x},t), \psi(\mathbf{x}',t)\} = 0, \tag{4.32b}$$

$$\{\psi^\dagger(\mathbf{x},t), \psi^\dagger(\mathbf{x}',t)\} = 0. \tag{4.32c}$$

After observing that the Chern-Simons field A is a pure gauge, *i.e.* $A_\mu(x) = \partial_\mu \Lambda(x)$, where $\Lambda(x) = \frac{e}{2\pi\kappa}\int d\mathbf{x}'\,\Theta(\mathbf{x}-\mathbf{x}')\,(\rho(x')-\rho_0)$, with ρ_0 a constant density shift, removal of A through the gauge transformation

$$A_\mu \mapsto A'_\mu = A_\mu - \partial_\mu \Lambda = 0 \tag{4.33a}$$

involves correspondingly a gauge redefinition of the field operators as

$$\psi(x) \mapsto \psi'(x) = e^{ie\Lambda(x)}\psi(x) = e^{i\alpha'\int d\mathbf{x}'\,\Theta(\mathbf{x}-\mathbf{x}')(\psi^\dagger(\mathbf{x}')\psi(\mathbf{x}')-\rho_0)}\psi(x), \tag{4.33b}$$

while

$$D_\mu \mapsto \partial_\mu. \tag{4.33c}$$

Thus, the action, Eq. (4.24), becomes that of free fields ψ', $\psi'^\dagger$

$$S' = \int d\mathbf{x}\left(i\psi'^\dagger\partial_0\psi' + \frac{1}{2m}\psi'^\dagger\Delta\psi'\right). \tag{4.34}$$

However, the new free fields do not obey standard anticommutation relations any longer. These are in fact replaced *e.g.* by the braiding relation

$$\psi'(\mathbf{x},t)\psi'(\mathbf{x}',t) = -e^{-i\alpha'\left(\Theta(\mathbf{x}-\mathbf{x}')-\Theta(\mathbf{x}'-\mathbf{x})\right)}\psi'(\mathbf{x}',t)\psi'(\mathbf{x},t). \tag{4.35}$$

Introducing a cut along the negative x axis and measuring angles from the positive x axis, so that all angles belong to $[-\pi,\pi)$ and Θ becomes single-valued, one has

$$\Theta(\mathbf{x}-\mathbf{x}') - \Theta(\mathbf{x}'-\mathbf{x}) = \begin{cases} \pi\,\mathrm{sgn}(x_2 - x'_2), & x_2 \neq x'_2, \\ \pi\,\mathrm{sgn}(x_1 - x'_1), & x_2 = x'_2. \end{cases} \tag{4.36}$$

It is then possible to show that Eq. (4.35) becomes

$$\psi'(\mathbf{x},t)\psi'(\mathbf{x}',t) = -q^{-1}\psi'(\mathbf{x}',t)\psi'(\mathbf{x},t), \tag{4.37a}$$

where $q = e^{i\alpha'\pi}$, for counterclockwise rotations of $\psi'(\mathbf{x}',t)$ around $\psi'(\mathbf{x},t)$, and

$$\psi'(\mathbf{x},t)\psi'(\mathbf{x}',t) = -q\psi'(\mathbf{x}',t)\psi'(\mathbf{x},t), \tag{4.37b}$$

for clockwise rotations. Therefore, the gauge transformation, Eqs. (4.33), is equivalent to transmuting fermions into anyons with statistical parameter α, and it can thus be considered as the two-dimensional generalization of the Jordan-Wigner transformation, which transmutes fermions into bosons in $d = 1$ dimensions.

4.2 Anyon exclusion statistics in arbitrary dimensions

As mentioned in the introduction, a conceptually different way of introducing fractional statistics, due to Haldane (1991), consists in generalizing Pauli exclusion principle. The addition of one particle to a system of N identical particles with a given statistical behaviour reduces by g the size of the subset of the Hilbert space available to the state of the new particle. This new 'statistical parameter', already defined in Eq. (4.2), obviously equals zero for bosons (no restriction to available states for an added particle), while $g = 1$ for fermions (*i.e.* Pauli exclusion principle). Correspondingly, the statistical weight for N identical particles occupying M possible quantum states is

$$W = \frac{(M+N-1)!}{N!\,(M-1)!} \tag{4.38a}$$

for bosons, and

$$W = \frac{M!}{N!\,(M-N)!} \tag{4.38b}$$

for fermions. One simple (albeit not unique: cf. Khare, 2005, and references therein) interpolation implying fractional exclusion statistics and Eq. (4.2) is (Haldane, 1991)

$$W = \frac{[M+(N-1)(1-g)]!}{N!\,[M-gN-(1-g)]!}, \tag{4.38c}$$

where one explicitly allows $0 \leq g \leq 1$. One advantage of fractional exclusion statistics over exchange statistics is evidently the fact that it is not restricted to dimensionality $d = 2$. Attempts at deriving Haldane's assumption from a microscopic theory have been made by Sutherland (1997) (see also Sutherland, 1971b,c,a, 1972), on the basis on an analogy with exactly solvable integrable systems in $d = 1$ dimensions. More recently, fractional excluson statistics has been derived from the quantum maximum entropy principle within the Wigner representation (Trovato and Reggiani, 2013; Trovato, 2014).

Although the statistical parameters α and g for fractional exchange and exclusion statistics, respectively, are conceptually distinct and obey quite different definitions, in the following, and for the sake of simplicity, we will use the same notation for both, *i.e.*

$$g \equiv \alpha, \tag{4.39}$$

given the fact that their values identify in the two extreme limits for bosons ($g = \alpha = 0$) and fermions ($g = \alpha = 1$). As already mentioned in the introduction, their relation is still far from fixed, despite several attempts in this direction (Murthy and Shankar, 1994; Speliotopoulos, 1997; Canright and Johnson, 1994).

Let ϵ_i denote the ith energy level, M_i its degeneracy, and N_i its occupation. Then Eq. (4.38c) generalizes to

$$W = \prod_i \frac{[M_i + (N_i - 1)(1-\alpha)]!}{N_i!\,[M_i - \alpha N_i - (1-\alpha)]!}. \tag{4.40}$$

Maximizing $\log W$ subject to the constraints $N = \sum_i N_i$ and $E = \sum_i N_i \epsilon_i$, Wu (1994) derived the statistical distribution for the occupation number $n = N_i/M_i$ as (see also Nayak and Wilczek, 1994)

$$n \equiv n_\alpha = \frac{1}{w(\zeta_i) + \alpha}, \tag{4.41}$$

where $\zeta_i = e^{(\epsilon_i - \mu)/k_\mathrm{B}T}$, and $w(\zeta)$ satisfies the functional equation (Wu, 1994)

$$w^\alpha (1+w)^{1-\alpha} = \zeta. \tag{4.42}$$

4.2.1 *Exclusons distribution function*

Wu's functional equation, Eq. (4.42) can be solved straightforwardly as $w(\zeta) = \zeta - 1$ for $\alpha = 0$, and as $w(\zeta) = \zeta$ for $\alpha = 1$, in which cases Eq. (4.41) reduces to the usual Bose-Einstein and Fermi-Dirac distributions, respectively, *viz.*

$$n_0(z) = \frac{1}{e^z - 1} \qquad (\text{BE: } \alpha = 0), \tag{4.43a}$$

$$n_1(z) = \frac{1}{e^z + 1} \qquad (\text{FD: } \alpha = 1), \tag{4.43b}$$

where $z = \beta(\epsilon - \mu)$. Other known cases in which Eq. (4.42) can be solved exactly include $\alpha = \frac{1}{2}$, for which it reduces to a quadratic equation, describing 'semions', characterized by the distribution function (Nayak and Wilczek, 1994)

$$n = \frac{1}{\sqrt{\frac{1}{4} + e^{2(\epsilon_i - \mu)/k_\mathrm{B}T}}}. \tag{4.44}$$

Moreover, the following duality property holds (Nayak and Wilczek, 1994; Rajagopal, 1995):

$$1 - \alpha n(\alpha; x) = \frac{1}{\alpha} n(1/\alpha; -x/\alpha), \tag{4.45}$$

where $x = (\epsilon - \mu)/k_{\mathrm{B}}T$. Other special cases were treated analytically by Joyce *et al.* (1996), who managed to relate Wu's function to special functions, and by Aoyama (2001), who expressed Wu's function in terms of the hypergeometric function when α is an integer reciprocal. In particular, Aoyama (2001) exploited the fact that a zero $\bar{z}$ of a complex function $f(z)$, $f(\bar{z}) = 0$ can be characterized as

$$\bar{z} = \frac{1}{2\pi i} \oint dz\, z \frac{f'(z)}{f(z)}, \tag{4.46}$$

where the integration is performed along a contour enclosing only $\bar{z}$. Performing the analytical continuation of $w(\zeta)$ in Eq. (4.42) to the complex plane, and setting $f(w) \equiv w^\alpha (1+w)^{1-\alpha} - \zeta$, one finds (Aoyama, 2001)

$$\bar{w} = \frac{\alpha}{2\pi i} \oint dw \frac{1 + \dfrac{1-\alpha}{\alpha} \dfrac{w}{1+w}}{1 - \zeta \dfrac{1}{1+w} \left(\dfrac{1+w}{w} \right)^\alpha}, \tag{4.47}$$

where the contour encloses $w = 0$.

An approximate form of the distribution function for exclusons had been earlier derived by March *et al.* (1993); March (1993, 1997) from collision theory, using the detailed balance hypothesis. Their result was that $n^{-1} = \zeta + a$, with a interpolating between Bose-Einstein ($a = -1$) and Fermi-Dirac ($a = +1$) statistics. Comparison with Wu's result, Eq. (4.42), then yields (March *et al.*, 1993; March, 1997)

$$a(\alpha, w) = w + \alpha - w^\alpha (1+w)^{1-\alpha} \approx 2\alpha - 1, \tag{4.48}$$

the latter approximation, independent of w, holding in the limit $w \gg 1$. Eq. (4.48) already correctly interpolates between the Bose-Einstein ($\alpha = 0$, $a = -1$) and Fermi-Dirac ($\alpha = 1$, $a = 1$) limits. Of course, as w becomes small, the dependence of a on w is significant, though $2\alpha - 1$ is not a bad approximation even at $w = 1$ (giving for $\alpha = \frac{1}{2}$, $a = 0$ instead of 0.09, and for $\alpha = \frac{1}{4}$, $a = -\frac{1}{2}$ rather than -0.44). Alternative expressions of the excluson distribution function have also been proposed by Dasnières de Veigy and Ouvry (1994, 1995).

At $T = 0$, and for any $0 < \alpha \leq 1$, assuming a continuous energy spectrum $\epsilon_i \mapsto \epsilon$, the average occupation number tends to a step function, as in the Fermi-Dirac limit (Wu, 1994):

$$n(\epsilon) = \begin{cases} 0, & \epsilon > E_{\mathrm{F}}, \\ 1/\alpha, & \epsilon < E_{\mathrm{F}}, \end{cases} \tag{4.49}$$

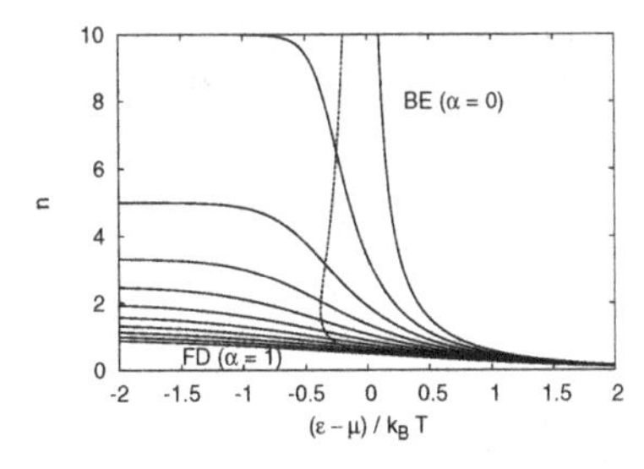

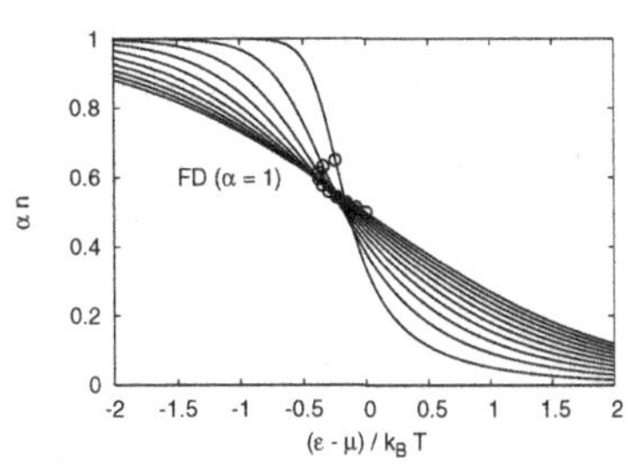

Fig. 4.1 (Left panel:) Distribution functions for anyon statistics, Eq. (4.41), at finite temperature $T \neq 0$, for equally spaced values of statistics parameter α ranging between $\alpha = 0$ (BE: Bose-Einstein) and $\alpha = 1$ (FD: Fermi-Dirac). The dashed line connects the inflection points of the distribution functions in the quasifermionic cases ($0 < \alpha \leq 1$), Eq. (4.51). Such points approximate the Fermi level, E_{F}. (Right panel:) Shows αn as a function of $\beta(\epsilon - \mu)$. Open circles mark the location of the inflection points, or 'quasi-Fermi levels'.

where the 'Fermi energy' E_{F} is implicitly defined by the requirement

$$\sum_{\epsilon_i < E_{\mathrm{F}}} M_i = \alpha N \tag{4.50}$$

(see also Eq. (4.74) below). At finite temperature, $n(\epsilon)$ becomes smoother, but is still a monotonously decreasing function, with an inflection point approximating E_{F} implicitly defined by the condition (Angilella *et al.*, 2006b)

$$w = -1 + \alpha + \sqrt{1 - \alpha + \alpha^2}. \tag{4.51}$$

Figure 4.1 shows Wu's distribution function $n(\epsilon)$ at a finite temperature T, for various values of the statistical parameter α.

For large values of $|\zeta|$, Joyce *et al.* (1996) show that Wu's function admits the expansion

$$w = \zeta + (\alpha - 1) + \alpha \sum_{m=1}^{\infty} \frac{(m\alpha - m)_m}{(m+1)!} \frac{(-1)^m}{\zeta^m}, \tag{4.52}$$

where $(z)_m = \Gamma(m+z)/\Gamma(z)$ denotes the Pochhammer symbol, and $|\zeta| \geq \alpha^\alpha (1-\alpha)^{1-\alpha}$. Correspondingly, Joyce *et al.* (1996) find

$$n = \frac{1}{w(\zeta) + \alpha} = \sum_{m=0}^{\infty} \frac{(m\alpha + \alpha - m)_m}{m!} \frac{(-1)^m}{\zeta^{m+1}}. \tag{4.53}$$

After performing the analytic continuation $\beta(\epsilon - \mu) \mapsto z \in \mathbb{C}$, the excluson distribution function, Eq. (4.41), exhibits some remarkable properties (Angilella *et al.*, 2006c). First of all, we observe that its BE and FD limits

$n_0(z)$ and $n_1(z)$, Eqs. (4.43), are meromorphic functions, with simple poles located along the imaginary axis at $z = (2k+1)i\pi$ and $z = 2ki\pi$ ($k \in \mathbb{Z}$), *i.e.* at odd and even multiples of $i\pi$, respectively. These facts have direct consequences in the Mittag-Leffler expansion of a modified hyperbolic tangent

$$\begin{aligned}\tanh_\alpha z &\equiv n_\alpha(-2z) - n_\alpha(2z) \\ &= \frac{1}{w(e^{-2z}) + \alpha} - \frac{1}{w(e^{2z}) + \alpha},\end{aligned} \tag{4.54}$$

which enters the definition of the pairing susceptibility in an anyonic model of superconductivity (Angilella *et al.*, 2006c). Such expansions for the two limiting cases of fermions and bosons read (Gradshteyn and Ryzhik, 1994), respectively,

$$\tanh_1 z = \tanh z = 2z \sum_{k=0}^{\infty} \frac{1}{z^2 + \left(k + \frac{1}{2}\right)^2 \pi^2}, \tag{4.55a}$$

$$-\tanh_0 z = \coth z = \frac{1}{z} + 2z \sum_{k=1}^{\infty} \frac{1}{z^2 + k^2\pi^2}. \tag{4.55b}$$

It should be noted that the origin of Bose-Einstein condensation for assemblies with $\alpha = 0$ (bosons) can be ultimately traced back to the presence of a single real pole ($z = 0$) in the expansion Eq. (4.55b) (Enz, 1992).

In the more general case of intermediate statistics ($0 < \alpha < 1$), neither expansion holds any longer. By direct inspection of Eq. (4.41), one finds that Wu's distribution function $n_\alpha(z)$ is characterized by infinitely many branching points of polar type, located in the complex plane at

$$z_k^\pm = \alpha \log \alpha + (1-\alpha)\log(1-\alpha) + (2k \pm \alpha)i\pi, \tag{4.56}$$

where $k \in \mathbb{Z}$. Around each singularity, Wu's distribution function behaves as (cf. Joyce *et al.*, 1996)

$$n_\alpha(z) \sim \mp \frac{i}{\sqrt{2\alpha(1-\alpha)}} \frac{1}{\sqrt{z - z_k^\pm}}, \qquad z \sim z_k^\pm, \tag{4.57}$$

as depicted in Fig. 4.2. Moving away from the FD limit ($\alpha = 1$), each simple pole for $n_1(z)$ located along the imaginary axis at odd multiples of $i\pi$ [$(2k+1)i\pi$, say; filled circles in Fig. 4.2] evolve into *two* square-root–like branching points of polar type for $n_\alpha(z)$ at z_k^+ and z_{k+1}^-, respectively. The same happens to $n_\alpha(-z)$, but now with singularity pairs located in the $\operatorname{Re} z > 0$ half plane (not shown in Fig. 4.2).

Making use of the above discussion, and of the definition Eq. (4.54), by analogy with Eq. (4.55a), Angilella *et al.* (2006c) then conjecture a Mittag-Leffler–like expansion of the 'modified' hyperbolic tangent, Eq. (4.54):

$$\tanh_\alpha z = \frac{1}{\alpha}\sum_{k=0}^{\infty}\left(\frac{1}{\sqrt{\left(z+\frac{1}{2}z_k^+\right)\left(z+\frac{1}{2}z_{k+1}^-\right)}} - \frac{1}{\sqrt{\left(z-\frac{1}{2}z_k^+\right)\left(z-\frac{1}{2}z_{k+1}^-\right)}}\right), \qquad (4.58)$$

where $\sqrt{z}$ denotes the principal branch of the square root function in the complex domain. Eq. (4.58) correctly reduces to Eq. (4.55a) in the limit $\alpha \to 1$, and has been verified within numerical accuracy for $0 < \alpha \leq 1$ against Eq. (4.54) over the real axis.

Moreover, by using Eq. (4.53) of Joyce *et al.* (1996) and Eq. (4.54) above, Angilella *et al.* (2006c) can express the generalized pairing susceptibility in the form of the following series

$$\frac{1}{2z}\tanh_\alpha \frac{z}{2} = \sum_{m=0}^{\infty}(-1)^m \frac{(m\alpha+\alpha-m)_m}{m!}\frac{\sinh[(m+1)z]}{z}, \qquad (4.59)$$

which is convergent for $0 \leq \alpha \leq 1$ and $\zeta_c < |e^z| < \zeta_c^{-1}$, with $\zeta_c = \alpha^\alpha(1-\alpha)^{1-\alpha}$ [implying $|\operatorname{Re} z| < -\alpha\log\alpha - (1-\alpha)\log(1-\alpha)$ along the real axis], and where $(n)_m = \Gamma(m+n)/\Gamma(n)$ denotes the Pochhammer symbol.

4.2.2 *Thermodynamic properties of exclusons*

In order to make contact with the ordinary Bose-Einstein and Fermi-Dirac limits, we shall now assume that $\epsilon_i = \hbar^2 k_i^2/(2m)$, and then treat $\mathbf{k}_i \mapsto \mathbf{k}$ as a continuous variable. Correspondingly, we shall write n_i as $n(\mathbf{k})$.

Within these assumptions, the general properties and some exact results of the thermodynamics of an assembly of non-interacting exclusons were derived to some extent by Wu (1994); Joyce *et al.* (1996); Iguchi (1997) (see also Iguchi, 1998; Iguchi and Sutherland, 2000; Su and Suzuki, 1998; Aoyama, 2001). In particular, by evaluating the particle density

$$\frac{N}{V} = \int \frac{d^d\mathbf{k}}{(2\pi)^d} n(\mathbf{k}) \qquad (4.60)$$

in dimensions $d = 2$, Wu (1994) was able to derive the relation between chemical potential μ, temperature T and particle density N/V explicitly as

$$\frac{\mu}{k_\mathrm{B}T} = \alpha\frac{\lambda^2 N}{V} + \log\left[1-\exp\left(-\frac{\lambda^2 N}{V}\right)\right], \qquad (4.61)$$

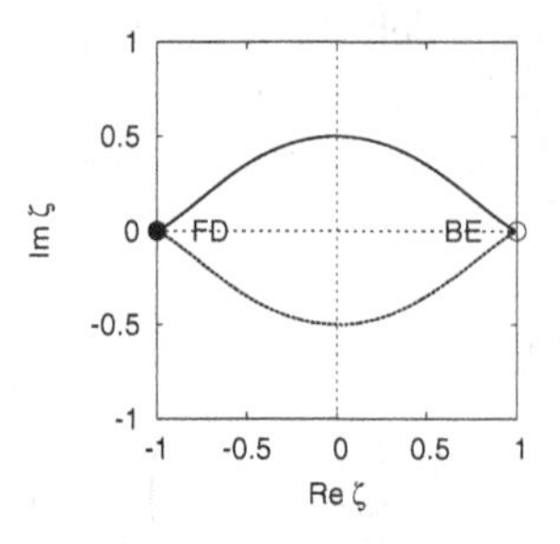

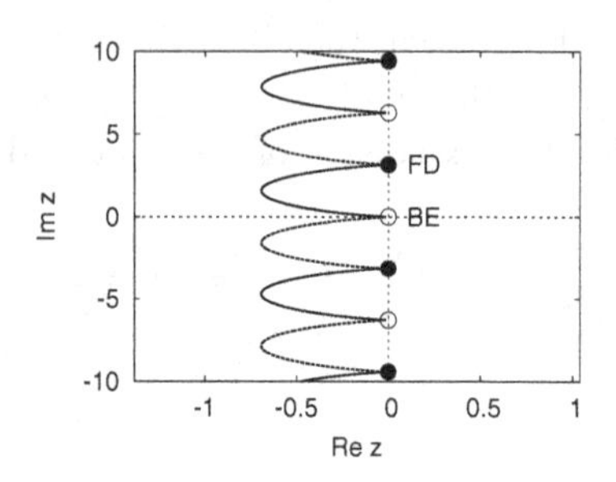

Fig. 4.2 Left panel shows the locus of the singularities of $n_\alpha(z)$ in the complex plane $\zeta = e^z$. The solid and dashed lines refer to the plus and minus signs, respectively, in Eq. (4.56). For each value of the statistical parameter $0 < \alpha < 1$, these singularities consist of two branching points of polar type, while for $\alpha = 0$ (BE, open circle) and $\alpha = 1$ (FD, filled circle) these merge into a single pole. Right panel shows the same singularities, but now in the complex plane z. Because of the (purely imaginary) periodicity of the exponential function, the two singularities in the ζ variable give rise to infinitely many such singularities as a function of z. One recognizes the familiar poles of $n_0(z)$ (BE, open circles) and $n_1(z)$ (FD, filled circles), Eqs. (4.43), along the imaginary axis in the two limiting cases $\alpha = 0$ and $\alpha = 1$, respectively.

where $\lambda = \hbar\sqrt{2\pi/(mk_{\rm B}T)}$ is the thermal wavelength. Performing the integration in Eq. (4.60) for a homogeneous system in arbitrary dimensions d (see also Iguchi, 1997), such a relation can be generalized implicitly as (Pellegrino *et al.*, 2007)

$$\Gamma\left(\frac{d}{2}\right)\frac{\lambda^d N}{V} = \int_{w_0}^{\infty} dw\, \frac{1}{w(1+w)}\bar{\epsilon}^{\frac{d-2}{2}}, \tag{4.62}$$

where $\Gamma(x)$ is Euler's function, $w_0 \equiv w_0(\mu/k_{\rm B}T)$ is a generalized inverse fugacity, implicitly defined by Wu's functional equation, Eq. (4.42), at $\epsilon = 0$, *i.e.*

$$w_0^\alpha (1+w_0)^{1-\alpha} = e^{-\mu/k_{\rm B}T}, \tag{4.63}$$

and

$$\bar{\epsilon} = \log\left[w^\alpha(1+w)^{1-\alpha}e^{\mu/k_{\rm B}T}\right] \equiv \log\frac{w^\alpha(1+w)^{1-\alpha}}{w_0^\alpha(1+w_0)^{1-\alpha}}. \tag{4.64}$$

Eliminating w_0 between Eqs. (4.62) and (4.63), Pellegrino *et al.* (2007) obtained the relation between the ratio $\mu/k_{\rm B}T$ of chemical potential to thermal energy and the scaled density $\lambda^d N/V$, parametrically. This is plotted in Fig. 4.3 for several values of the statistical parameter α, and $d = 1, 2, 3$. We explicitly note that the dilute limit $\lambda^d N/V \ll 1$ corresponds to $w_0 \gg 1$ ($\mu/k_{\rm B}T \to -\infty$). We thus recover a monotonic relation, as

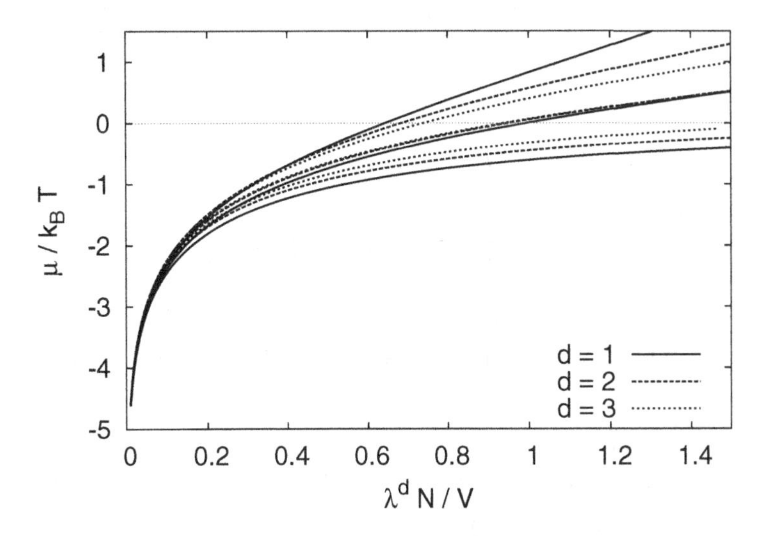

Fig. 4.3 Chemical potential *vs* scaled density, Eq. (4.62), for $\alpha = 0,\ 0.5,\ 1$ (bottom to top) and $d = 1, 2, 3$.

expected, which for $\alpha = 0$ (Bose-Einstein limit) leads to an infinitesimally small negative chemical potential, in the limit of large density.

Other notable thermodynamic properties of an excluson gas include the thermodynamic potential (Wu, 1994; Joyce *et al.*, 1996; Iguchi, 1997)

$$\Omega = -pV = -k_{\mathrm{B}}T \sum_{\mathbf{k}} \log\left(\frac{1 + w_{\mathbf{k}}}{w_{\mathbf{k}}}\right), \tag{4.65}$$

where p denotes pressure, V a d-dimensional volume, and $w_{\mathbf{k}}$ is Wu's function evaluated at $\epsilon_{\mathbf{k}}$, and the statistics-independent relation for the internal energy E,

$$pV = \frac{2}{d}E. \tag{4.66}$$

Converting integrals over d-dimensional momenta into integrals over energy, and making use of the d-dimensional density of states for free particles with a quadratic dispersion

$$\rho(\epsilon) = g_s \left(\frac{2\pi m}{h^2}\right)^{d/2} \frac{V}{\Gamma(d/2)} \epsilon^{(d-2)/2}, \tag{4.67}$$

where $g_s = 2s+1$ is the spin degeneracy factor, Joyce *et al.* (1996) find

$$\frac{\beta E}{dN/2} = \frac{1}{x} I_{d/2}(\alpha, \eta), \tag{4.68a}$$

$$x \equiv \frac{\lambda^d}{g_s} \frac{N}{V} = I_{(d-2)/2}(\alpha, \eta), \tag{4.68b}$$

$$I_n(\alpha, \eta) = \frac{1}{\Gamma(n+1)} \int_0^\infty \frac{t^n}{w(e^{t-\eta}) + \alpha} dt, \tag{4.68c}$$

where $\eta = \beta\mu$. Joyce *et al.* (1996) can then find closed expressions for the heat capacity at constant volume, C_V, and for the heat capacity at constant pressure, C_p, as

$$\frac{C_V}{dNk_\text{B}/2} = \frac{d+2}{2x} I_{d/2}(\alpha, \eta) - \frac{dx}{2I_{(d-2)/2}(\alpha, \eta)}, \tag{4.69a}$$

$$\frac{C_p}{C_V} = \frac{d+2}{dx^2} I_{d/2}(\alpha, \eta) I_{(d-2)/2}(\alpha, \eta). \tag{4.69b}$$

Joyce *et al.* (1996) quote the following asymptotic expansion at large η of their integral, Eq. (4.68c):

$$I_n(\alpha, \eta) \sim \frac{\eta^{n+1}}{\alpha\Gamma(n+2)} \left[1 + \alpha(n+1) \sum_{j=1}^\infty \binom{n}{j} \frac{C_j(\alpha)}{\eta^{j+1}} \right], \tag{4.70}$$

where

$$C_j(\alpha) = \int_0^\infty dz\, z^j \left[\frac{1}{w(e^z) + \alpha} + (-1)^j \left(\frac{1}{w(e^{-z}) + \alpha} - \frac{1}{\alpha} \right) \right], \tag{4.71}$$

with $C_0(\alpha) = 0$, for arbitrary α (Delves *et al.*, 1997). In the Fermi-Dirac limit, $\alpha = 1$, this essentially reduces to the well-known Sommerfeld expansion, for which correspondingly one has

$$C_{2m}(1) = 0, \tag{4.72a}$$

$$C_{2m-1}(1) = \frac{1}{m}(2^{2m-1} - 1)|B_{2m}|\pi^{2m}, \tag{4.72b}$$

where B_m is the mth Bernoulli number. Such expansions allow Joyce *et al.* (1996) to derive the low-temperature asymptotic formulas ($0 < \alpha \leq 1$,

$k_BT \ll E_F$)

$$\frac{\mu}{E_F} \sim 1 - \alpha E_F \frac{\rho'(E_F)}{\rho(E_F)} C_1(\alpha)\tau^2 + O(\tau^3) \tag{4.73a}$$

$$\sim 1 - \alpha(d-2)\left[\frac{1}{2}C_1(\alpha)\tau^2 + \frac{1}{8}(d-4)C_2(\alpha)\tau^3 - \frac{1}{48}(d-6)\left(3\alpha(d-2)C_1^2(\alpha) - (d-4)C_3(\alpha)\right)\tau^4 + O(\tau^5)\right]; \tag{4.73b}$$

$$\frac{d+2}{d}\frac{E}{NE_F} \sim 1 + \frac{\alpha}{2}(d+2)C_1(\alpha)\tau^2 + \alpha(d-2)\left[\frac{1}{4}(d+2)C_2(\alpha)\tau^3 - \frac{1}{16}(d+2)\left(3\alpha(d-2)C_1^2(\alpha) - (d-4)C_3(\alpha)\right)\tau^4 + O(\tau^5)\right]; \tag{4.73c}$$

$$\frac{C_V}{dNk_B} \sim \frac{2}{d}\frac{E_F}{N}\rho(E_F)C_1(\alpha)\tau + O(\tau^2) \tag{4.73d}$$

$$\sim \alpha C_1(\alpha)\tau + (d-2)O(\tau^2); \tag{4.73e}$$

$$\frac{C_p}{C_V} \sim 1 + 2\alpha C_1(\alpha)\tau^2 + \alpha(d-2)\left[\frac{3}{2}C_2(\alpha)\tau^3 - \frac{1}{2}\left(\alpha(3d-4)C_1^2(\alpha) - (d-4)C_3(\alpha)\right)\tau^4 + O(\tau^5)\right], \tag{4.73f}$$

where $\tau = k_BT/E_F$, and the pseudo-Fermi energy E_F, defined in general by Eq. (4.50), for a free anyon gas with a quadratic dispersion relation in d dimensions, and a density of states given by Eq. (4.67), takes the explicit expression

$$E_F = \left(\frac{h^2}{2\pi m}\right)\left[\Gamma\left(\frac{d+2}{2}\right)\frac{\alpha}{g_s}\frac{N}{V}\right]^{2/d}. \tag{4.74}$$

Explicit expressions for the entropy, the magnetization, the magnetic susceptibility, and the entropy have been derived by Wu (1994), but for a quite specific model, characterized by a single energy level $\epsilon = \hbar\omega_c/2$, as is *e.g.* the case if all anyons are in the lowest Landau level in the presence of a magnetic field, at very low temperatures. Similar results were derived by Nayak and Wilczek (1994), with explicit reference to semions ($\alpha = \frac{1}{2}$). Weakly interacting exclusons have been studied by Iguchi (1998), thereby generalizing Landau theory of Fermi liquids to Haldane-Wu liquids (see also

Iguchi and Sutherland, 2000). Slightly more general dispersion relations were considered by Su and Suzuki (1998), leading to essentially the same results, which however allowed to derive the virial expansion of the equation of state for an anyon gas (see also Murthy and Shankar, 1994; Khare, 2005).

4.3 Statistical correlations in an excluson gas

Statistical correlations have well-known consequences in the energy spectrum and density distribution of quantum fluids, even in the case of a non-interacting gas. One limit is provided by a fermion gas, where Pauli exclusion principles prevents any two particles to occupy the same quantum state. In particular, no two fermions with the same spin may reside at the same position in space. As a result, each fermion is surrounded by an 'exchange hole', a region in which the density of same-spin fermions is smaller than average (Wigner and Seitz, 1933; Slater, 1934a,b). In the case of charged fermions, such as electrons, within an exchange hole the positive background charge is not exactly cancelled. It is the interaction of each electron with the positive charge of the exchange hole that gives rise to the exchange energy (Giuliani and Vignale, 2005). The situation is quite the opposite for bosons, where statistical correlations are attractive and may be thought of being ultimately responsible of Bose-Einstein condensation.

In the case of intermediate or fractional statistics, Pellegrino *et al.* (2007, 2008) raised therefore the natural question of how many particles an excluson effectively 'sees' around itself. In other words, one is led to consider the effect of Haldane's generalized exclusion principle, Eq. (4.2), on the exchange hole which an excluson 'digs' around itself. Pellegrino *et al.* (2007, 2008) then studied a suitably defined generalization of the pair correlation function between exclusons in arbitrary dimensions d ($d = 1, 2, 3$), as a function of temperature T and particle density. A general result, applying to all non-bosonic values of the fractional parameter ($0 < \alpha \leq 1$), is that the pair correlation function is characterized by Friedel-like oscillations, which become more pronounced for increasing α or density, or for decreasing T.

In this context, earlier results include the seminal work of Sutherland (1971b), for a system of fermions or bosons in $d = 1$ interacting through a singular potential $\sim 1/r^2$, a problem which can be mapped into anyons (Forte, 1992). The evaluation of a pairwise correlation function for anyons (*i.e.* particles obeying fractional *exchange* statistics) in $d = 2$ at $T = 0$ was also analyzed by Gutierrez (2004), in connection with an intensity interferometry gedanken experiment, which may prove useful in the context of quantum computing. The question is also intimately related to that of finding Uhlenbeck's statistical interparticle potential (Uhlenbeck and Gropper, 1932), which has been studied for the two-anyons system in $d = 2$ (Huang, 1995). A somewhat related study (Sen *et al.*, 2008) considered the two-particle kernel in the lowest Landau level of a quantum Hall system, but now in $d = 2$ and within the context of exchange fractional statistics.

4.3.1 *Pair correlation function for non-interacting exclusons*

The pair correlation function $g(\mathbf{r})$ between two indistinguishable, hard-core particles is defined as the normalized probability of finding them at a relative distance $\mathbf{r}$ (Slater, 1934a; Giuliani and Vignale, 2005). In second quantization, and neglecting spin, one has

$$g(\mathbf{r}) = \frac{\langle \Psi^\dagger(\mathbf{r})\Psi^\dagger(0)\Psi(0)\Psi(\mathbf{r})\rangle}{n(\mathbf{r})n(0)}, \tag{4.75}$$

where Ψ, $\Psi^\dagger$ are quantum field operators, $n(\mathbf{r}) = \langle \Psi^\dagger(\mathbf{r})\Psi(\mathbf{r})\rangle$ is the single-particle probability density at position $\mathbf{r}$, and $\langle \cdots \rangle$ denotes a quantum statistical average associated with the equilibrium distribution of the particle assembly under study (see below). In the case of a homogeneous system, one obviously has $n(\mathbf{r}) = n(0)$ and $g(\mathbf{r}) \equiv g(r)$.

In order to reduce the four-point average in Eq. (4.75) to simpler quantities by means of something analogous to Wick's theorem, one needs to specify the commutation properties of the excluson field operators under exchange. As emphasized in the introduction, the relation between fractional exclusion and exchange statistics is not completely settled. Within fractional exchange statistics in reduced dimensionality ($d \leq 2$), one may make recourse to Eqs. (4.37) (cf. Forte, 1992; Ha, 1996, in $d = 2$ and $d = 1$, respectively), which of course reduce to the familiar commutation and anticommutation relations for bosons ($\alpha = 0$) and fermions ($\alpha = 1$), respectively. Following Pellegrino *et al.* (2007), we postulate that standard quantum Bose-Einstein or Fermi-Dirac statistics are weakly violated as prescribed by

$$\begin{aligned}
[\Psi(\mathbf{r}), \Psi(\mathbf{r}')]_\alpha &\equiv \Psi(\mathbf{r})\Psi(\mathbf{r}') - \cos(\alpha\pi)\Psi(\mathbf{r}')\Psi(\mathbf{r}) \\
&= 0, && (4.76a)\\
[\Psi^\dagger(\mathbf{r}), \Psi^\dagger(\mathbf{r}')]_\alpha &= 0, && (4.76b)\\
[\Psi(\mathbf{r}), \Psi^\dagger(\mathbf{r}')]_\alpha &= \delta(\mathbf{r} - \mathbf{r}'). && (4.76c)
\end{aligned}$$

Eqs. (4.76) again reduce to the familiar relations for bosons and fermions, in the appropriate limits. Following Greenberg (1991) for the generalization of Wick's theorem corresponding to Eqs. (4.76), one finds

$$\langle\Psi^\dagger(\mathbf{r})\Psi^\dagger(0)\Psi(0)\Psi(\mathbf{r})\rangle = \langle\Psi^\dagger(\mathbf{r})\Psi(\mathbf{r})\rangle\langle\Psi^\dagger(0)\Psi(0)\rangle + \cos(\alpha\pi)\langle\Psi^\dagger(\mathbf{r})\Psi(0)\rangle\langle\Psi^\dagger(0)\Psi(\mathbf{r})\rangle. \quad (4.77)$$

Correspondingly, the pair correlation function for a homogeneous assembly of exclusons reads

$$g(r) = 1 + \cos(\alpha\pi)\frac{|\langle\Psi^\dagger(r)\Psi(0)\rangle|^2}{n^2(0)}, \quad (4.78)$$

which correctly reduces to the fermion limit ($\alpha = 1$), with $g(r) \leq 1$, manifestly (Giuliani and Vignale, 2005).

For translationally invariant systems, one may use a plane wave expansion to find

$$g(r) = 1 + \cos(\alpha\pi)\left|\frac{\tilde{n}(r)}{\tilde{n}(0)}\right|^2, \quad (4.79)$$

where

$$\tilde{n}(r) = \int \frac{d^d\mathbf{k}}{(2\pi)^d} e^{-i\mathbf{k}\cdot\mathbf{r}} n(\mathbf{k}) \quad (4.80)$$

is the Fourier transform in d dimensions of the single-particle distribution function $n(\mathbf{k})$ for exclusons in equilibrium at temperature T, Eq. (4.41). Performing the integration required in Eq. (4.80) for an isotropic system in d dimensions, Pellegrino *et al.* (2007) find

$$\lambda^d \tilde{n}(x) = x^{-\nu} \int_{w_0}^\infty dw \, \frac{1}{w(1+w)} \bar{\epsilon}^{\nu/2} J_\nu\left(2x\bar{\epsilon}^{1/2}\right), \quad (4.81)$$

where $x = \sqrt{\pi}r/\lambda$, $\nu = (d-2)/2$, $\lambda = \hbar\sqrt{2\pi/(mk_\mathrm{B}T)}$ is again the thermal wavelength, and $J_\nu(z)$ is a Bessel function of first kind. Making use of the asymptotic properties of the Bessel functions (Gradshteyn and Ryzhik, 1994), it can be shown that Eq. (4.81) correctly reduces to Eq. (4.62) for the particle density in the limit $x \to 0$.

4.3.1.1 *Recursion formulas for the local excluson density at different dimensionalities*

Eq. (4.81) yields $\tilde{n}_\nu(x, w_0) \equiv \tilde{n}(x)$ as a function of scaled interparticle distance x and generalized inverse fugacity w_0, at a given statistical parameter α and reduced dimensionality $\nu = (d-2)/2$. By differentiating Eq. (4.81) with respect to x, and making use of the appropriate recursion formula for the Bessel functions (Gradshteyn and Ryzhik, 1994), Pellegrino *et al.* (2007) obtain

$$\frac{\partial \tilde{n}_\nu}{\partial x} + 2x\tilde{n}_{\nu+1} = 0. \tag{4.82}$$

Similarly, by differentiating the same Eq. (4.81) with respect to w_0, and taking into account Eq. (4.64), one obtains

$$\frac{\partial \tilde{n}_\nu}{\partial w_0} + \left(\frac{d\bar{\epsilon}}{dw}\right)_0 \tilde{n}_{\nu-1} + \rho_\nu(w_0) = 0, \tag{4.83}$$

where

$$\left(\frac{d\bar{\epsilon}}{dw}\right)_0 = \frac{\alpha + w_0}{w_0(1+w_0)}, \tag{4.84}$$

and

$$\rho_\nu(w) = \frac{1}{\Gamma(\nu+1)} \frac{1}{w(1+w)} \bar{\epsilon}^\nu \tag{4.85}$$

is the density of states in the w variable, in $d = 2\nu + 2$ dimensions, cf. Eq. (4.67). Eqs. (4.82) and (4.83) are then differential recursion formulas relating the local excluson density at different dimensionalities. Combining Eqs. (4.82) and (4.83), and noting that $\rho_{\nu+1}(w_0) = \lim_{w\to w_0} \rho_{\nu+1}(w) = 0$ for $d = 1, 2, 3$, one obtains (Pellegrino *et al.*, 2007)

$$\frac{\partial^2 \tilde{n}_\nu}{\partial w_0 \partial x} = 2x \frac{\alpha + w_0}{w_0(1+w_0)} \tilde{n}_\nu, \tag{4.86}$$

which is a hyperbolic partial differential equation for the local excluson density, similar to the Klein-Gordon equation, but with a variable 'mass' term.

4.3.2 *Friedel oscillations in the excluson pair correlation function at $T \neq 0$*

The fact that Wu's distribution function is characterized by an inflection point, Eq. (4.51), is responsible for the appearance of oscillations in its Fourier transform in all pseudo-fermion cases. This is apparent in Fig. 4.4, showing $g(x)$ in $d = 2$ for $w_0 = 0.1$ (corresponding to a relatively large particle density), and for $\alpha = 0 - 1$ (Pellegrino *et al.*, 2007). Close to and in the fermion limit ($0.5 \lesssim \alpha \leq 1$), the pair correlation function exhibits a correlation 'hole' around $x = 0$, whose depth decreases with decreasing α and eventually vanishes as $\alpha \to \frac{1}{2}$. In the same range of values of the

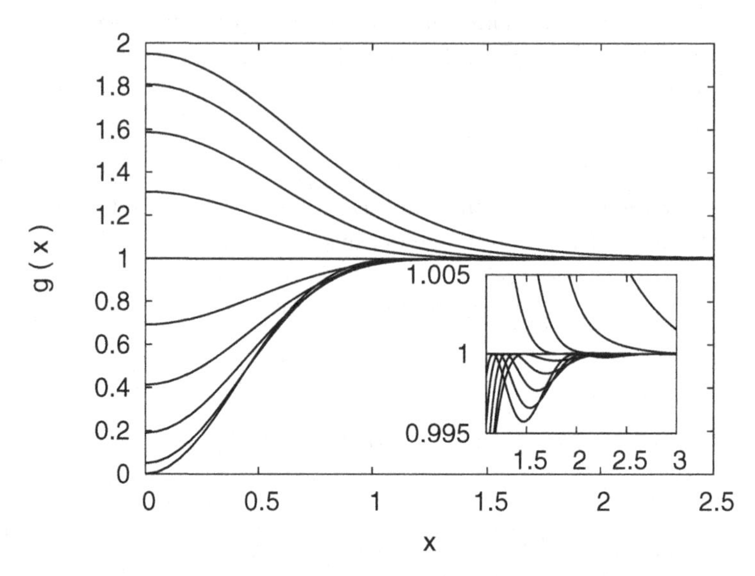

Fig. 4.4 Pair correlation function $g(x)$ (in scaled units), Eq. (4.79), in $d = 2$, for $w_0 = 0.1$ (dense limit) and $\alpha = 0 - 1$ (top to bottom). Inset shows Friedel oscillations, which are present only close to the FD limit $(0.5 < \alpha \leq 1)$.

statistical parameter, $g(x)$ is characterized by damped Friedel-like oscillations (cf. inset of Fig. 4.4), which are more pronounced close to the fermion limit. These oscillations tend to disappear as one approaches the bosonic limit, $0 \leq \alpha \lesssim 0.5$, where the pair correlation function displays a monotonically decreasing behavior, with correlations 'piling up' at $x = 0$, in contrast with the fermion limit. These oscillations are generic also in $d = 1$ and $d = 3$. However, a monotonic behaviour (*i.e.* no oscillations) is recovered as w_0 increases, corresponding to a decrease in the particle density (Pellegrino *et al.*, 2007). The latter observation suggested to use the criterion $w_0 = w_*$, where w_* denotes the inflection point, as given by Eq. (4.51), to mark the onset of such Friedel-like oscillations, as a function of density. This is depicted in Fig. 4.5, where the plane of scaled density $\lambda^d N/V$ *vs* statistical parameter α is divided into two regions by the line implicitly defined by the equation $w_* = w_0$. At a fixed value of the statistical parameter α, say, one might think of increasing the fermionic-like correlations of a given excluson assembly (*i.e.* sharpening the Friedel-like oscillations in the pair correlation function) by suitably increasing its scaled density $\lambda^d N/V$, *i.e.* increasing density N/V or reducing temperature T.

Within this picture, one may also estimate the wavelength Λ of the

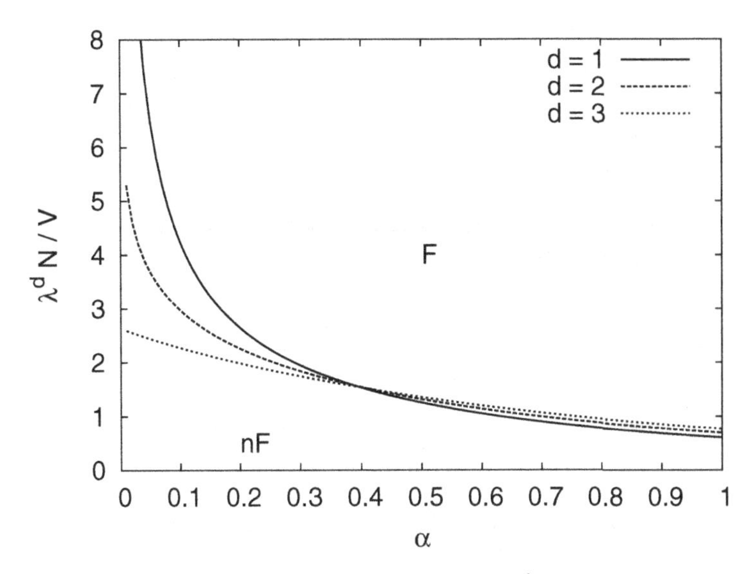

Fig. 4.5 Phase diagram in the plane of scaled density $\lambda^d N/V$ *vs* statistical parameter α. The phase diagram features regions where pronounced Friedel-like oscillations occur in the pair correlation function (F), and where such oscillations are absent, or beyond graphical resolution (nF). Different lines refer to various dimensionalities ($d = 1, 2, 3$).

oscillations in the pair correlation function with the approximate condition $2\sqrt{\epsilon_*}\Lambda \approx 2\pi$, which straightforwardly leads to

$$\Lambda^2 \approx \frac{2\pi^2\hbar^2}{m\epsilon_*}, \tag{4.87}$$

where all dimensions have been restored. Clearly, as $\epsilon_* \to 0$ as $\alpha \to 0$, one recovers the absence of Friedel-like oscillations in the bosonic limit.

4.3.3 *Friedel oscillations in the excluson pair correlation function at $T = 0$*

At $T = 0$ and for each non-zero value of the statistical parameter α ($0 < \alpha \leq 1$), Wu's distribution function Eq. (4.41) becomes a step function, given by Eq. (4.49), as in Fermi statistics. On the contrary, in the bosonic case ($\alpha = 0$) the distribution function behaves like a Dirac delta function, $n(\epsilon) = N\delta(\epsilon)$. This enabled Pellegrino *et al.* (2008) to derive the analytical expression of the exclusons pair correlation function in arbitrary dimension

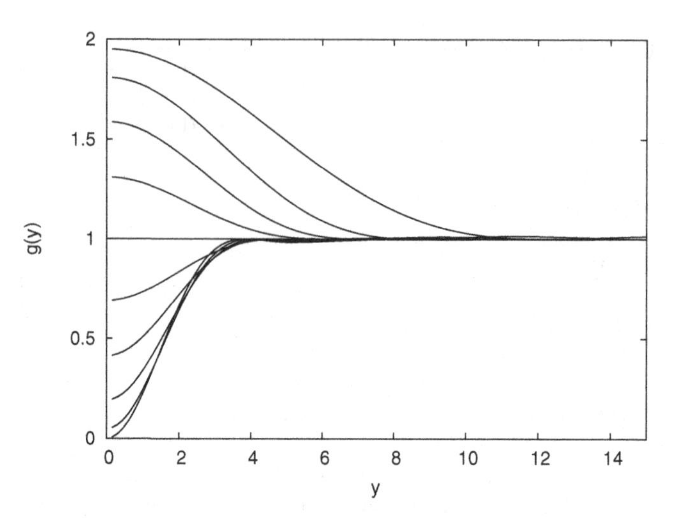

Fig. 4.6 Pair correlation function $g(y)$ (in scaled units), Eq. (4.90), in $d = 2$ for $\alpha = 0-1$ (α increases from top to bottom).

d at $T = 0$ as:

$$g(r) = 1 + \cos(\alpha\pi)\Gamma\left(\frac{d}{2}\right)\left(\frac{2}{k_{\mathrm{F},\alpha}r}\right)^d J^2_{d/2}(k_{\mathrm{F},\alpha}r), \tag{4.88}$$

where $J_{d/2}(z)$ is a Bessel function of first kind of order $d/2$ (Gradshteyn and Ryzhik, 1994), and $k_{\mathrm{F},\alpha}$ denotes the wavevector implicitly defined by:

$$C_d\left(\frac{k_{\mathrm{F},\alpha}}{2\pi}\right)^d = \alpha\frac{N}{V}, \tag{4.89}$$

where $C_d = \pi^{d/2}/\Gamma(d/2)$ is the volume of the unit sphere in d dimensions. In the fermion case ($\alpha = 1$), $\hbar k_{\mathrm{F},\alpha}$ coincides with the Fermi momentum.

Introducing the dimensionless variable, common to all values of α, defined as $y = k_{\mathrm{F}}r$, where $k_{\mathrm{F}} \equiv k_{\mathrm{F},\alpha=1}$ is the Fermi wavevector, Eq. (4.89) can be rewritten as

$$g(y) = 1 + \cos(\alpha\pi)\Gamma\left(\frac{d}{2}\right)\frac{2^d}{\alpha y^d}J^2_{d/2}(\alpha^{1/d}y). \tag{4.90}$$

Fig. 4.6 then shows $g(y)$ at $T = 0$ in $d = 2$ for $\alpha = 0-1$. In the fermionic regime ($\frac{1}{2} < \alpha \leq 1$), the pair correlation function exhibits a *hole* centred at $y = 0$, whose depth decreases with decreasing α, before disappearing at $\alpha = \frac{1}{2}$. On the other hand, for $0 < \alpha < \frac{1}{2}$, $g(y)$ shows rather a *hump* centred at $y = 0$, whose height decreases with increasing α. In the bosonic limit ($\alpha = 0$), the pair correlation function $g(y)$ is a constant, $g(y) = 2$

(Giuliani and Vignale, 2005). Similarly to the case $T \neq 0$, damped Friedel-like oscillations characterize the pair correlation function for each non-zero value of the statistical parameter α, except for the case $\alpha = \frac{1}{2}$ (Pellegrino *et al.*, 2008).

Making use of Eq. (4.88) and of the asymptotic properties of the Bessel function $J_\nu(z)$, in the limit $z \gg 1$, the wavelength Λ of the Friedel-like oscillations in the pair correlation function can be derived as:

$$\Lambda = \frac{2\pi}{k_{\mathrm{F},\alpha}} = \sqrt{\pi}\left(\alpha\Gamma\left(d/2\right)\frac{N}{V}\right)^{-\frac{1}{d}}. \tag{4.91}$$

For each dimensionality d, at $T = 0$ and at fixed density N/V, Λ increases with decreasing α, with $\Lambda \to \infty$ as $\alpha \to 0$ (bosonic limit).

On the basis of Eq. (4.91), one would find a finite wavelength Λ also for $\alpha = \frac{1}{2}$. However, the absence of Friedel-like oscillations in the pair correlation function in the so-called semionic case ($\alpha = \frac{1}{2}$) is not a consequence of the exclusion statistics in momentum space, but of the exchange properties implied by Eqs. (4.76) (Pellegrino *et al.*, 2007).

In addition to Friedel-like oscillations, generically due to the inflection point, Eq. (4.51), in Wu's distribution function, one also expects an exponential damping of the pair distribution function, with characteristic decay length $\lambda^2 k_{\mathrm{F},\alpha}$, where λ denotes the thermal wavelength. Accordingly, Pellegrino *et al.* (2008) then derived an approximate form of the pair correlation function at finite temperature, consisting in multiplying $g(r)$, which is exact at $T = 0$, by a damping factor with a characteristic decay length:

$$L = \lambda^2 k_{\mathrm{F},\alpha}. \tag{4.92}$$

This results in an approximate correlation function be defined as:

$$g(r) = 1 + e^{-y/Y}\cos(\alpha\pi)\frac{2^d\Gamma(d/2)}{k_{\mathrm{F},\alpha}r^d}J^2_{d/2}(k_{\mathrm{F},\alpha}r). \tag{4.93}$$

Similarly, at $T = 0$ the pair correlation function can be rewritten by means of a change of variable,

$$g(y) = 1 + e^{-y/Y}\cos(\alpha\pi)\frac{2^d\Gamma(d/2)}{\alpha y^d}J^2_{d/2}(\alpha^{1/d}y), \tag{4.94}$$

where the dimensionless quantity Y, defined as

$$Y = \alpha^{1/d}\lambda^2 k_{\mathrm{F}}^2, \tag{4.95}$$

acts as a characteristic decay length. Again, one recovers damped, Friedel-like oscillations in the quasi-Fermi regime, and humps in the boson regime.

It is clear from Eq. (4.95) that under the same conditions (*i.e.* temperature, density, dimensionality), the damping of the Friedel-like oscillations depends explicitly on the parameter α. Indeed, the value of the scaling variable Y increases with increasing α. Near the bosonic limit, the characteristic decay length is so short that the oscillations are completely damped out, whereas close to the fermionic limit the characteristic decay length is so large that the first Friedel-like oscillations survive. Pictorially, one can state that the ideal anyonic liquids losing Friedel-like oscillations in their pair correlation function at low T, are those which have smaller volume within the generalized Fermi surface.

4.4 Concluding remarks

The possible role of fractional statistics in the temperature dependence of some thermodynamic properties, such as the specific heat, of collective excitations, such as phonons or magnons, of two-dimensional materials, such as graphene or MgB_2, have been investigated by Angilella *et al.* (2006b).

Angilella *et al.* (2006c) also discussed a generalized BCS model for anyonic superconductivity. Specifically, they derived the ratio of the zero temperature superconducting gap and critical temperature, $\Delta(0)/k_{\rm B}T_c$, as a function of the statistical parameter α. This was found to increase monotonically from $\alpha = 0$ (bosons, where $\Delta(0)/k_{\rm B}T_c = 0$), to $\alpha = 1$ (fermions, where it reaches the BCS standard value).

Possible experimental realizations of anyonic fluids in reduced dimensional systems include a one-dimensional optical lattice, whose description would be based on the Hubbard model, but where fractional statistics would also play a prominent role (Keilmann *et al.*, 2011; Longhi and Della Valle, 2012; Greschner and Santos, 2015; Tang *et al.*, 2015). In particular, several competing ordered phases have been predicted to occur in this system. (See also Guan *et al.*, 2013, for a general review on the experimental realization of Fermi gases in $d = 1$ dimension.)

Evidence of fractional exclusion statistics has also been predicted to be detectable in spectroscopic measurements involving quantum Hall (or Laughlin) droplets (Cooper and Simon, 2015).

The connection between excluson statistics and topological matter in $d = 2$ dimensions has been the object of a study by Hu *et al.* (2014), which elucidates to some extent the structure of the Hilbert space in such systems, as well as several thermodynamic properties of their quasiparticle excitations. These results may be relevant to establish the occurrence of fractional Chern insulators, having properties similar to those of the fractional quantum Hall effect, in some two-dimensional organometallic materials (Li *et al.*, 2014).

Chapter 5

Superconductivity and superfluidity

5.1 Superconductivity

Superconductivity is perhaps one of the most spectacular manifestations of quantum mechanics at the macroscopic scale. In its most basic version, by superconductivity one means ideal electrical conductance (*zero* dc resistance), along with perfect diamagnetism (complete expulsion of an external magnetic field), below a critical temperature T_c. While the former effect was historically the first to be discovered (in elemental Hg below $T_c = 4.1$ K (Kamerlingh Onnes, 1911a,b,c)), it is probably perfect diamagnetism, or the Meissner-Ochsenfeld effect (Meissner and Ochsenfeld, 1933), which is most characteristic of superconductivity.

Ginzburg and Landau (1950) provided an effective albeit phenomenological theory, which enables to classify superconductivity as a phase transition characterized by an order parameter, *viz.* a field ψ (possibly non-homogeneous or time-dependent) which is exactly zero above, and different from zero below, the critical temperature T_c. Ginzburg-Landau theory allows to encompass superconductivity within a wide class of phase transitions, also including magnetism and superfluidity (see Chapter 9.1.3 and Sec. 5.2, respectively), almost universally described by this same theory. From a more general point of view, for such phase transitions it is usually true that, while the relevant thermodynamic potentials or the Hamiltonian are invariant with respect to a specific group of symmetry transformations at any temperature T, the ground state below T_c is only invariant with respect to a (smaller) subgroup of such transformations. This state of affairs is termed a *spontaneous symmetry breaking (SSB)*. For instance, in the case of magnetism, a system of many magnetic moments (atomic or molecular spins, say) is rotationally invariant above the Curie temperature

T_C, whereas a preferential direction develops as soon as a finite magnetization $\mathbf{M}$ sets in below T_C. In this case, it is the magnetization that plays the role of the order parameter ψ of Ginzburg-Landau theory.

The existence of an order parameter also for a superconductor later allowed Anderson (1963) to recognize the spontaneous breaking of a specific symmetry associated with the superconducting material interacting with the electromagnetic field. In the case of superconductivity, this turned out to be gauge invariance. From a phenomenological point of view, the SSB of gauge invariance effectively 'screens' a photon, *i.e.* the mediator of the electromagnetic interaction, within the superconductor, and endows it with a finite range λ. This is the same as the penetration length, *viz.* the distance from the surface of a superconducting bulk beyond which the magnetic field falls exponentially to zero. Such a penetration length makes its natural appearance in the Maxwell equations for the electromagnetic field, modified to take into account of the interaction with a superconducting medium, and allows to derive the phenomenological London equation. It is remarkable that the same mechanism based on SSB is at the core of Higgs (1964); Englert and Brout (1964) theory, describing the generation of mass for elementary particles, and is in fact dubbed the Anderson-Higgs mechanism, in certain (condensed matter oriented) contexts.[1]

5.1.1 *BCS theory*

Despite the success of Ginzburg-Landau theory to describe most of the phenomenology connected with superconductivity, it was not before three decades after its experimental discovery that a truly microscopic, quantum theory of superconductivity became available. This is the celebrated Bardeen-Cooper-Schrieffer, or BCS theory (Bardeen *et al.*, 1957). Together with Bloch's theory of band structure in crystalline solids and with Landau's theory of Fermi liquids, BCS theory is probably the strongest cornerstone of modern condensed matter theory.

BCS start by considering the Frölich (1954) Hamiltonian with a pairing

[1] It should be emphasized, however, that the paradigm entailed by Ginzburg-Landau theory, the existence of an order parameter, and the breakdown of a specific symmetry, does not exhaust all possible phase transitions, specifically those of quantum origin. Notable exceptions fall within the class of topological transitions, whereof the quantum Hall effects, and the recently discovered topological insulators are important examples of current and rapidly increasing interest. (See Thouless, 1998, for a review.)

interaction

$$H = H_0 + H_{\text{int}} = \sum_{\mathbf{k}\sigma} \xi_{\mathbf{k}} c^\dagger_{\mathbf{k}\sigma} c_{\mathbf{k}\sigma} + \sum_{\mathbf{k}\mathbf{k}'\sigma} V_{\mathbf{k}\mathbf{k}'} c^\dagger_{\mathbf{k}\sigma} c^\dagger_{-\mathbf{k},-\sigma} c_{-\mathbf{k}',-\sigma} c_{\mathbf{k}'\sigma}. \tag{5.1}$$

Here, $c^\dagger_{\mathbf{k}\sigma}$ ($c_{\mathbf{k}\sigma}$) creates (respectively, destroys) a particle in a Bloch state labelled by wavenumber $\mathbf{k}$, to be restricted within the first Brillouin zone (1BZ), and spin projection $\sigma \in \{\uparrow, \downarrow\}$ along a specified direction (say, z), with energy dispersion $\epsilon_{\mathbf{k}}$ and $\xi_{\mathbf{k}} = \epsilon_{\mathbf{k}} - \mu$ (μ is the chemical potential). The matrix element $V_{\mathbf{k}\mathbf{k}'}$ denotes the interaction potential, which within BCS theory is mediated by phonons, and will specified below.

Before giving an albeit self-contained account of BCS solution, it is instructive to briefly review the so-called Cooper problem (cf. March *et al.*, 1995).

5.1.1.1 *The Cooper problem*

This consists in attacking the inherently many-body problem of superconductivity from a much more simplified viewpoint. Namely, consider a single pair of electrons out of a well-formed Fermi sea. The simplifying assumption is that an arbitrarily small attractive interaction (of an as yet unspecified nature) is at work only between those two electrons (1 and 2, say), while it is 'switched off' between the two electrons and the remaining ones, and among the latter ones. So to speak, all the remaining electrons participate to the overall system as non-interacting 'spectators'. On the other hand, their Fermion nature makes them occupy all states below the Fermi level E_F, which is therefore well-defined. The problem may then be stated as the eigenvalue equation for the reduced Hamiltonian for the state $\psi(\mathbf{r}_1, \mathbf{r}_2)$ of the pair, namely

$$\left(-\frac{\hbar^2}{2m} \nabla_1^2 - \frac{\hbar^2}{2m} \nabla_2^2 + v(\mathbf{r}_1, \mathbf{r}_2) \right) \psi(\mathbf{r}_1, \mathbf{r}_2) = E \psi(\mathbf{r}_1, \mathbf{r}_2). \tag{5.2}$$

In the unperturbed case ($v = 0$), the ground state is obviously that of two free electrons, with energy $E = 2E_F$ and wavefunction given by the product of two plane waves normalized within a volume V,[2]

$$\psi^0_{\mathbf{k}}(\mathbf{r}_1,\mathbf{r}_2) = \frac{1}{V} e^{i\mathbf{k}\cdot(\mathbf{r}_1-\mathbf{r}_2)}. \tag{5.3}$$

Here, $|\mathbf{k}| > k_{\mathrm{F}}$, such restriction being an immediate consequence of the Pauli exclusion principle enforced by all the 'spectator' electrons below the Fermi level. Setting $E_k = 2\epsilon_{\mathbf{k}} = \hbar^2 k^2/m$, we now look for a solution of Eq. (5.2) in the form of a superposition of those unperturbed states, which form a basis:

$$\psi(\mathbf{r}_1,\mathbf{r}_2) = \sum_{|\mathbf{k}'|>k_{\mathrm{F}}} \alpha_{\mathbf{k}'} \frac{1}{V} e^{i\mathbf{k}'\cdot(\mathbf{r}_1-\mathbf{r}_2)}, \tag{5.4}$$

or, in second quantization,

$$\psi(\mathbf{r}_1,\mathbf{r}_2) = \sum_{|\mathbf{k}'|>k_{\mathrm{F}}} \alpha_{\mathbf{k}'} e^{i\mathbf{k}'\cdot\mathbf{r}_1} e^{-i\mathbf{k}'\cdot\mathbf{r}_2} c^\dagger_{\mathbf{k}'\uparrow} c_{-\mathbf{k}'\downarrow} |0\rangle, \tag{5.5}$$

where $|0\rangle$ denotes the unperturbed Fermi sea ground state. Substituting in the Schrödinger equation for the pair wavefunction, Eq. (5.2), multiplying by $\psi^{0*}_{\mathbf{k}}$ and integrating with respect to $\mathbf{r}_1$ and $\mathbf{r}_2$ over the volume V, one obtains

$$(E_{\mathbf{k}} - E)\alpha_{\mathbf{k}} = - \sum_{|\mathbf{k}'|>k_{\mathrm{F}}} \alpha_{\mathbf{k}'} v_{\mathbf{k},\mathbf{k}'}, \tag{5.6}$$

where

$$v_{\mathbf{k},\mathbf{k}'} \equiv v(\mathbf{k},-\mathbf{k},\mathbf{k}',-\mathbf{k}') = \frac{1}{V^2} \int d^3\mathbf{r}_1 d^3\mathbf{r}_2 e^{-i\mathbf{k}\cdot(\mathbf{r}_1-\mathbf{r}_2)} v(\mathbf{r}_1,\mathbf{r}_2) e^{i\mathbf{k}'\cdot(\mathbf{r}_1-\mathbf{r}_2)}. \tag{5.7}$$

We further impose that such a potential be different from zero and attractive only within an energy shell of width δ around the Fermi energy, *viz.*

$$v_{\mathbf{k},\mathbf{k}'} = -\frac{v}{V}, \text{if } E_F < \tfrac{1}{2}E_{\mathbf{k}}, \tfrac{1}{2}E_{\mathbf{k}'} < E_F + \delta, \text{ and 0 otherwise}, \tag{5.8}$$

with $v > 0$. Writing $v_{\mathbf{k},\mathbf{k}'}$ as

$$v_{\mathbf{k},\mathbf{k}'} = -\frac{v}{V}\theta_{\mathbf{k}}\theta_{\mathbf{k}'}, \tag{5.9}$$

---

[2]We are here assuming, as is customary, that the total momentum of the pair is null, $\mathbf{Q} = \mathbf{0}$. This implies, for the fully many-particle problem (Sec. 5.1.1.2), that the superconducting quantum condensate of Cooper pairs be translationally invariant. However, a spatially modulated density of Cooper pairs has been predicted in the so-called Fulde-Ferrell-Larkin-Ovchinnikov (FFLO) phase (Fulde and Ferrell, 1964; Larkin and Ovchinnikov, 1964), and observed in an ultracold ^{6}Li gas (Liao *et al.*, 2010) and, more recently, in the $Bi_2Sr_2CaCu_2O_{8+x}$ layered superconductor, by means of scanning tunneling microscopy (Hamidian *et al.*, 2016).

with $\theta_{\mathbf{k}} = 1$ if $E_F < \frac{1}{2}E_{\mathbf{k}} < E_F + \delta$ and 0 otherwise, shows that this is a particular case of a *separable potential.* Inserting Eq. (5.8) into Eq. (5.2), one finds

$$\alpha_{\mathbf{k}} = \frac{v\theta_{\mathbf{k}}}{E_{\mathbf{k}} - E} \frac{1}{V} \sum_{\mathbf{k}'}{}' \alpha_{\mathbf{k}'}\theta_{\mathbf{k}'} = \text{const} \times \frac{\theta_{\mathbf{k}}}{E_{\mathbf{k}} - E}, \tag{5.10}$$

where $\sum_{\mathbf{k}'}' \equiv \sum_{E_F < \frac{1}{2}E_{\mathbf{k}'} < E_F + \delta}$, which fixes the functional form of the pair wavefunction. In order to obtain information on the eigenvalue E, one sums again over $\mathbf{k}$, to obtain

$$\frac{1}{v} = \frac{1}{V} \sum_{\mathbf{k}}{}' \frac{1}{E_{\mathbf{k}} - E}, \tag{5.11}$$

which is the *Bethe-Goldstone equation* (in this approximation). A graphical solution of Eq. (5.11) (cf. March *et al.*, 1995) shows that, apart from a continuum of unbound states with energy E lying between $2E_F$ and $2(E_F + \delta)$, there always exists a solution with energy eigenvalue $E_0 < 2E_F$, corresponding to a bound or paired state, regardless of the strength of the pairing interaction, provided it to be positive. Denoting $\Delta = E_F - \frac{1}{2}E_0$ the gap energy per particle, and passing to the continuum limit, one finds

$$\Delta \approx \delta e^{-2/(\rho_F v)}, \tag{5.12}$$

with ρ_F the single-particle density of states at the Fermi level. The non-analytic dependence of Δ on the pairing potential v contained in Eq. (5.12) shows that Cooper pairing is a non-perturbative effect. This feature will indeed be retained in the fully many-body solution provided by BCS theory (cf. Sec. 5.1.1.2), whose nature is in fact variational. A more general version of the Bethe-Goldstone equation for the Cooper problem, Eq. (5.11), including an anisotropic, but again separable, potential, has been discussed by Angilella *et al.* (2004), who emphasized its possible relevance for the cuprates and other 'exotic' superconductors.

5.1.1.2 *Mean-field solution to the BCS Hamiltonian*

On the basis of the experimental evidence that $T_c \propto M^{-1/2}$, where M is the mass of the atomic isotope of the bulk superconducting material, Bardeen *et al.* (1957) surmised that the effective electron-electron interaction should have a phononic origin. They started therefore with Frölich Hamiltonian, Eq. (5.1), and considered a variational wavefunction allowing for (Cooper) pairs, *viz.*

$$|\text{BCS}\rangle = \prod_{\mathbf{k}} \left(u_{\mathbf{k}} + v_{\mathbf{k}} c^{\dagger}_{\mathbf{k}\uparrow} c^{\dagger}_{-\mathbf{k}\downarrow} \right) |0\rangle. \tag{5.13}$$

Here, $b_{\mathbf{k}}^\dagger = c_{\mathbf{k}\uparrow}^\dagger c_{-\mathbf{k}\downarrow}^\dagger$ is a pair creation operator, $u_{\mathbf{k}}$ and $v_{\mathbf{k}}$ are (complex) variational parameters, and $|\mathrm{BCS}\rangle$ is therefore a coherent superposition of states, each containing a given number of Cooper pairs. In terms of the pair creation operator, it may be indeed shown that

$$|\mathrm{BCS}\rangle = \mathrm{const} \times \exp\left(\sum_{\mathbf{k}} \frac{v_{\mathbf{k}}}{u_{\mathbf{k}}} b_{\mathbf{k}}^\dagger\right) |0\rangle. \tag{5.14}$$

The total number of Cooper pairs in the BCS trial state is thus undetermined: the macroscopic number of electrons provides one with a reservoir allowing for fluctuations in the particle number. It is also straightforward to show that, while

$$[b_{\mathbf{k}}^\dagger, b_{\mathbf{k}'}^\dagger] = 0, \tag{5.15a}$$

$$[b_{\mathbf{k}}, b_{\mathbf{k}'}] = 0, \tag{5.15b}$$

one has instead

$$[b_{\mathbf{k}}, b_{\mathbf{k}'}^\dagger] = \delta_{\mathbf{k}\mathbf{k}'}(1 - n_{\mathbf{k}\uparrow} - n_{-\mathbf{k}\downarrow}) \neq \delta_{\mathbf{k}\mathbf{k}'}, \tag{5.15c}$$

where $n_{\mathbf{k}\sigma} = c_{\mathbf{k}\sigma}^\dagger c_{\mathbf{k}\sigma}$ is the number operator in state $(\mathbf{k}, \sigma)$. Therefore, Cooper pairs are not strictly bosons (one cannot create two Cooper pairs in the same state), and one rather speaks of *hard core bosons.*

Historically, BCS characterized their variational ground-state via a variational estimate of $u_{\mathbf{k}}$, $v_{\mathbf{k}}$, through the minimization of the free energy (Bardeen *et al.*, 1957) (cf. Leggett, 1975, for a review). Another, although completely equivalent, route is that followed by Bogoliubov (1958); Valatin (1958), who employed a canonical transformation to diagonalize the BCS Hamiltonian, within the mean-field approximation (cf. Fetter and Walecka, 2004; Rickayzen, 1965; Jones and March, 1986b; March *et al.*, 1995, for a review).

Since the superconducting ground state is characterized by the presence of Cooper pairs, it is natural to assume that the expectation value of the pair number operator is nonzero in the superconducting state ($T < T_c$), and zero in the normal state ($T > T_c$):

$$\langle c_{-\mathbf{k}\downarrow} c_{\mathbf{k}\uparrow}\rangle \begin{cases} = 0, & T > T_c, \\ \neq 0, & T < T_c. \end{cases} \tag{5.16}$$

Therefore, it serves as an order parameter for the superconducting transition, in the sense of Landau. A mean-field treatment of the BCS Hamiltonian then consists of writing the pair operators in terms of fluctuation operators with respect to their average values in the ground state, *viz.*

$$b_{\mathbf{k}} = c_{-\mathbf{k}\downarrow} c_{\mathbf{k}\uparrow} \equiv \langle c_{-\mathbf{k}\downarrow} c_{\mathbf{k}\uparrow} \rangle + \delta b_{\mathbf{k}}, \tag{5.17a}$$

$$b^\dagger_{\mathbf{k}} = c^\dagger_{\mathbf{k}\uparrow} c_{-\mathbf{k}\downarrow} \equiv \langle c^\dagger_{\mathbf{k}\uparrow} c^\dagger_{-\mathbf{k}\downarrow} \rangle + \delta b^\dagger_{\mathbf{k}}. \tag{5.17b}$$

Substituting into the BCS Hamiltonian and neglecting terms quadratic in the fluctuations in the interaction part, one has

$$H = \sum_{\mathbf{k}\sigma} \xi_{\mathbf{k}} c^\dagger_{\mathbf{k}\sigma} c_{\mathbf{k}\sigma} - \sum_{\mathbf{k}} (\Delta_{\mathbf{k}} c^\dagger_{\mathbf{k}\uparrow} c^\dagger_{-\mathbf{k}\downarrow} + \text{H.c.}) + \text{const}, \tag{5.18}$$

where we have introduced the *energy gap*

$$\Delta_{\mathbf{k}} = - \sum_{\mathbf{k}'} V_{\mathbf{k}\mathbf{k}'} \langle c_{-\mathbf{k}'\downarrow} c_{\mathbf{k}'\uparrow} \rangle. \tag{5.19}$$

The BCS Hamiltonian within the mean-field approximation is now manifestly a quadratic (Hermitean) form, that can be easily diagonalized by means of the canonical transformation (Bogoliubov, 1958; Valatin, 1958)

$$\gamma_{\mathbf{k}\uparrow} = u_{\mathbf{k}} c_{\mathbf{k}\uparrow} - v_{\mathbf{k}} c^\dagger_{-\mathbf{k}\downarrow}, \tag{5.20a}$$

$$\gamma^\dagger_{-\mathbf{k}\downarrow} = v^*_{\mathbf{k}} c_{\mathbf{k}\uparrow} + u^*_{\mathbf{k}} c^\dagger_{-\mathbf{k}\downarrow}, \tag{5.20b}$$

where

$$|u_{\mathbf{k}}|^2 = \frac{1}{2} \left(1 + \frac{\xi_{\mathbf{k}}}{E_{\mathbf{k}}} \right), \tag{5.21a}$$

$$|v_{\mathbf{k}}|^2 = \frac{1}{2} \left(1 - \frac{\xi_{\mathbf{k}}}{E_{\mathbf{k}}} \right) \tag{5.21b}$$

are BCS *coherence factors*, with $E^2_{\mathbf{k}} = \xi^2_{\mathbf{k}} + |\Delta_{\mathbf{k}}|^2$, and the energy gap is self-consistently determined by the *gap equation*

$$\Delta_{\mathbf{k}} = - \sum_{\mathbf{k}'} V_{\mathbf{k}\mathbf{k}'} \frac{\Delta_{\mathbf{k}'}}{2E_{\mathbf{k}'}} \tanh \left(\frac{\beta E_{\mathbf{k}'}}{2} \right). \tag{5.22}$$

An approximate solution of the above gap equation close to $T = 0$ for an isotropic phonon-mediated attractive interaction yields $\Delta(T = 0) \approx 2\hbar\omega_D e^{-1/(\lambda\rho_F)}$ in the weak-coupling limit ($\lambda\rho_F \ll 1$, λ being the electron-phonon coupling, and ω_D the Debye frequency), in qualitative agreement with the Cooper problem solution, Eq. (5.12). One also finds that the momentum distribution for the superconducting state is given by

$$n_{\mathbf{k}} = 1 - \frac{\xi_{\mathbf{k}}}{E_{\mathbf{k}}} \tanh \left(\frac{\beta E_{\mathbf{k}}}{2} \right). \tag{5.23}$$

5.1.2 *Richardson model*

The success of BCS theory also relies in the fact that it applies to *macroscopic* superconducting samples. The relative size of the samples is implicitly taken into account into the mean-field approximation, in that the number of Cooper pairs is assumed to be relatively *large,* and therefore number fluctuations can be safely neglected. Actually, that the BCS theory is indeed an approximation can be seen directly from the form of the BCS trial wavefunction for the superconducting ground state, Eq. (5.13), which is in fact a coherent superposition of states with different numbers of electrons, therefore with an undefined number of Cooper pairs: only its thermodynamic average is well defined (and relatively large). This is in contrast with the fact that the BCS Hamiltonian, Eq. (5.1), does commute with the total number operator, and therefore the *exact* ground state must have a well defined number of electrons and pairs.

Deviations from BCS theory are therefore to be expected when number fluctuations start to become important, *viz.* close to T_c and/or for sample size down to the nanoscale size. Although the theory of fluctuations in superconductors was given birth almost at the same time as Ginzburg-Landau theory (Ginzburg, 1960; Levanyuk, 1959), it was not before sufficiently small superconducting samples, such as aluminum nanograins (Black *et al.*, 1996), or superconductors with sufficiently large critical temperatures T_c's and sufficiently small coherence lengths ξ, such as the cuprate superconductors (Bednorz and Müller, 1986), became available that it was possible to test such a theory experimentally, and consider the relevance of fluctuations both on thermodynamic and dynamical (*i.e.* transport) properties (see Larkin and Varlamov, 2002, 2005, for a review).

Within the context of nuclear theory, the BCS Hamiltonian, Eq. (5.1), also describes pairing between nucleons with an effective nucleon-nucleon interaction $V_{\mathbf{kk}'} \equiv v$, and is in fact called *pairing model* in that context. Soon after BCS theory, Richardson (1963) derived a set of algebraic equations which yield all eigenvalues and eigenstates of the pairing model. Such equations are in principle *exact,* although their solution obviously imply some numerical approximation.

Richardson (1963) starts by observing that the interaction part in the BCS Hamiltonian, Eq. (5.1), acts only on doubly-occupied $\mathbf{k}$-states. In other words, k-states which are occupied by a single electron do not take part in the dynamics. The k-labels of these are therefore good quantum numbers, and may be used to describe the overall ground state. Here,

$k \equiv (\mathbf{k}, \sigma)$.

Let $\Omega = \{k\colon |k\rangle \text{is singly occupied}\}$ denote the set of singly-occupied k-states (or a subset thereof), and let M denote the number of such states. Then, the generic eigenstate of the BCS Hamiltonian H can be written as

$$|N; M\rangle = \prod_{k\in\Omega} c_k^\dagger |N\rangle, \tag{5.24}$$

where

$$|N\rangle = \sum_{k_1,\ldots k_N \notin \Omega} \alpha_{k_1,\ldots k_N} b_{k_1}^\dagger \ldots b_{k_N}^\dagger |0\rangle \tag{5.25}$$

is a generic state with exactly N pairs of electrons in the states $(\mathbf{k}_1, \uparrow)$, $(-\mathbf{k}_1, \downarrow)$, $\ldots(\mathbf{k}_N, \uparrow)$, $(-\mathbf{k}_N, \downarrow)$. Due to the 'hard core boson' condition specified above, $(b_k^\dagger)^2 = 0$, one can restrict the summation to different momenta only, $k_1 \neq k_2 \neq \ldots \neq k_N$. Furthermore, $[b_k^\dagger, b_{k'}^\dagger] = 0$, and therefore the pairing amplitudes $\alpha_{k_1,\ldots k_N}$ can be chosen as symmetric with respect to permutations of the indices $k_1, \ldots k_N$. The eigenstate $|N; M\rangle$ therefore contains a well-defined number $2N+M$ of electrons, whereof N are paired, and M are in singly-occupied states. The latter states are unaffected by the interaction part of the Hamiltonian, and therefore contribute with a constant term $E(M) = \sum_{k\in\Omega} \epsilon_k$ to the total energy. For this reason, one can safely neglect those electrons, and assume $\Omega = \emptyset$ $(M = 0)$ in what follows, and refer directly to $|N\rangle$ as given by Eq. (5.25), without loss of generality.

Since one is assuming $k_1 \neq k_2 \neq \ldots \neq k_N$, one can omit the term $n_{\mathbf{k}\uparrow} + n_{-\mathbf{k}\downarrow}$ in Eq. (5.15c). In other words, the operators b_k, $b_{k'}$ do act as boson operators within the restricted Hilbert space in Eq. (5.25), as long as we create at most one hard core boson in a given state k (this restriction is of course not required for true bosons). One explicitly finds

$$\begin{aligned}
& b_k b_{k_1}^\dagger b_{k_2}^\dagger \ldots b_{k_N}^\dagger |0\rangle \\
&= [b_k, b_{k_1}^\dagger] b_{k_2}^\dagger \ldots b_{k_N}^\dagger |0\rangle + b_{k_1}^\dagger [b_k, b_{k_2}^\dagger] \ldots b_{k_N}^\dagger |0\rangle + \ldots + b_{k_1}^\dagger b_{k_2}^\dagger \ldots [b_k, b_{k_N}^\dagger] |0\rangle \\
&= \delta_{kk_1} b_{k_2}^\dagger \ldots b_{k_N}^\dagger |0\rangle + \delta_{kk_2} b_{k_1}^\dagger \ldots b_{k_N}^\dagger |0\rangle + \ldots + \delta_{kk_{N-1}} b_{k_1}^\dagger b_{k_2}^\dagger \ldots b_{k_{N-1}}^\dagger |0\rangle.
\end{aligned} \tag{5.26}$$

In order to understand the procedure leading to Richardson's equations, let us consider the simplest nontrivial case, *viz.* that of $N = 2$ Cooper pairs

$$|N = 2\rangle = \sum_{p\neq q} \alpha_{pq} b_p^\dagger b_q^\dagger |0\rangle, \tag{5.27}$$

with $\alpha_{pq} = \alpha_{qp}$. Operating with H on both sides, assuming an isotropic interaction for the sake of simplicity, and making use of Eq. (5.26), one finds

$$H|N=2\rangle = \sum_{p\neq q} \alpha_{pq} \left((2\epsilon_p + 2\epsilon_q) b_p^\dagger b_q^\dagger - v \sum_{k\neq q} b_k^\dagger b_q^\dagger - v \sum_{k\neq p} b_p^\dagger b_k^\dagger \right) |0\rangle. \tag{5.28}$$

Swapping indices as $k \leftrightarrow p$ in the first inmost summation, and as $k \leftrightarrow q$ in the second one, one may cast the eigenvalue equation $H|N=2\rangle = E|N=2\rangle$ as a system of linear algebraic equations for the amplitudes α_{pq}:

$$(2\epsilon_p + 2\epsilon_q)\alpha_{pq} - v\sum_{k\neq q} \alpha_{kq} - v \sum_{k\neq p} \alpha_{pk} = E\alpha_{pq}. \tag{5.29}$$

The constraint in the summations is reminiscent of the fact that no more than one Cooper pair is allowed in a given pairing state. This does not prevent the existence of the solution, which may be sought in the (naturally symmetric) form

$$\alpha_{pq} = \alpha_p^{(1)}\alpha_q^{(2)} + \alpha_q^{(1)}\alpha_p^{(2)}, \tag{5.30}$$

with $E = E_1 + E_2$, correspondingly. The summations can be extended to all states, if an extra term is subtracted to Eq. (5.29). One thus finds

$$\begin{aligned} &\left(\alpha_q^{(2)}(2\epsilon_p - E_1)\alpha_p^{(1)} + \alpha_q^{(1)}(2\epsilon_p - E_2)\alpha_p^{(2)}\right) + \{p \leftrightarrow q\} \\ &= v\left(\alpha_q^{(2)}\sum_k \alpha_k^{(1)} + \alpha_q^{(1)}\sum_k \alpha_k^{(2)} + \{p\leftrightarrow q\}\right) - 2v\left(\alpha_p^{(1)}\alpha_p^{(2)} + \alpha_q^{(1)}\alpha_q^{(2)}\right) \end{aligned} \tag{5.31}$$

It is actually this extra term, *i.e.* last term in the right-hand side in Eq. (5.31), that introduces a sort of 'exchange' correlation between the amplitudes in the two Cooper pair states. Without the extra term, the solution of Eq. (5.31) would read as for the Cooper problem, Sec. 5.1.1.1, as

$$\alpha_p^{(1)} = \frac{1}{2\epsilon_p - E_1}, \tag{5.32a}$$

$$\alpha_p^{(2)} = \frac{1}{2\epsilon_p - E_2}, \tag{5.32b}$$

with E_i a root of

$$\sum_k \frac{1}{2\epsilon_k - E_i} = \frac{1}{v}. \tag{5.32c}$$

This solution would correspond to two *independent* pairs.

The inclusion of the pair-to-pair correlation term in Eq. (5.31) does not alter the form of the solutions, Eqs. (5.32a) and (5.32b), for the amplitudes $\alpha_p^{(1)}$ and $\alpha_p^{(2)}$, but does modify the structure of the equation for the eigenvalue, which becomes now nonlinear, although retaining an algebraic character. Indeed, from Eqs. (5.32a) and (5.32b), it follows that

$$\alpha_p^{(1)}\alpha_p^{(2)} = \frac{\alpha_p^{(2)} - \alpha_p^{(1)}}{E_2 - E_1}, \tag{5.33}$$

with which Eq. (5.31) can be rewritten as

$$\begin{aligned} &\alpha_q^{(2)}\left((2\epsilon_p - E_1)\alpha_p^{(1)} - v\sum_k \alpha_k^{(1)} + \frac{2v}{E_2 - E_1}\right) + \{p \leftrightarrow q\} \\ &+ \alpha_q^{(1)}\left((2\epsilon_p - E_2)\alpha_p^{(2)} - v\sum_k \alpha_k^{(2)} - \frac{2v}{E_2 - E_1}\right) + \{p \leftrightarrow q\} = 0. \end{aligned} \tag{5.34}$$

Eqs. (5.32a) and (5.32b) are still solutions, but now the pair energies E_1 and E_2 are solutions of the *coupled* system of (nonlinear) algebraic equations

$$\sum_k \frac{1}{2\epsilon_k - E_1} = \frac{1}{v} + \frac{2}{E_2 - E_1}, \tag{5.35a}$$

$$\sum_k \frac{1}{2\epsilon_k - E_2} = \frac{1}{v} + \frac{2}{E_1 - E_2}. \tag{5.35b}$$

In the more general case of an arbitrary number N of Cooper pairs, the Richardson equations can be derived in a similar, albeit somewhat more complicated, way. The amplitude $\alpha_{k_1\ldots k_N}$ is now a (globally) symmetric tensor in the given indices, and one may make use of the Ansatz

$$\alpha_{k_1\ldots k_N} = \sum_P \alpha_{k_{P(1)}}^{(1)} \cdots \alpha_{k_{P(N)}}^{(N)} = \sum_P \prod_{i=1}^N \alpha_{k_{P(i)}}^{(i)} \tag{5.36}$$

to determine it, where P is a generic permutation of the given indices. After some straightforward calculation, the Richardson equations read

$$\begin{aligned} \sum_P \sum_{i=1}^N \Bigg[&\left(\prod_{j\neq i}^N \alpha_{k_{P(j)}}^{(j)}\right)\left((2\epsilon_{k_{P(i)}} - E_i)\alpha_{k_{P(i)}}^{(i)} - v\sum_p \alpha_p^{(i)}\right) \\ &+ v\sum_{\ell\neq i}^N \left(\prod_{j\neq i,\ell}^N \alpha_{k_{P(j)}}^{(j)}\right)\alpha_{k_{P(\ell)}}^{(i)}\alpha_{k_{P(\ell)}}^{(\ell)}\Bigg] = 0. \end{aligned} \tag{5.37}$$

Their solutions can be presented again in the form

$$\alpha_k^{(i)} = \frac{1}{2\epsilon_k - E_i}, \tag{5.38}$$

from which it follows that

$$\alpha_k^{(i)} \alpha_k^{(j)} = \frac{\alpha_k^{(j)} - \alpha_k^{(i)}}{E_j - E_i}. \tag{5.39}$$

This allows to decouple the eigenvalue equation as

$$\sum_P \sum_{i=1}^N \left(\prod_{j \neq i}^N \alpha_{k_{P(j)}}^{(j)} \right) \times \left((2\epsilon_{k_{P(i)}} - E_i) \alpha_{k_{P(i)}}^{(i)} - v \sum_p \alpha_p^{(i)} + 2v \sum_{\ell \neq i} \frac{1}{E_\ell - E_i} \right) = 0, \tag{5.40}$$

where now the pair energies are solutions of the coupled system of (nonlinear) algebraic equations

$$\sum_k \frac{1}{2\epsilon_k - E_i} = \frac{1}{v} + 2 \sum_{\ell \neq i} \frac{1}{E_\ell - E_i} \equiv \frac{1}{v_i}. \tag{5.41}$$

The last equation shows, in particular, that the pairing potential effectively seen by the two electrons in the pairing state i is renormalized by the energy spectrum of all other pairs.

We refer to Dukelsky *et al.* (2004) for a recent review of the Richardson model, and its numerical solution both in condensed matter and in nuclear systems. In particular, the numerical solution of the Richardson model with application to ultrasmall superconducting grains has been obtained by means of the Lanczos algorithm by Mastellone *et al.* (1998). The effect of the pairing correlations on the thermodynamic properties of small superconductors with a fixed number of Cooper pairs has been studied by Gambacurta and Lacroix (2012), with applications also to nuclear systems.

5.1.3 *Geminals for pairing correlations*

The success of the BCS variational ground state, Eq. (5.13), relies on the fact that it embeds, since the outstart, pair correlations. That pair correlations should be included in any appropriate description of a many-body ensemble was apparent also in the early beginnings of quantum chemistry. In particular, somehow foreshadowing modern density functional theory

(DFT), Coulson (1960) stated that "It has frequently been pointed out that a conventional many-electron wave function tells us more than we need to know. (...) There is an instinctive feeling that matters such as electron correlation should show up in the two-particle density matrix" (also quoted in Coleman, 1963) (cf. Chapter 1). This prompted the interest of both theoretical quantum chemists and, more recently, several solid-state theorists, towards two-particle functions, or *geminals*,[3] which generalize the concept of orbitals, *i.e.* single-particle functions, with applications to molecules and other correlated condensed phases.

The main idea dates back at least to Lewis (1916), and consists in assuming electron pair states, rather than single-particle orbitals, as the fundamental building blocks for the electronic wave function. One then starts by considering an antisymmetrized product of geminals (APG), which in second-quantization reads

$$|\Psi_{\mathrm{APG}}\rangle = \prod_{p=1}^{P} b_p^\dagger |\theta\rangle, \tag{5.42a}$$

where

$$b_p^\dagger = \sum_{k=1}^{2K} \sum_{\ell=1}^{2K} c_{p;k\ell} a_k^\dagger a_\ell^\dagger \tag{5.42b}$$

is a geminal creation operator. Here, $a_k^\dagger$ creates an electron in the kth spin-orbital, there are K spatial orbitals (and $2K$ spin-orbitals), and $P = N/2$ electron pairs, where N is the number of electrons outside the vacuum reference state. In Eqs. (5.42), $|\theta\rangle$ denotes the vacuum with respect to the creation of geminals, *i.e.* it does not need to be a zero-electron state. For example, an odd number of electrons can be treated by choosing a state with an odd number of electrons as the vacuum. From the point of view of quantum chemistry applied *e.g.* to molecules, Eqs. (5.42) are appealing in that geminals describe the formation or breaking of chemical bonds.

One may explicitly verify that the following commutation relations hold

$$[b_{p'}, b_p^\dagger] = \delta_{pp'} - 4 \sum_{k\ell\ell'} c_{p';k\ell'}^* c_{p;k\ell} a_\ell^\dagger a_{\ell'}, \tag{5.43}$$

showing that the geminal operators are ladder operators for composite bosons. The analogy of Eqs. (5.42) with Eq. (5.13), and of Eq. (5.43) with Eq. (5.15), is apparent, and demonstrates the possible relevance of

---

[3]From *gemini*, Latin for 'twins'.

the geminals formalism also for condensed phases, such as superconductivity and superfluidity (see Coleman and Yukalov, 2000, for a review). The basic differences are that (i) the BCS trial ground state includes pairs of electrons in itinerant states (Bloch waves), while geminals create pairs of electrons in (possibly localized) one-particle spin-orbitals, and (ii) the BCS trial ground state is not number-conserving.

By performing a change of basis, the complex expansion coefficients $c_{p;k\ell} \in \mathbb{C}$ in Eqs. (5.42), with $c_{p;k\ell} = -c_{p;\ell k}$, can be written in their equivalent 'natural' form, in which the orbital-pairing scheme is explicit (Löwdin, 1955)

$$|\Psi_{\mathrm{APG}}\rangle = \prod_{p=1}^{P} \left(\sum_{k=1}^{K} c_{p;k} a^{\dagger}_{p;2k-1} a^{\dagger}_{p;2k} \right) |\theta\rangle. \tag{5.44}$$

In the case of an atom or a molecule containing an even number N of electrons, it has been demonstrated that the energy is an exact functional of a single geminal, thereby laying the basis for a 'geminal functional theory' (Smith, 1966; Coleman, 1997; Mazziotti, 2000). It soon became evident that geminals, although appealing from the theoretical point of view, were numerically intractable, even through several modifications, such as orthogonalization. However, quite recent advances in the field are reviving the concept of geminals and its computational manageability (Limacher *et al.*, 2013).

The relevance of reduced density matrices for ordered phases, such as superconductors and superfluids, has been studied by Coleman and Yukalov (1996). These authors first discuss the consequences of a phase transition on the eigenvalues of the density matrices, through the concept of *order index.* They start by generalizing the concept of nth order reduced density matrix (already introduced in Chapter 1)[4] to embrace temperature as

$$\rho_n(\mathbf{r}_1, \dots \mathbf{r}_n; \mathbf{r}'_1, \dots \mathbf{r}'_n) = \langle \psi^\dagger(\mathbf{r}'_1) \dots \psi^\dagger(\mathbf{r}'_n) \psi(\mathbf{r}_n) \dots \psi(\mathbf{r}_1) \rangle \tag{5.45}$$

at equilibrium, where ψ ($\psi^\dagger$) are quantum field annihilation (creation) operators, and the brackets imply averaging with respect to the quantum statistical operator, $\langle A \rangle = \mathrm{Tr}(e^{-\beta H} A) / \mathrm{Tr}(e^{-\beta H})$. The generic element ρ_n of the nth order reduced density matrix in Eq. (5.45) can be viewed as the kernel of an integral operator $\varphi \mapsto \rho_n \circ \varphi$ defined as

$$\rho_n \circ \varphi(\mathbf{r}_1, \dots \mathbf{r}_n) = \int d^3\mathbf{r}'_1 \dots d^3\mathbf{r}'_n \, \rho_n(\mathbf{r}_1, \dots \mathbf{r}_n; \mathbf{r}'_1, \dots \mathbf{r}'_n) \varphi(\mathbf{r}'_1, \dots \mathbf{r}'_n). \tag{5.46}$$

[4]Here, we follow the notation and the normalization convention as in Coleman (1963); Coleman and Yukalov (1996), which differs from that of Chapter 1.

One may then consider the eigenvalue problem[5]

$$\rho_n \circ \varphi_{ni} = \gamma_{ni}\varphi_{ni} \tag{5.47}$$

and define the (spectral) norm of ρ_n as its largest eigenvalue (see Yukalov and Yukalova, 2013, for a review)

$$\| \rho_n \| = \sup_i \gamma_{ni}. \tag{5.48}$$

Note that $\gamma_{ni} \geq 0$, since ρ_n is positive semi-definite. The *order index* is then defined as (Coleman and Yukalov, 1991, 1992, 1993)

$$\alpha_n = \lim_{N\to\infty} \frac{\log \| \rho_n \|}{\log N}. \tag{5.49}$$

In the thermodynamic limit, this implies that the largest eigenvalue of ρ_n scales as $\sim N^{\alpha_n}$. It has been demonstrated (see Coleman, 1963, and references therein) that the order indices satisfy the inequalities

$$\begin{aligned} 0 \leq \alpha_n \leq \frac{n}{2}, &\quad \text{(fermions)}, \\ 0 \leq \alpha_n \leq n, &\quad \text{(bosons)}. \end{aligned} \tag{5.50}$$

Coleman and Yukalov (1996) then define a characteristic index as

$$\alpha_n^0 = \begin{cases} n/2, & \text{if } n \text{ is even}, \\ (n-1)/2, & \text{if } n \text{ is odd}\ , \end{cases} \tag{5.51}$$

which enables them to classify *off-diagonal order* as (Coleman and Yukalov, 1993)

$$\begin{aligned} \alpha_n = 0, &\quad \text{short-range or absence of order}, \\ 0 < \alpha_n < \alpha_n^0, &\quad \text{even mid-range order}, \\ \alpha_n = \alpha_n^0, &\quad \text{even long-range order}, \end{aligned} \tag{5.52}$$

for both fermions and bosons, and

$$\begin{aligned} \alpha_n^0 < \alpha_n < n, &\quad \text{total mid-range order}, \\ \alpha_n = n, &\quad \text{total long-range order}, \end{aligned} \tag{5.53}$$

[5]It should be remarked that, in the case $n = 1$, ρ_n reduces to the one-particle density matrix $\rho_1 \equiv \rho$, which, with the normalization set out in Chapter 1, is explicitly defined as $\rho(\mathbf{r}, \mathbf{r}') = N \int d^3\mathbf{r}_2 \dots d^3\mathbf{r}_N\, \Psi^*(\mathbf{r}, \mathbf{r}_2, \dots \mathbf{r}_N)\Psi(\mathbf{r}', \mathbf{r}_2, \dots \mathbf{r}_N)$. Then, the eigenvectors of the one-particle density matrix define the *natural orbitals*, $\varphi_i(\mathbf{r})$ say, and its eigenvalues the associated *occupation numbers*, n_i (Löwdin, 1955). Making use of the spectral theorem, one may then express the one-particle density matrix as $\rho = \sum_i n_i |\varphi_i\rangle\langle\varphi_i|$. One finds $0 \leq n_i \leq 1$, and $\sum_i n_i = N$, with $n_i = 0$ or 1 for free fermions. These two conditions are equivalent to the *N-representability* of the density matrix, *i.e.* the fact that it must be a (Hermitean) matrix that can be obtained from an ensemble of (fermionic) many-particle wave functions (Coleman, 1963; Löwdin, 1955).

for bosons.

In order to illustrate the role and relevance of the order index, Coleman and Yukalov (1996) discuss the case of the superconducting transition of an ensemble of N spin one-half electrons. For a uniform system, the field operator can be expanded in plane waves as

$$\psi_\sigma(\mathbf{r}) = \sum_{\mathbf{k}} c_{\mathbf{k}\sigma}\varphi_{\mathbf{k}\sigma}(\mathbf{r}), \tag{5.54}$$

where $\varphi_{\mathbf{k}\sigma}(\mathbf{r}) = \mathcal{V}^{-1/2}e^{i\mathbf{k}\cdot\mathbf{r}}\chi_\sigma$, and $\sigma \in \{\uparrow,\downarrow\}$. The spin-resolved momentum distribution $n_{\mathbf{k}\sigma} = \langle c^\dagger_{\mathbf{k}\sigma}c_{\mathbf{k}\sigma}\rangle$ may then be related to the one-particle density matrix $\rho_1(\mathbf{r},\mathbf{r}') = \sum_\sigma \langle \psi^\dagger_\sigma(\mathbf{r}')\psi_\sigma(\mathbf{r})\rangle$ as (cf. footnote 5 on page 89)

$$\rho_1(\mathbf{r},\mathbf{r}') = \sum_{\mathbf{k}} n_{\mathbf{k}} \left(\varphi^*_{\mathbf{k}\uparrow}(\mathbf{r})\varphi_{\mathbf{k}\uparrow}(\mathbf{r}') + \varphi^*_{-\mathbf{k}\downarrow}(\mathbf{r})\varphi_{-\mathbf{k}\downarrow}(\mathbf{r}')\right), \tag{5.55}$$

where we have assumed $n_{\mathbf{k}\sigma} = \frac{1}{2}n_{\mathbf{k}}$ for both spin projections, and $n_{-\mathbf{k}} = n_{\mathbf{k}}$. In the superconducting state, $n_{\mathbf{k}}$ is given by Eq. (5.23), while in the normal state it reduces to the Fermi distribution function, $f_{\mathbf{k}}$ say. Making use of the occupation number for the superconducting state within BCS theory, Eq. (5.23), one finds that

$$\| \rho_1 \| = \sup_{\mathbf{k}} n_{\mathbf{k}} \leq 1. \tag{5.56}$$

Thus, the first-order index vanishes

$$\alpha_1 = 0. \tag{5.57}$$

Coming now to the $n = 2$ reduced density matrix, it is well known (Gutzwiller, 1962; Coleman, 1964) that, for a BCS-like Hamiltonian, thanks to the relationship between density matrices and Green's functions (Blöchl *et al.*, 2013), one has

$$\begin{aligned}\rho_2(\mathbf{r}_1,\mathbf{r}_2;\mathbf{r}'_1,\mathbf{r}'_2) &= \rho_1(\mathbf{r}_1,\mathbf{r}'_1)\rho_1(\mathbf{r}_2,\mathbf{r}'_2) \\ &\quad + \rho_1(\mathbf{r}_1,\mathbf{r}'_2)\rho_1(\mathbf{r}_2,\mathbf{r}'_1) - \chi^*(\mathbf{r}_1,\mathbf{r}_2)\chi(\mathbf{r}'_1,\mathbf{r}'_2),\end{aligned} \tag{5.58}$$

where χ is related to the 'anomalous' Green's function (Valatin, 1961; Gutzwiller, 1962) as

$$\begin{aligned}\chi(\mathbf{r}'_1,\mathbf{r}'_2) &= \sum_{\mathbf{k}} \left(n_{\mathbf{k}}(1-n_{\mathbf{k}}) - f_{\mathbf{k}}(1-f_{\mathbf{k}})\right)^{1/2} \\ &\quad \times \left(\varphi_{\mathbf{k}\uparrow}(\mathbf{r}'_1)\varphi_{-\mathbf{k}\downarrow}(\mathbf{r}'_2) - \varphi_{-\mathbf{k}\downarrow}(\mathbf{r}'_1)\varphi_{\mathbf{k}\uparrow}(\mathbf{r}'_2)\right),\end{aligned} \tag{5.59}$$

where $f_{\mathbf{k}}$ is Fermi distribution, while $n_{\mathbf{k}}$ is the BCS occupation number, Eq. (5.23). Since $\| \chi \|^2 \sim N$ for a superconductor in the thermodynamic limit (Yang, 1962) (and zero in the normal state), it follows that (Coleman and Yukalov, 1996)

$$\alpha_2 = 1. \tag{5.60}$$

More generally, it has been shown that

$$\alpha_{2n} = \alpha_{2n+1} = n\alpha_2. \tag{5.61}$$

Thus superconductivity falls into the category that Coleman and Yukalov (1996) have termed even long-range order. However, even though a non-zero order parameter, Eqs. (5.16) and (5.19), implies non-zero order indices, corresponding to even long-range order, this does not rule out the possibility of an anisotropic gap, characterized by nodal loci in the first Brillouin zone, as is the case of the high-T_c cuprates. In this context, it is relevant to mention that the study of the occupation numbers of natural orbitals and geminals has been brought into contact with pairing correlation effects, such as the reduction of the Fermi line, as observed *e.g.* in the cuprates by angle-resolved photoemission spectroscopy (Barbiellini, 2014).

We will come back on the relevance of order indices for characterizing phase transitions later in Section 5.2.5, after a general introduction on superfluidity.

5.2 Superfluidity

5.2.1 *General introduction to the Bose-Einstein condensation and superfluidity*

Superfluidity is a macroscopic manifestation of the phenomenon of Bose-Einstein condensation (BEC). This was discovered by Einstein in the early 1920s (Einstein, 1924, 1925) (see Rickayzen, 1965; Minguzzi *et al.*, 2003, for a review) in generalizing Bose statistics, originally formulated for photons, to material particles ('Bosons') with integer spin. Since their collective wave function must be completely symmetrical with respect to the interchange of all quantum numbers of any two of them, the occupancy of any single quantum state is not *a priori* restricted (as is the case for 'Fermions', particles with half-integer spin, obeying Fermi-Dirac statistics). In fact, it turns out that all particles of an ideal Bose gas condense in the ground-state at $T = 0$, such a condensation starting at a critical temperature T_c, determined by the condition of vanishing of the chemical potential, which can be expressed as

$$n\lambda_{\rm dB}^3(T = T_c) = \zeta(\tfrac{3}{2}) \simeq 2.612. \tag{5.62}$$

Here, n is the particle number density, and $\lambda_{\rm dB} = \sqrt{(2\pi\hbar^2)/(mk_{\rm B}T)}$ is the de Broglie thermal wavelength. Eq. (5.62) implies that the linear scale at which quantum effects are essential becomes comparable with the interparticle mean separation at temperature T_c, thereby enabling quantum interference among the individual particles wave functions to conjure towards the formation of a single, coherent quantum state.

Denoting by n_0 the particle number density in the ground state, for an ideal Bose gas one finds that the condensate fraction increases with decreasing temperature below T_c as

$$\frac{n_0}{n} = 1 - \left(\frac{T}{T_c}\right)^{\frac{3}{2}}, \tag{5.63}$$

with $n_0/n = 1$ at exactly $T = 0$. This value is reduced due to correlations in a Bose liquid with hard-core interactions, where (Bogoliubov, 1947)

$$\left.\frac{n_0}{n}\right|_{T=0} = 1 - \frac{8}{3\sqrt{\pi}}\sqrt{na^3}. \tag{5.64}$$

Here, a is the hard-core diameter, and na^3 is a dimensionless measure of the particle packing, assumed to be small ($na^3 \ll 1$). Neutron inelastic scattering at high energy and momentum transfers (Glyde *et al.*, 2000) as well as numerical simulations within the Diffusion Monte Carlo (DMC) method (Ceperley, 1995; Moroni *et al.*, 1997) allowed to estimate a condensate fraction at $T = 0$ of $n_0/n = 7.26\%$ (Minguzzi *et al.*, 2003).

An important qualification of the quantum many-body nature of a Bose-Einstein condensate was provided by Penrose (1951); Penrose and Onsager (1956), who studied the particle-particle correlations in low-temperature Boson assemblies, by addressing the one-body density matrix (cf. Chapter 1)

$$\rho(\mathbf{r},\mathbf{r}') = \langle\Psi^\dagger(\mathbf{r})\Psi(\mathbf{r}')\rangle. \tag{5.65}$$

This is defined as a suitable quantum average of an annihilation and a creation quantum field operators at arbitrary locations in space, with respect to the quantum ground state, here characterized by BEC. In a homogeneous system, $\rho(\mathbf{r},\mathbf{r}') \equiv \rho(\mathbf{r}-\mathbf{r}')$ is a translationally invariant quantity, and its Fourier transform yields the momentum distribution $n(\mathbf{p})$. For $T > T_c$, the momentum distribution is a smooth function of $\mathbf{p}$ at small momenta, therefore

$$\rho(\mathbf{r}-\mathbf{r}') \to 0, \quad \text{as } |\mathbf{r}-\mathbf{r}'| \to \infty. \tag{5.66}$$

On the other hand, at $T \leq T_c$ the zero-momentum state gets macroscopically occupied, and one has to write $n(\mathbf{p}) = N_0\delta(\mathbf{p}) + \tilde{n}(\mathbf{p})$, with $\tilde{n}(\mathbf{p})$ a smooth function of $\mathbf{p}$ at small momenta.[6] Fourier transforming back to real space, one finds that the one-body density matrix does not vanish at large distances, but rather behaves as

$$\rho(\mathbf{r}-\mathbf{r}') \to n_0, \quad \text{as } |\mathbf{r}-\mathbf{r}'| \to \infty. \tag{5.67}$$

This is often referred to as off-diagonal long-range order (ODLRO), as it involves values of the one-body density matrix far away from the condition $\mathbf{r} = \mathbf{r}'$. ODLRO is usually associated with the occurrence of an order parameter, which is readily identified with the condensate fraction, n_0/n, in the case of superfluidity.

More specifically, Leggett (1999), following earlier ideas of Ginzburg and Landau (1950), related the generally inhomogeneous and time-dependent condensate number density $n_0(\mathbf{r},t)$ to the macroscopic quantum wave function of the condensate as $\Psi_0(\mathbf{r},t) = \sqrt{n_0(\mathbf{r},t)}\exp[i\phi(\mathbf{r},t)]$. Here, $\phi(\mathbf{r},t)$ is the phase of the condensate wave function, a local function itself of coordinates and time. Its spatial gradient can be related to the superfluid velocity as $\mathbf{v}_s(\mathbf{r},t) = (\hbar/m)\nabla\phi(\mathbf{r},t)$. One consequence of superfluidity is that such velocity be irrotational, while a consequence of the fact that the wave function Ψ_0 be single-valued, a part from inessential phase factors equal to an integer times 2π, is that the line integral of the velocity field along a closed contour must be quantized as

$$\oint \mathbf{v}_s \cdot d\boldsymbol{\ell} = s\frac{h}{m}, \tag{5.68}$$

with s an integer (Onsager, 1949a,b), with $s = 0$ for a simply-connected superfluid sample, while a value $s > 0$ signals the presence of vortices. Indeed, experiments confirmed that vortex lines are quantized in superfluid He II and, more recently, in Bose-Einstein condensed alkali atoms (see Minguzzi *et al.*, 2003, and references therein).

Superfluidity breaks down when v_s exceeds a critical value, v_c say. The Landau criterion establishes that viscous flow sets back in when it becomes energetically favourable for elementary excitations to be created Rickayzen (1965); March and Tosi (2002). Let $\epsilon(\mathbf{p})$ be the dispersion relation of such elementary excitations, with $\mathbf{p}$ their momentum with respect to the (super)fluid. If relative motion sets in at a relative velocity $\mathbf{v}$, this is modified into $\epsilon(\mathbf{p})+\mathbf{v}\cdot\mathbf{p}$, which is negative when $\mathbf{p}$ is antiparallel to $\mathbf{v}$, and v exceeds the critical value $v_c = \min_p \epsilon(\mathbf{p})/p$. Such a minimum does in fact exist in

[6] Higher moments of $\tilde{n}(\mathbf{p})$ do exhibit anomalies as well, and these are also governed by the condensate fraction, *e.g.* $\lim_{p\to 0} p\tilde{n}(\mathbf{p}) = \frac{1}{2}n_0 mc/\hbar$, where c is the superfluid sound velocity (Gavoret and Nozières, 1964).

the excitation spectrum of He II, as has been verified by means of neutron spectroscopy, and the elementary excitations corresponding to it have been termed 'rotons' by Feynman (Rickayzen, 1965; March and Tosi, 2002).

5.2.2 *Experimental manifestations of superfluidity and BEC*

Superfluidity was first observed in ^{4}He (Kapitza, 1938; Allen and Misener, 1938a,b). Below its boiling point, liquid ^{4}He behaves as an ordinary fluid with low viscosity, until at $T_c = 2.17$ K the transition to a new phase (He II) is signalled by the vanishing of viscosity, and by the occurrence of an anomaly in the temperature dependence of the specific heat (λ point) (see Angilella *et al.*, 2003, for a review on liquid-liquid transitions). Several effects (fountain effect, second sound, thermal superconductivity) were observed and ascribed to the lack of viscosity in this truly quantum fluid (see Lynton, 1964, for an early review). In particular, the temperature dependence contained in Eq. (5.63) for the condensate fraction was demonstrated by Andronikashvili (1946), who measured the effective density of the normal component as a function of temperature from the period of torsional oscillations (and hence the moment of inertia) of a pile of thin metal disks, which were closely spaced to ensure that the normal fluid in the interstices would be dragged along while the superfluid remained stationary.

The overall phenomenology of superfluid He II could be accounted for within the so-called two-fluid model (Tisza, 1938; Landau, 1941). Within such a model, He II is described as a mixture of two fluids: a normal fluid, characterized by Newtonian viscosity, and a superfluid, which is responsible of all the anomalous effects. Later, London identified superfluidity in He II as a manifestation of BEC in a real material.

More recently, instances of BEC have been experimentally observed in atomic condensates (Anderson *et al.*, 1995) (see also Leggett, 2001; Inguscio, 2003; Angilella *et al.*, 2006a, for a review). These are low density ($n \sim 10^{13}$ cm^{-3}) assemblies of alkali atoms at ultracold temperatures ($T \sim 0.1$ μK) confined in optical and magnetic traps. Below their BE condensation temperature, practically all atoms occupy the same quantum level. This implies complete quantum coherence, which has been in fact verified in interference, and Raman- and Bragg-scattering experiments.

5.2.3 *Superfluidity without BEC: the two-dimensional Bose fluid*

5.2.3.1 *Microscopic definition of superfluid density*

In Sec. 5.2.1 we related the condensate density n_0 to the condensate wave function as $n_0(\mathbf{r}, t) = |\Psi_0(\mathbf{r}, t)|^2$. This relation qualifies n_0 as an appropriate order parameter, in the sense that a nonzero value thereof characterizes the phase with ODLRO. However, a different and, in general, unrelated 'order parameter' can be introduced for a superfluid, namely the superfluid density n_s (Josephson, 1966). This can be barely defined within linear response theory as the response to a transverse current probe (Minguzzi *et al.*, 2003). The superfluid density is at the basis of any hydrodynamic description of superfluidity, it may serve to express the superfluid free energy density as $F_s = \frac{1}{2}mn_s v_s^2$, and is of course the quantity actually measured in experiments. On the other hand, any microscopic theory of superfluidity usually starts with the condensate density (Bogoliubov, 1947). As a matter of fact, experiments (and theory) confirm that $n_0 \propto n_s \sim (T - T_c)^{\frac{2}{3}}$ close to the critical temperature for bulk superfluidity.[7] In a seminal paper, it was Josephson (1966) who first addressed, on phenomenological grounds, the actual relation between the condensate and the superfluid densities. It turns out that the two are related via an identity involving the long-wavelength limit of the static single-particle propagator $G_{11}(\mathbf{q}, \omega = 0)$, *viz.* (Josephson, 1966)

$$n_s = - \lim_{\mathbf{q}\to 0} \frac{mn_0}{q^2 G_{11}(\mathbf{q}, \omega = 0)}. \tag{5.69}$$

Eq. (5.69) poses severe constraints on the structure of the interparticle self-energy at low momenta (Talbot and Griffin, 1983), and has been recently proved within diagrammatic perturbation theory (Holzmann and Baym, 2007) (see also Taylor, 2008, and references therein). Their proof generalizes the earlier finding of Bogoliubov (1947), Gavoret and Nozières (1964), and Hohenberg and Martin (1965) that $n_s = mn_0$ at $T = 0$.

5.2.3.2 *Superfluidity without a condensate in $d = 2$: role of phase fluctuations*

In dimensions $d < 3$, Mermin-Wagner theorem rules out the spontaneous breaking of any continuous symmetry and the establishment of ODLRO at

[7] The same proportionality, although possibly with a different critical exponent, applies for other second-order phase transitions in $d = 3$.

any finite temperature T (Mermin and Wagner, 1966). This is basically due to the enhancement of thermal fluctuations due to the reduced dimensionality. In particular, this excludes the occurrence of BEC in $d = 2$. This specific result was also proved, following an independent argument by Bogoliubov, by Hohenberg (1967). In the case of neutral Bosons interacting via a hard-core repulsion, or charged Bosons interacting coulombically (in a neutralizing background), the following inequality holds for the non-singular part of the low-momentum distribution function (see also Minguzzi *et al.*, 2003; March, 2012):

$$\tilde{n}(\mathbf{p}) \geq -\frac{1}{2} + \frac{mk_{\mathrm{B}}T}{p^2}\frac{n_0}{n}. \tag{5.70}$$

The requirement that the integral of the total distribution $n(\mathbf{p})$ be finite (and fixed by the total number of particles) implies therefore that $n_0 = 0$ (Hohenberg, 1967).

The inequality Eq. (5.70) has been generalized by Magro and Ceperley (1994) to a Bose system interacting via a $\ln r$ repulsion as

$$\tilde{n}(\mathbf{p}) \geq -\frac{1}{2} + \frac{1}{S(\mathbf{p})}\frac{n_0}{n}. \tag{5.71}$$

Here, $S(\mathbf{p})$ is the static structure factor (cf. Chapter 7), which in the long-wavelength limit behaves as $S(p) \sim p^2/\omega_{\mathrm{pl}}$, where $\omega_{\mathrm{pl}} = \sqrt{2\pi ne^2/m}$ is the plasma frequency. Plasma density fluctuations are therefore responsible of the destruction of a condensate at any $T > 0$ in such a two-dimensional Bose system.

Although a condensate is absent in any interacting Bose fluid in $d = 2$ at $T \neq 0$, the system shows *algebraic* off-diagonal long-range order, *i.e.* the one-body density matrix behaves asymptotically as

$$\rho(\mathbf{r}) \sim \frac{1}{r^\alpha}, \quad r \to \infty. \tag{5.72}$$

This can be understood by saying that, while thermal fluctuations prevent the onset of a nonzero condensate density n_0, phase correlations can be developed at sufficiently low temperature. Of course, a real finite Bose system in $d = 2$ might show condensation if its size is smaller than the phase correlation length.

This state of affairs is not dissimilar to the Berezinskii-Kosterlitz-Thouless (BKT) transition (Berezinskii, 1971, 1972; Kosterlitz and Thouless, 1973), which is a notable example of topological phase transition (in reduced dimensionality) with no condensate ($n_0 = 0$), but with phase coherence ($n_s \neq 0$).[8] Indeed, in such system, in the diluted limit and at low temperature, the superfluid density can be estimated using the formula due to Landau

$$\frac{n_s}{n} = 1 - \frac{1}{nk_{\rm B}T} \sum_{\mathbf{k}\neq 0} \frac{k^2}{2m} \frac{e^{\hbar\omega_{\mathbf{k}}/k_{\rm B}T}}{[e^{\hbar\omega_{\mathbf{k}}/k_{\rm B}T} - 1]^2}, \tag{5.73}$$

where $\omega_{\mathbf{k}}$ is the dispersion relation of the superfluid collective excitations (Minguzzi *et al.*, 2003).

5.2.4 *BEC of magnons*

Given the inverse relationship between the critical temperature T_c for Bose-Einstein condensation (BEC) and the mass m of the bosons taking part in such phase transition, as is implicit in Eq. (5.62), there has recently been a quest for boson quasiparticles having a sufficiently small effective mass, so that BEC can take place at a relatively high temperature. However, their density should also be tunable, and their lifetime should be larger than that of other dissipative processes occurring in the system, so that BEC can be experimentally accessible.

In this context, the elementary excitations of a magnetic systems or magnons seem to fulfill such requirements, and indeed BEC of magnons has received considerable interest in the recent past, both theoretically (Bugrij and Loktev, 2007; Bunkov and Volovik, 2010; Bugrij and Loktev, 2013), and experimentally (Demokritov *et al.*, 2006; Demidov *et al.*, 2007; Dzyapko *et al.*, 2007a; Demidov *et al.*, 2008). In particular, BEC of magnons has been observed at room temperature in a thin film of yttrium iron garnet (YIG), under parallel pumping (Demokritov *et al.*, 2006). This corresponds to applying a radio-frequency magnetic field, which excites certain magnon modes in the sample. The absorbed energy can either be distributed through other magnon modes, or transferred to phonon modes, via magnon-magnon or magnon-phonon interactions, respectively. For ferromagnetic materials such as YIG, it turns out that the relaxation processes of the former kind take place in a considerably shorter time (~ 1 ns) than

[8] A notable example of a phase transition characterized, on the contrary, by the presence of a condensate ($n_0 \neq 0$) but without phase coherence ($n_s \neq 0$) is given by the disordered Bose glass, which is one of the possible solutions of the Bose-Hubbard model (Gersch and Knollman, 1963; Giamarchi and Schulz, 1988; Fisher *et al.*, 1989). Recently, this model has been revived to study cold Boson atoms in an optical lattice (see Yao *et al.*, 2014, and references therein). See also Chapter 8.

those involving phonons ($\sim 1\ \mu$s). Hence, magnons can be effectively considered as decoupled from phonons over a time range of the order of 1 μs. Therefore, it has been possible to observe magnon BEC, by applying pulsed microwave magnetic field of relatively short durations (~ 100 ns), compared with the natural relaxation time of the system (Demokritov *et al.*, 2006).

The experimental findings of Demokritov *et al.* (2006) have been described quantitatively by Kar *et al.* (2012). They start by considering the experimental apparatus of Demokritov *et al.* (2006), which can be approximated as a thin slab of ferromagnetic material of dimensions $L_x \sim 2$ mm, $L_y \sim 5\ \mu$m, and $L_z \sim 20$ mm, with $L_y \ll L_x \lesssim L_z$. Therefore, spin waves can propagate in the xz plane with wave vector $\mathbf{k}$, while they are standing waves in the transverse y direction, with $k_y = n_\perp \pi / L_y$, and $n_\perp$ an integer. The magnon Hamiltonian then reads (Kar *et al.*, 2012)

$$H_0 = \sum_{\mathbf{k},n_\perp} \hbar\omega_{\mathbf{k},n_\perp} c^\dagger_{\mathbf{k},n_\perp} c_{\mathbf{k},n_\perp}, \tag{5.74}$$

where $\omega_{\mathbf{k},n_\perp}$ is the magnon dispersion relation, given by

$$\omega^2_{\mathbf{k},n_\perp} = \gamma^2 \left(H + Dk^2 + D\frac{\pi n_\perp^2}{L_y^2} + 4\pi M(1 - F_\mathbf{k}) \sin^2\theta \right) \times \left(H + Dk^2 + D\frac{\pi n_\perp^2}{L_y^2} + 4\pi M F_\mathbf{k} \right). \tag{5.75}$$

Here, γ is the gyromagnetic ration, $D = 2JSa^2/\hbar\gamma$ is the exchange stiffness, with J being the nearest-neighbour exchange constant and a the lattice parameter, H is the (static) magnetic field, M is the magnetization, θ is the angle formed by the wave vector $\mathbf{k}$ with the z-axis, and $F_\mathbf{k} = (1 - e^{-kL_y})/kL_y$. For a sufficiently thick film ($L_y \gtrsim 1\ \mu$m), the magnon dispersion relation presents a minimum for $kL_y \geq 1$ at a k value given by $k_0^3 \approx \pi M/DL_y$, with

$$\omega_{k_0,\min} \approx \gamma \left(H + 3\left(\frac{\pi M}{DL_y}\right)^{2/3} D \right). \tag{5.76}$$

When a 'pumping' microwave magnetic field $h(t) = he^{-i\omega_P t}$ is applied, with both H and h along z (*i.e.* in the plane of the thin film), then the Hamiltonian, Eq. (5.74), will be subjected to the additional perturbing and time-dependent term

$$H_P(t) = \frac{\hbar}{2} \sum_{\mathbf{k},n_\perp} h\rho_{\mathbf{k},n_\perp} e^{-i\omega_P t} c^\dagger_{\mathbf{k},n_\perp} c^\dagger_{-\mathbf{k},n_\perp} + \text{H.c.}, \tag{5.77}$$

where

$$\rho_{\mathbf{k},n_\perp} = \gamma\omega_M[(1 - F_{\mathbf{k}})\sin^2\theta - F_{\mathbf{k}}]/4\omega_{\mathbf{k},n_\perp} \tag{5.78}$$

is the coupling of the pumping field with the $(\mathbf{k}, -\mathbf{k})$ magnon pair, having frequency $\omega_{\mathbf{k},n_\perp}$, and $\omega_M = \gamma 4\pi M$. Due to the pumping perturbation, Eq. (5.77), the occupation numbers of particular magnon modes will increase. Solving the Heisenberg equation for the spin wave operators with $H = H_0 + H_P(t)$, one finds the average number of magnons per unit volume to be

$$\langle n_{\mathbf{k},n_\perp}(t)\rangle = \langle n_{\mathbf{k},n_\perp}(0)\rangle \exp(2\lambda_{\mathbf{k},n_\perp} t), \tag{5.79}$$

where $\langle n_{\mathbf{k},n_\perp}(0)\rangle$ is the average magnon density at thermal equilibrium, and

$$\lambda_{\mathbf{k},n_\perp} = \sqrt{(h\rho_{\mathbf{k},n_\perp})^2 - \Delta\omega^2_{\mathbf{k},n_\perp}} - \eta_{\mathbf{k},n_\perp}, \tag{5.80}$$

with $\eta_{\mathbf{k},n_\perp} \equiv 1/(2\tau_{\mathbf{k},n_\perp})$ the magnon relaxation rate, and $\Delta\omega_{\mathbf{k},n_\perp} = \omega_{\mathbf{k},n_\perp} - \omega_P/2$. It follows that energy from the pumping field is more efficiently transferred to magnon modes with frequency close to $\omega_P/2$.

Kar *et al.* (2012) estimate the number density of the pumped magnons at time t as

$$\Delta N_{P1} = \frac{N_T\langle n(0)\rangle\Delta}{2\pi\gamma D}\left[\exp\left((2h\rho - 2\eta - \Delta^2/3h\rho)t\right) - 1\right], \tag{5.81}$$

where η is the reference value of the magnon relaxation rate, $\langle n(0)\rangle$ is the number of thermal magnons of any mode in the dominant group, $h\rho \approx \eta$, and $\pm\Delta$ determine the range for $\Delta\omega_{\mathbf{k},n_\perp}$ as $\Delta^2 = (h\rho)^2 - \eta^2$. ΔN_{P1} can be further decoupled as $\Delta N_{P1} = \Delta N_M + \Delta N_{\text{con}}$, where

$$\Delta N_M = \frac{1}{4\pi D}\frac{k_B T}{\hbar\gamma} \times \sum_{n_\perp} \log \frac{1 - \exp\left(-\frac{\hbar\gamma}{k_B T}\left(H + 2\pi M + D(n_\perp\pi/L_y)^2\right)\right)}{1 - \exp\left(-\frac{\hbar\gamma}{k_B T}\left(2\pi M - 3D(\pi M/L_y D)^{2/3} + D(n_\perp\pi/L_y)^2\right)\right)} \tag{5.82}$$

is the portion distributed over the whole spectrum according to the Bose-Einstein distribution function with the maximum value at the chemical potential, whereas ΔN_{con} corresponds to coherent magnons which have condensed at the minimum magnon energy.

Bose-Einstein condensation occurs when the density of pumped magnons exceeds ΔN_M, which provides an estimate for the rise-time for magnon condensation as (Kar *et al.*, 2012)

$$t_R = \frac{1}{2h\rho - 2\eta - \Delta^2/3h\rho} \log\left(1 + \frac{2\pi\gamma D\Delta N_M}{\langle n(0)\rangle N_T\Delta}\right). \tag{5.83}$$

Taking reasonable values for the various parameters, as in the available experiments for magnons in YIG, Kar *et al.* (2012) can estimate $t_R = 300 - 900$ ns for a static field $H = 1000$ Oe and a microwave field $h = 16.1 - 14.9$ Oe, correspondingly. These values compare favourably with the experimental values of $22 - 19.56$ Oe of Dzyapko *et al.* (2007b), and are below the magnon-phonon relaxation time. Kar *et al.* (2012) also studied the case of a pumping magnetic field perpendicular to the film, and found a microwave field of the same order of available experimental results (Dzyapko *et al.*, 2009).

5.2.5 *Mesoscopic condensation: relevance of pairing correlations*

Earlier in Section 5.1.3 we have discussed the relevance of pairing correlations for the onset of off-diagonal long range order, such as in superconductivity, as signalled by geminals within density matrix theory (cf. Chapter 1) (Coleman, 1963; Coleman and Yukalov, 2000; Yukalov and Yukalova, 2013). After having introduced the phenomenon of superfluidity, we come back on that topic, by mentioning that the order index (already introduced in Sec. 5.1.3) would indicate the occurrence of superfluidity, as it changes continuously from zero to a finite value as a function of the parameters in a model.

An explicit model allowing for superfluidity for particular values of the parameters was proposed by Coleman and Yukalov (1996), who considered the Hamiltonian for a Bose system given by ($\hbar = 1$)

$$H = \int d^3\mathbf{r}\, \psi^\dagger(\mathbf{r}) \left(-\frac{1}{2m}\nabla^2 - \mu\right) \psi(\mathbf{r}) + \frac{\epsilon N^\delta}{\sqrt{\rho}}(\psi_0^\dagger e^{i\varphi} + \psi_0 e^{-i\varphi}), \tag{5.84}$$

where ϵ and $0 \leq \varphi \leq 2\pi$ are real parameters, $\psi(\mathbf{r}) = \psi_0 + \chi(\mathbf{r})$ is the Bose field, with ψ_0 is for the condensate and $\chi(\mathbf{r})$ for its fluctuations, $\rho = N/V$ is the average particle density, and $\delta < 1$ ($\mu \neq 0$) is the actual parameter of the model. Since for a uniform system one has $\langle\psi_0\rangle = \eta =$ const, and $\psi_0^\dagger\chi(\mathbf{r})\rangle = 0$, one finds the reduced single-particle matrix as

$$\rho_1(\mathbf{r}, \mathbf{r}') = \langle\psi_0^\dagger\psi_0\rangle + \langle\chi^\dagger(\mathbf{r})\chi(\mathbf{r})\rangle. \tag{5.85}$$

Let us pass to the momentum representation, with

$$\varphi_0(\mathbf{r}) = \frac{1}{\sqrt{V}}, \tag{5.86a}$$

$$\varphi_{\mathbf{k}}(\mathbf{r}) = \frac{1}{\sqrt{V}} e^{i\mathbf{k}\cdot\mathbf{r}}, \tag{5.86b}$$

so that

$$\psi_0(\mathbf{r}) = \frac{1}{\sqrt{V}} a_0, \tag{5.87a}$$

$$\chi(\mathbf{r}) = \sum_{\mathbf{k}\neq 0} \varphi_{\mathbf{k}}(\mathbf{r}) a_{\mathbf{k}}. \tag{5.87b}$$

The Hamiltonian, Eq. (5.84), then becomes

$$H = \sum_{k\neq 0} \left(\frac{k^2}{2m} - \mu \right) a_{\mathbf{k}}^\dagger a_{\mathbf{k}} - \mu a_0^\dagger a_0 + \frac{\epsilon N^\delta}{\sqrt{N}} (a_0^\dagger e^{i\varphi} + a_0 e^{-i\varphi}). \tag{5.88}$$

This Hamiltonian can be diagonalized by means of the canonical transformation

$$a_0 = b_0 + \frac{\epsilon N^\delta}{\mu\sqrt{N}} e^{i\varphi}, \tag{5.89a}$$

$$a_{\mathbf{k}} = b_{\mathbf{k}}, \quad \mathbf{k} \neq 0, \tag{5.89b}$$

which leads to

$$H = \sum_{k\neq 0} \left(\frac{k^2}{2m} - \mu \right) b_{\mathbf{k}}^\dagger b_{\mathbf{k}} - \mu b_0^\dagger b_0 + \frac{\epsilon^2}{\mu^2} N^{2\delta - 1}. \tag{5.90}$$

Correspondingly, the reduced single-particle matrix takes the form

$$\rho_1(\mathbf{r}, \mathbf{r}') = \frac{N_0}{V} + \sum_{k\neq 0} N_{\mathbf{k}} \varphi_{\mathbf{k}}(\mathbf{r}) \varphi_{\mathbf{k}}^*(\mathbf{r}'), \tag{5.91}$$

where $N_0 = \langle a_0^\dagger a_0 \rangle$ and $N_{\mathbf{k}} = \langle a_{\mathbf{k}}^\dagger a_{\mathbf{k}} \rangle$ specify the occupancy of the natural orbitals φ_0 and $\varphi_{\mathbf{k}}$, respectively. Since for a Bose gas one has $\mu \leq 0$ (Reichl, 2009), as $\mu \to 0^-$ one finds $N \to \infty$, which implies Bose-Einstein condensation. On the other hand, for a fixed value $\mu \lneqq 0$, one finds

$$\langle a_0^\dagger a_0 \rangle = \frac{1}{e^{-\beta\mu} - 1} + \frac{\epsilon^2}{\mu^2} N^{2\delta - 1}, \tag{5.92}$$

which, in the thermodynamic limit ($N \to \infty$), yields

$$\langle a_0^\dagger a_0 \rangle = \begin{cases} (e^{-\beta\mu} - 1)^{-1}, & \delta < \frac{1}{2}, \\ (e^{-\beta\mu} - 1)^{-1} + \epsilon^2 \mu^{-2}, & \delta = \frac{1}{2}, \\ \epsilon^2 \mu^{-2} N^{2\delta - 1}, & \delta > \frac{1}{2}. \end{cases} \tag{5.93}$$

In order to evaluate the order index α_1, defined by Eq. (5.49), we need to calculate the norm $\| \rho_1 \|$. This turns out to be finite for all N for $\mu < 0$, if $\delta \leq \frac{1}{2}$, whereas $\| \rho_1 \| = \epsilon^2 \mu^{-2} N^{2\delta - 1}$, if $\delta > \frac{1}{2}$. Therefore, the first order index for the model Hamiltonian, Eq. (5.84), equals

$$\alpha_1 = \begin{cases} 0, & \delta \leq \frac{1}{2}, \\ 2\delta - 1, & \delta > \frac{1}{2}, \end{cases} \tag{5.94}$$

showing that α_1 depends continuously on the parameter δ.

In order to check whether the condition $\delta > \frac{1}{2}$ does indeed correspond to long-range order, Coleman and Yukalov (1996) study the Heisenberg equation of motion for the condensate operator b_0, which reads

$$i \frac{db_0}{dt} = [b_0, H] = -\mu b_0. \tag{5.95}$$

Taking the statistical average, one finds

$$i \frac{d\langle b_0 \rangle}{dt} = -\mu \langle b_0 \rangle. \tag{5.96}$$

But now, at equilibrium $\langle b_0 \rangle = \text{const}$, therefore the left-hand side vanishes, and since we have assumed $\mu \neq 0$, one concludes that also $\langle b_0 \rangle$ vanishes. From the canonical transformation Eq. (5.89a) one then obtains

$$\langle a_0 \rangle = \frac{\epsilon N^\delta}{\mu \sqrt{N}} e^{i\varphi}, \tag{5.97}$$

which for the order parameter implies

$$\eta = \langle \psi_0 \rangle = \sqrt{\frac{\rho}{N}} \langle a_0 \rangle = \sqrt{\rho} \frac{\epsilon}{\mu N^{1-\delta}} e^{i\varphi}, \tag{5.98}$$

and in the thermodynamic limit

$$\lim_{N \to \infty} \langle \psi_0 \rangle = 0 \quad (\delta < 1). \tag{5.99}$$

A vanishing order parameter in the thermodynamic limit implies no long-range order, therefore, according to the classification of Coleman and Yukalov (1996) [Eq. (5.53)], the case $0 < \alpha_1 < 1$ corresponds to total mid-range order, or *mesoscopic order.*

Since $\langle \psi_0^\dagger \psi_0 \rangle = \rho \epsilon^2 \mu^{-2} N^{\alpha_1 - 1}$ for $\delta > \frac{1}{2}$, and since the distance $|\mathbf{r} - \mathbf{r}'|$ between two particles scales as $V^{1/3}$, with $V = N/\rho$, further analysis of the reduced single-particle density matrix demonstrates that

$$\rho_1(\mathbf{r}, \mathbf{r}') \simeq \frac{\epsilon^2 \rho^{\alpha_1}}{\mu^2} |\mathbf{r} - \mathbf{r}'|^{-3(1-\alpha_1)}. \tag{5.100}$$

Similarly, for the reduced n-particle density matrix, one finds

$$\rho_n(\mathbf{r},\ldots\mathbf{r};\mathbf{r}',\ldots\mathbf{r}') \sim |\mathbf{r}-\mathbf{r}'|^{-3(n-\alpha_n)}, \tag{5.101}$$

asymptotically as $|\mathbf{r}-\mathbf{r}'| \to \infty$, where $\alpha_n = n\alpha_1$. Therefore, the order index may be used in conjunction with the order parameter to signal the occurrence of off-diagonal long-range order, and it also characterizes the behaviour of the off-diagonal terms of the reduced density matrices.

It would be interesting to investigate the relevance of the concept of order index for Bose-Einstein condensation in an optical lattice (Morsch and Oberthaler, 2006; Yukalov, 2013).

5.2.6 *An exactly solvable model for an interacting Boson gas in one dimension*

A major contribution to the many-body study of interacting Bosons was made by Lieb and Liniger (1963) and by Lieb (1963). These authors (LL) dealt with one-dimensional Bose particles with repulsive δ-function interactions. All the eigenfunctions were obtained explicitly, while the energies were given by the solution of a transcendental equation.

Thus, the starting point for the discussion below is the Schrödinger equation (in suitable units) of N particles with the interaction specified above

$$\left[-\sum_i \frac{\partial^2}{\partial x_i^2} + 2c\sum_{ij}\delta(x_i - x_j)\right]\Psi = E\Psi, \tag{5.102}$$

where following LL the amplitude of the δ-function is written as $2c$. The region of space to be considered below is $0 \le x_i \le L$ $(i = 1,\ldots N)$. The wave function $\Psi = \Psi(x_1,\ldots x_N)$ determined below following LL satisfies periodic boundary conditions in each variable. Below, the entire focus is on the repulsive form of the interaction, so that $c \ge 0$.

LL point out first that a δ-function potential is equivalent to the boundary condition

$$\frac{\partial\Psi}{\partial x_{j+1}} - \frac{\partial\Psi}{\partial x_j} = c\Psi, \quad \text{at } x_j = x_{j+1}. \tag{5.103}$$

The above condition amounts to say that Ψ is continuous whenever two particles touch, but there is a jump of $2c$ in the derivative of Ψ. The original Schrödinger equation is therefore equivalent to a Helmholtz equation, with mixed boundary conditions. Here, all the focus is on symmetric (Bose) wave functions.

Lieb and Liniger (1963) seek then Ψ constructed according to the following Ansatz. Let $\{k\} \equiv \{k_1, \dots k_N\}$ be an ordered set of N numbers, and P a generic permutation thereof. LL then define Ψ as

$$\Psi(x_1, \dots x_N) = \sum_P a(P) P \exp\left(i \sum_{j=1}^N k_j x_j \right), \tag{5.104}$$

the summation extending over all permutations of $\{k\}$, and where $a(P)$ are coefficients to be determined, in order for Ψ to satisfy the Schrödinger equation, Eq. (5.102), with $E = \sum_{j=1}^N k_j^2$. Such coefficients would reduce to $a(P) = (-1)^P$ for a one-component Fermi function.

The problem, as LL point out, is quite similar to the one-dimensional Heisenberg model of ferromagnetism with nearest-neighbour interactions, first solved by Bethe (see Orbach, 1958, and earlier references therein). LL show then first that the total momentum must be an integral multiple of $2\pi/L$. Also, LL stress that for any state with momentum p there exists a state with momentum

$$p' = p + 2\pi n_0 \rho, \tag{5.105}$$

with $\rho = N/L$. Conversely, LL point out that for any state with momentum $|p| \geq \pi\rho$, there is a corresponding state having momentum satisfying

$$-\pi\rho < p \leq \pi\rho. \tag{5.106}$$

One thus has to consider only states in the interval specified in Eq. (5.106). All other states are obtained by the 'Umklapp'

$$k_i' = k_i + 2\pi n_0 / L \tag{5.107}$$

for an appropriate integer n_0. This integer n_0 is thus one of the quantum numbers characterizing an energy eigenstate. The energy of the new state is

$$E' = \sum_{j=1}^N (k_j')^2 = E + \frac{4\pi n_0 \rho}{N} p + \frac{1}{N} (2\pi n_0 \rho)^2, \tag{5.108}$$

which LL point out is a consequence of Galilean invariance.

5.2.6.1 *Summary of ground-state energy*

LL next note that, from dimensional arguments, that the ground-state energy can always be expressed in the form

$$E_0 = N\rho^2 \ell(N, cL), \tag{5.109}$$

where ℓ is a dimensionless function of the arguments shown. However, as E_0 is an extensive variable, ℓ can depend only on intensive variables. The only dimensionless intensive variable in the present problem is

$$\gamma = c/\rho. \tag{5.110}$$

Therefore, the ground-state energy must take the form

$$E_0 = N\rho^2 e(\gamma). \tag{5.111}$$

5.2.6.2 *Other ground-state properties*

There are other ground-state properties that can be related to the central quantity $e(\gamma)$ introduced above. Thus, again following LL, the chemical potential is given by

$$\mu = \frac{\partial E_0}{\partial N} = \rho^2 \left(3e - \gamma \frac{de}{d\gamma}\right). \tag{5.112}$$

The potential energy v per particle is also of some interest, and can be obtained from

$$v = \frac{c}{N}\frac{\partial E_0}{\partial c} = \rho^2 \gamma \frac{de}{d\gamma}. \tag{5.113}$$

Combining Eqs. (5.112) and (5.113) readily yields

$$\mu = 3\rho^2 e - v. \tag{5.114}$$

Finally, the kinetic energy per particle $t = T/N$ is given by

$$t = \frac{T}{N} = \frac{E_0}{N} - v = \rho^2 \left(e - \gamma \frac{de}{d\gamma}\right) = \rho^2 e - v. \tag{5.115}$$

5.2.6.3 *Transcendental equation for* $e(\gamma)$

As discussed fully by LL, $e(\gamma)$ is related to a function $g(x)$ by

$$e(\gamma) = \frac{\gamma^3}{\lambda^3} \int_{-1}^{1} g(x) x^2 dx, \tag{5.116}$$

where $g(x)$ is the solution of the inhomogeneous Fredholm integral equation of the second kind

$$1 + 2\lambda \int_{-1}^{1} \frac{g(x)dx}{\lambda^2 + (x-y)^2} = 2\pi g(y), \tag{5.117}$$

subject to the condition

$$\gamma \int_{-1}^{1} g(x)dx = \lambda. \tag{5.118}$$

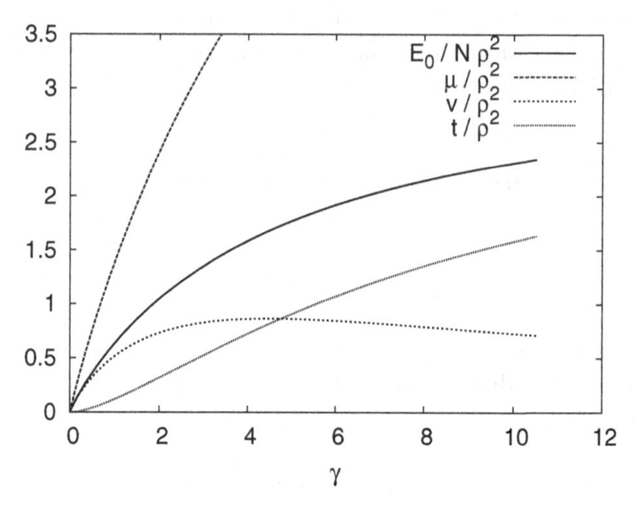

Fig. 5.1 Properties of the ground state, as a function of $\gamma = c/\rho$. Plotted quantities are $E_0/N\rho^2 \equiv e(\gamma)$, Eq. (5.111), μ/ρ^2, Eq. (5.112), v/ρ^2, Eq. (5.113), and t/ρ^2, Eq. (5.115). Redrawn after Lieb and Liniger (1963).

LL show furthermore that $e(\gamma)$ is an analytic function of γ, except at $\gamma = 0$, and can therefore be evaluated numerically with confidence. For any fixed $\lambda > 0$ (the case $\lambda = 0$ leading to $\gamma = e(\gamma) = 0$), one starts by solving Eq. (5.117) (*e.g.*, iteratively). Equation (5.118) can then be employed to determine the corresponding value of γ, thus showing that γ is a monotonic function of λ. Finally, Eq. (5.116) yields the corresponding value of $e(\gamma)$. Figure 5.1 shows our numerical results for $e(\gamma)$ *vs* γ, along with several other properties of the ground-state that can be derived in terms of these quantities.

5.2.6.4 *Excited states*

Having discussed the ground-state in considerable detail, we turn, but now more briefly, to treat the excitation spectrum, following closely Lieb (1963). Below, we stress Lieb's important conclusion that the energy spectrum of the above model contains two separate branches. While the Bogoliubov's perturbation theory carried out by Lieb and Liniger (1963) correctly accounts at weak potentials for one of the two branches [$\epsilon_+(p)$, cf. Eq. (5.121) below], the second branch is only obtained with the exact procedure of Lieb (1963), and is defined only for momenta p such that $|p| \leq \pi\rho$.

Let us denote by $K = K(\gamma)$ the absolute value of largest k_i in $\{k\}$, so

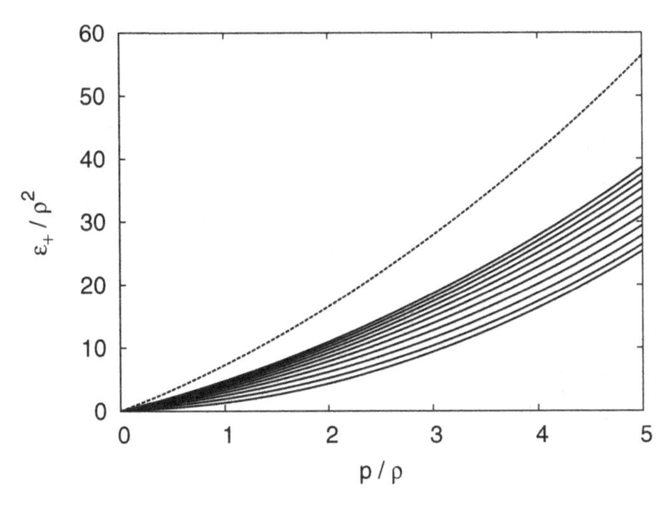

Fig. 5.2 Energy spectrum $\epsilon_+(p)$ of type I elementary excitations, Eq. (5.121), for various values of the interaction parameter γ, increasing from bottom to top. Dashed line is the limiting dispersion for hard-core repulsion ($\gamma = \infty$), Eq. (5.120). Redrawn after Lieb (1963).

that $-K \leq k_i \leq K$. The numbers k_i need not be distributed uniformly within this range. Moreover, it turns out that K is an intensive quantity, with $K(0) = 0$ and $K(\infty) = \pi\rho$, and more generally (Lieb and Liniger, 1963)

$$K = \rho\frac{\gamma}{\lambda} = \rho\left(\int_{-1}^{1} g(x)dx\right)^{-1}. \tag{5.119}$$

In the hard-core case ($\gamma = \infty$), the Bose gas within the LL model can be mapped onto a non-interacting one-component Fermi gas (albeit with a completely antisymmetric wave function). The elementary excitations of such a system can then be described as a 'particle-hole pair'. In other words, an elementary excitation can be constructed by promoting a particle from a state with momentum q ($|q| < K$), to a state with momentum k ($|k| > K$). The energy of such an excitation would then be given by $\epsilon(q, k) = k^2 - q^2 > 0$, and its momentum by $p(k, q) = k - q$. The drawback of such a description is that both quantities would necessitate of two variables (q and k, each with different constraints), and it is not obvious how this description would be continued to finite coupling ($\gamma < \infty$), and whether the dispersion relation $\epsilon = \epsilon(p)$ would remain single-valued (it actually turns out to be *not* the case).

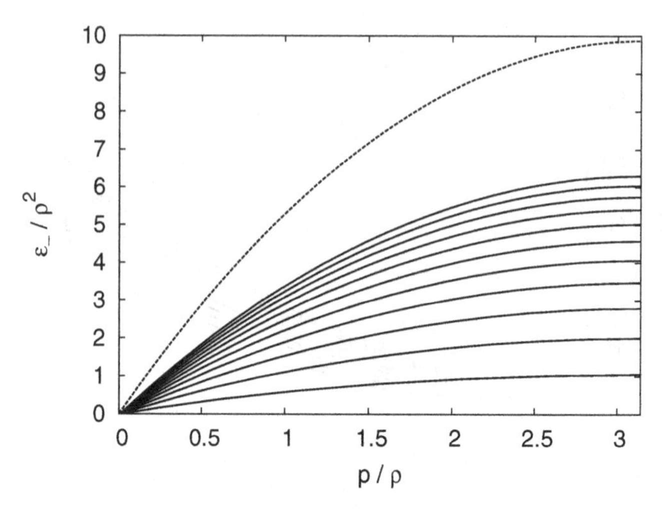

Fig. 5.3 Energy spectrum $\epsilon_-(p)$ of type II elementary excitations, Eq. (5.121), for various values of the interaction parameter γ, increasing from bottom to top. Dashed line is the limiting dispersion for hard-core repulsion ($\gamma = \infty$), Eq. (5.120). Redrawn after Lieb (1963).

Lieb (1963) circumvents both difficulties by defining two types of elementary excitations as follows.

Type I excitations correspond to promoting a particle from K to $q > K$ (or from $-K$ to $q < -K$). This corresponds to a momentum $p = q - K$, if $q > K$ (or $p = q + K$, if $q < -K$). In both cases, the energy of such an excitation is $\epsilon_+ = q^2 - K^2$.

Type II excitations correspond to promoting a particle from $0 < q < K$ to $K + 2\pi/L$ (or from $-K < q < 0$ to $-K - 2\pi/L$). This corresponds to a momentum $p = K - q$, if $0 < q < K$ (or $p = -K - q$, if $-K < q < 0$). In both cases, the energy of such an excitation is $\epsilon_- = K^2 - q^2$.

The two branches of the excitation spectrum, in the hard core limit, therefore obey the dispersion relation

$$\epsilon_\pm(p) = 2\pi\rho|p| \pm p^2, \tag{5.120}$$

the upper (lower) sign applying to type I (type II) excitations. In analogy to the non-interacting Fermi gas, type I excitations correspond to 'particle' states, while type II excitations correspond to 'holes'.

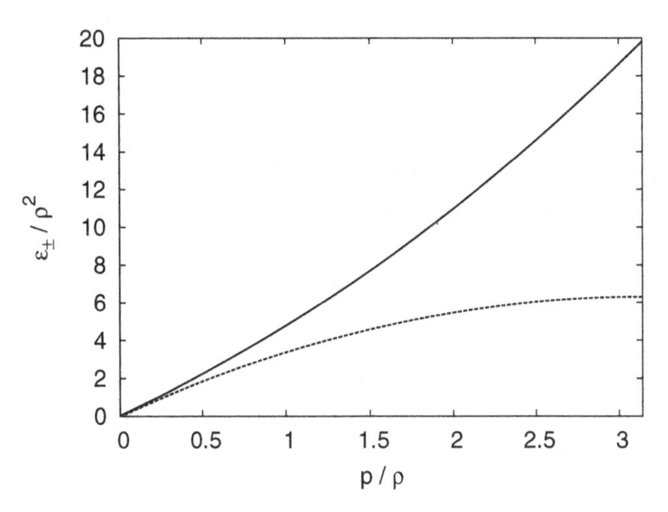

Fig. 5.4 Energy spectra $\epsilon_\pm(p)$ of type I (solid line) and type II (dashed line) elementary excitations, Eq. (5.121), for $\gamma = 10.518$, showing the same slope at $q = 0$.

This distinction is maintained also for a generic interaction strength ($\gamma < \infty$), but now the dispersion relations deviate quantitatively from their hard-core limit, Eq. (5.120). As Lieb (1963) shows, the dispersion relation of the two branches of the elementary excitation spectrum can be found implicitly by eliminating the auxiliary parameter q from

$$\epsilon_\pm = \mp\mu \mp q^2 + 2\int_{-K}^{K} kJ_\pm(k)dk, \tag{5.121a}$$

$$p = \pm q + \int_{-K}^{K} J_\pm(k)dk, \tag{5.121b}$$

$$2\pi J_\pm(k) = \mp\pi \mp \theta(q-k) + 2\rho\gamma \int_{-K}^{K} dk' \frac{J_\pm(k')}{\rho^2\gamma^2 + (k-k')^2}, \tag{5.121c}$$

where $\theta(q) = -2\,\mathrm{atan}[q/(\rho\gamma)]$. Our numerical results, reproducing those of Lieb (1963), are reported in Fig. 5.2, for type I excitations $[\epsilon_+(p)]$, and in Fig. 5.3, for type II excitations $[\epsilon_-(p)]$. It is to be noted that type II excitations exist only for $|p| < \pi\rho$, whereas no restriction applies to the momentum of type I excitations ($-\infty < p < \infty$).

As Lieb emphasizes, no energy gap is associated with either branch, and the two spectra have a common slope at $p = 0$ (Fig. 5.4). This implies that they propagate sound at the same velocity (Feynman, 1953, 1954)

$$v_s = \lim_{p\to 0} \frac{\partial\epsilon_\pm(p)}{\partial p}. \tag{5.122}$$

5.2.7 Quench dynamics of a one-dimensional interacting Bose gas

The experimental realization of ultracold atoms down to the nano-Kelvin range in laser or magnetic traps (see Inguscio, 2003; Angilella *et al.*, 2006a, and references therein) has stimulated the theoretical study of their rapid evolution out of equilibrium (Bloch *et al.*, 2008). These systems are formed by trapping a gas of atoms using counterpropagating laser waves, or within magnetic traps, or a combination of these techniques. The atomic gases are then cooled evaporatively, and are kept isolated from any thermal bath. One is then interested in the thermalization of such isolated quantum systems. In addition to the atomic species, one can also tune the system by applying suitable external magnetic or electric fields, thereby simulating the occurrence of periodic lattices, not only in three dimensions, but also with reduced dimensionality $d = 1$ or 2, and ideally without disorder. On the other hand, one can introduce the desired degree of disorder in a 'clean' or controlled way, by inserting two or more atomic species into the system.

The rapid evolution of these systems out of equilibrium can be modelled by *quench dynamics, viz.* the (rapid) evolution of a system, initially prepared in some initial state $|\Phi_0\rangle$, when this is subjected to the action of a given Hamiltonian H at $t \geq 0$. From the theoretical point of view, the study of quench dynamics is particularly interesting when H is integrable (Cazalilla *et al.*, 2011). Such is the case of the Bose-Hubbard model, the XXZ model, the Sine-Gordon model, and the Lieb-Liniger model (Sec. 5.2.6). In particular, the latter model is relevant for a continuum gas of bosons in a one-dimensional trap, which interact through δ-functions. The resulting Schrödinger equation has been derived and solved exactly by Lieb and Liniger (1963); Lieb (1963), and presented in Eq. (5.102) above.

Following Iyer and Andrei (2012); Iyer *et al.* (2013), the Hamiltonian corresponding to the Lieb-Liniger model can be written as

$$H = \int [\partial b^\dagger(x)\partial b(x) + cb^\dagger(x)b(x)b^\dagger(x)b(x)]d^N x, \tag{5.123}$$

where $b(x)$ is a bosonic field, $x = \{x_1, \dots x_N\}$, and c is the interaction strength. As discussed in Sec. 5.2.6, the model is integrable, and its N-particles eigenstates can be presented in the form of a Bethe Ansatz as (Iyer and Andrei, 2012; Iyer *et al.*, 2013)

$$|k\rangle = \int dy \prod_{i<j} \mathcal{S}_y[Z_{ij}^y(k_i - k_j) \prod_j e^{ik_j y_j}]b^\dagger(y_j)|0\rangle \tag{5.124}$$

in terms of the momenta $\{k\}$. Here, $\mathcal{S}_y$ is a symmetrizer, and the factor $Z_{ij}^y(z) = [z - ic\,\mathrm{sgn}(y_i - y_j)]/(z - ic)$ incorporates the S matrix, whose elements are implicitly defined as $S_{ij}(k_i - k_j) = (k_i - k_j + ic)/(k_i - k_j - ic)$, and describe the scattering of two bosons with momenta k_i, k_j. The corresponding eigenenergy can be written as $E_k = \sum_{j=1}^N k_j^2$ [cf. Eq. (5.108)]. The momenta are real-valued for repulsion ($c > 0$), or are complex conjugate pairs (corresponding to bound states) for attraction ($c < 0$).

Iyer and Andrei (2012) start by considering a generic (symmetric) initial state

$$|\Phi_0\rangle = \int d^N x \Phi_0(x_1, \ldots x_N) \prod_{j=1}^N b^\dagger(x_j)|0\rangle, \tag{5.125}$$

and study its quench dynamics (*i.e.* time evolution at $t > 0$, after the application of the Hamiltonian H at $t = 0$) as

$$|\Phi_0, t\rangle = e^{-iHt}|\Phi_0\rangle = \sum_k e^{-iE_k t}|k\rangle\langle k|\Phi_0\rangle. \tag{5.126}$$

Iyer and Andrei (2012) then circumvents the difficulty arising in performing the summation over all momenta ('identity', in terms of eigenstates of H), by means of a generalization of the Yudson representation (Yudson, 1985, 1988) (see Iyer *et al.*, 2013, for more details). This amounts to express such summations as the contour integrations

$$|\Phi_0\rangle = \int d^N x\, d^N y\, d^N k\, \theta(x)\Phi_0(x) \times \left(\mathcal{S}_y \prod_j e^{ik_j(y_j - x_j)} \prod_{i<j} Z_{ij}^y(k_i - k_j) \right) b^\dagger(y_j)|0\rangle, \tag{5.127}$$

where $\theta(x) = \theta(x_1 > x_2 > \ldots x_N)$, and the contours are fixed by the nature of the interaction, *viz.* all k_j have to be integrated along the real line, in the repulsive case, while all k_j have to be integrated along the line $i(|c| + 0^+)$, in the attractive case. It can be shown that each k_j integration produces a factor $\delta(y_j - x_j)$, thus proving the validity of Eq. (5.127) (Iyer *et al.*, 2013).

Iyer and Andrei (2012) finally consider the relevant example of the quenching of a system of bosons initially trapped in a deep periodic trap (Mott state):

$$|\Phi_{\mathrm{latt}}\rangle = \prod_j \frac{1}{(2\pi\sigma^2)^{1/4}} \int d^N x e^{-[x_j + (j-1)a]/2\sigma^2} b^\dagger(x_j)|0\rangle, \tag{5.128}$$

where a is the lattice step, and $\sigma \ll a$ is the envelope width. In the case of two particles, the evolved state at a finite time t can be given the closed expression in the repulsive case (Iyer and Andrei, 2012)

$$|\Phi_{\text{latt}}, t\rangle = \frac{1}{4\pi i t} \int d^N x \, d^N y \, e^{-[x_1^2 + (x_2+a)^2]/\sigma^2} e^{i[(y_1-x_1)^2+(y_2-x_2)^2]/4t} \times \left[1 - c\sqrt{\pi i t}\theta(y_2 - y_1) e^{i\alpha^2(t)/8t} \operatorname{erf}\left(\frac{i-1}{4}\frac{i\alpha(t)}{\sqrt{t}}\right)\right] b^\dagger(y_1) b^\dagger(y_2)|0\rangle, \tag{5.129}$$

where $\alpha(t) = 2ct - i(y_1 - x_1) - i(y_2 - x_2)$. The function erf must be replaced by erfc in the attractive case. While the repulsive and noninteracting cases are almost indistinguishable, the attractive case develops oscillations in the average density of the condensate, with period $\sim 1/c^2$. This may be ascribed to the competition between diffusion of the bosons and the action of the attractive potential. Exact integrations beyond the two particles case require numerical methods, although the Yudson representation affords an exact analysis of the asymptotics of the quenched states (Iyer and Andrei, 2012; Iyer *et al.*, 2013).

This same analytical procedure has been applied by Liu and Andrei (2014) to extract the quench dynamics of the anisotropic XXZ Heisenberg model on the infinite line. Further aspects of the non-equilibrium dynamics and thermodynamics of the Lieb-Liniger and related models have been studied by Goldstein and Andrei (2013, 2014a,b,c,d, 2015).

Chapter 6

Exact results for an isolated impurity in a solid

6.1 Derivation of the expression of the Dirac density matrix from Schrödinger equation

The concept of density matrix (DM), already introduced in Chapter 1, proves useful also for the description of the problem of an isolated impurity, such as a localized defect or a vacancy, in a metal [March and Murray (1961); see also March and Murray (1960a)]. Indeed, its diagonal element coincides with the electron density $\rho(\mathbf{r}, \mathbf{r}_0; k)$, which is the quantity of central interest here.

Following March and Murray (1961), we start by deriving an expression for $\rho(\mathbf{r}, \mathbf{r}_0; k)$, starting from Schrödinger's equation for a single particle with energy $E = k^2/2$ in a potential V, which we rewrite as

$$\nabla^2\psi + k^2\psi = 2V\psi. \tag{6.1}$$

Making use of the Green function $\cos kr/r$ for its associated homogeneous equation, the general integral of Eq. (6.1) may be written in the form

$$\psi(\mathbf{r}) = -\frac{1}{2\pi}\int d\mathbf{r}' V(\mathbf{r}')\psi(\mathbf{r}')\frac{\cos k|\mathbf{r}-\mathbf{r}'|}{|\mathbf{r}-\mathbf{r}'|} + \int d\mathbf{r}'\alpha(\mathbf{r}')\frac{\sin k|\mathbf{r}-\mathbf{r}'|}{|\mathbf{r}-\mathbf{r}'|}. \tag{6.2}$$

Here, one recognizes in the first contribution a particular integral of the complete Equation (6.1), while the second contribution is the general integral of the associated homogeneous equation, with $\alpha(\mathbf{r}')$ to be determined by the boundary conditions.

Introducing the quantity

$$\sigma(\mathbf{r}, \mathbf{r}_0; k) = \frac{2\pi^2}{k}\frac{\partial\rho(\mathbf{r}, \mathbf{r}_0; k)}{\partial k}, \tag{6.3}$$

one may then reconstruct ρ as

$$\rho(\mathbf{r}, \mathbf{r}_0; k) = \int_0^k \frac{k}{2\pi^2}\sigma(\mathbf{r}, \mathbf{r}_0; k')dk'. \tag{6.4}$$

It may be shown that $\sigma(\mathbf{r},\mathbf{r}_0;k)$ is related to the average wave function product for energy $E=k^2/2$, and hence satisfies Eq. (6.1). In the case of free electrons, one explicitly finds

$$\rho_0(\mathbf{r},\mathbf{r}_0;k)=\frac{k^3}{2\pi^2}\frac{j_1(k|\mathbf{r}-\mathbf{r}_0|)}{k|\mathbf{r}-\mathbf{r}_0|}, \tag{6.5}$$

where $j_1(\rho)=(\sin\rho-\rho\cos\rho)/\rho^2$ denotes the spherical Bessel function of order one. Correspondingly,

$$\sigma_0(\mathbf{r},\mathbf{r}_0;k)=\frac{\sin k|\mathbf{r}-\mathbf{r}_0|}{|\mathbf{r}-\mathbf{r}_0|}. \tag{6.6}$$

Replacing $\sigma(\mathbf{r},\mathbf{r}_0;k)$ for $\psi(\mathbf{r})$ in the solution to the Schrödinger's equation, Eq. (6.2), one finds

$$\sigma(\mathbf{r},\mathbf{r}_0;k)=-\frac{1}{2\pi}\int d\mathbf{r}'V(\mathbf{r}')\sigma(\mathbf{r}',\mathbf{r}_0;k)\frac{\cos k|\mathbf{r}-\mathbf{r}'|}{|\mathbf{r}-\mathbf{r}'|}$$
$$+\int d\mathbf{r}'\alpha(\mathbf{r}',\mathbf{r}_0)\frac{\sin k|\mathbf{r}-\mathbf{r}'|}{|\mathbf{r}-\mathbf{r}'|}, \tag{6.7}$$

where $\alpha(\mathbf{r}',\mathbf{r}_0)$ is now to be determined under the condition that $\sigma(\mathbf{r},\mathbf{r}_0;k)$ be symmetrical with respect to the interchange of $\mathbf{r}$ and $\mathbf{r}_0$. This is now an integral equation, which may be solved by iterations in powers of V. Although such a form of the solution is perturbative in nature, it will afford the derivation of exact results, to be discussed in Sec. 6.2. Writing

$$\sigma(\mathbf{r},\mathbf{r}_0;k)=\sum_{j=0}^{\infty}\sigma_j(\mathbf{r},\mathbf{r}_0;k), \tag{6.8}$$

with σ_0 given by Eq. (6.6), and inserting σ_0 in the right-hand side of Eq. (6.7), one finds

$$\sigma_1(\mathbf{r},\mathbf{r}_0;k)=-\frac{1}{2\pi}\int d\mathbf{r}_1V(\mathbf{r}_1)\frac{\sin k(|\mathbf{r}_0-\mathbf{r}_1|+|\mathbf{r}_1-\mathbf{r}|)}{|\mathbf{r}_0-\mathbf{r}_1|\cdot|\mathbf{r}_1-\mathbf{r}|}. \tag{6.9}$$

Similarly, following this procedure, at order j one finds

$$\sigma_j(\mathbf{r},\mathbf{r}_0;k)=\int\prod_{\ell=1}^{j}\left(-\frac{d\mathbf{r}_\ell}{2\pi}V(\mathbf{r}_\ell)\right)\frac{\sin k\left(\sum_{\ell=1}^{j+1}s_\ell\right)}{\prod_{\ell=1}^{j+1}s_\ell}, \tag{6.10}$$

where $s_\ell=|\mathbf{r}_\ell-\mathbf{r}_{\ell-1}|$, with $\mathbf{r}_{j+1}\equiv\mathbf{r}$. Writing $\rho=\sum_j\rho_j$, inserting Eq. (6.10) into Eq. (6.4) and integrating, one finds (March and Murray, 1961)

$$\rho_j(\mathbf{r},\mathbf{r}_0;k)=\frac{k^2}{2\pi^2}\int\prod_{\ell=1}^{j}\left(-\frac{d\mathbf{r}_\ell}{2\pi}V(\mathbf{r}_\ell)\right)\frac{j_1\left(k\sum_{\ell=1}^{j+1}s_\ell\right)}{\prod_{\ell=1}^{j+1}s_\ell}. \tag{6.11}$$

The same expression has also been derived by March and Murray (1961) using a different route, involving the Bloch equation for the canonical density matrix.

6.2 Friedel sum rule and integrated density of states

The scattering process of a plane wave $e^{i\mathbf{k}\cdot\mathbf{r}}$ by an isotropic impurity described by potential $V(r)$ produces an outgoing wave which is the superposition of the incoming plane wave, plus a spherical wave times a scattering factor $f(\theta)$ depending on the scattering angle θ (Jones and March, 1986a)

$$\psi = e^{ikz} + f(\theta)\frac{e^{ikr}}{r}. \tag{6.12}$$

The corresponding solution of the Schrödinger equation $\psi(\mathbf{r})$ can then be analyzed in terms of spherical harmonics as

$$\psi(\mathbf{r}) = \sum_{\ell=0}^{\infty}(2\ell+1)i^{\ell}c_{\ell}R_{\ell}(r)P_{\ell}(\cos\theta), \tag{6.13}$$

where $P_\ell(x)$ are Legendre polynomials, c_ℓ are suitable complex numbers, and $R_\ell(r)$ is the real normalized solution of the radial Schrödinger equation

$$\frac{d^2}{dr^2}(rR_\ell) + 2\left[\frac{k^2}{2} - V(r) - \frac{\ell(\ell+1)}{2r^2}\right]rR_\ell = 0. \tag{6.14}$$

Their solutions in the free case are $R_\ell(r) \propto j_\ell(kr)$, apart from an inessential normalization factor, where $j_\ell(\rho)$ is a Bessel spherical harmonic. Therefore, they behave asymptotically for $r \to \infty$ as

$$j_\ell(kr) \to \frac{1}{kr}\sin\left(kr - \frac{\ell\pi}{2}\right). \tag{6.15}$$

Such a behaviour must be obtained also by the solutions of the complete equation, Eq. (6.14), since $V(r) \to 0$, when $r \to \infty$, apart from a phase shift η_ℓ:

$$R_\ell(r) \to \frac{1}{kr}\sin\left(kr + \eta_\ell - \frac{\ell\pi}{2}\right). \tag{6.16}$$

Scattering theory off an isolated impurity then teaches that most properties of the outgoing wave can be expressed in terms of series involving such phase shifts, and that only the very first few terms have often to be retained, in order to get good numerical approximation (Jones and March, 1986a).

In particular, the scattering amplitude, which essentially describes the outgoing wave, can be expressed as

$$f(\theta) = \frac{1}{k}\sum_{\ell=0}^{\infty}(2\ell+1)e^{i\eta_\ell}\sin\eta_\ell P_\ell(\cos\theta). \tag{6.17}$$

Their value is fixed by the scattering potential through (Jones and March, 1986b)

$$\sin\eta_\ell = -k\int_0^\infty j_\ell(kr)V(r)R_\ell(r)r^2 dr. \tag{6.18}$$

The s-wave scattering length a was already introduced in Eq. (3.1), along with the effective range r_s of the interaction, and is defined through the expansion

$$k\cot\eta_{\ell=0} = -\frac{1}{a} + \frac{1}{2}r_s k^2 + \dots. \tag{6.19}$$

The Friedel sum rule (1952) relates the phase shifts to the total charge Ze ($e > 0$) of a charged impurity with potential $V(r) = Ze/r$ in an electron gas (Jones and March, 1986b). Consider now a sufficiently large sphere centred on the impurity, with radius R. The total screening charge displaced into this sphere is related to the difference between the probability densities associated with incoming and outgoing waves, integrated within this sphere, *i.e.*

$$\frac{1}{\mathcal{V}}\int^R d^3\mathbf{r}\left[|\psi(\mathbf{r})|^2 - |e^{i\mathbf{k}\cdot\mathbf{r}}|^2\right] = \frac{4\pi}{\mathcal{V}}\sum_{\ell=0}^{\infty}(2\ell+1)\int_0^R dr\, r^2\left[R_\ell^2(r;k) - j_\ell^2(kr)\right], \tag{6.20}$$

where we have made explicit the (parametric) dependence on k also on the radial part as $R_\ell(r) \equiv R_\ell(r;k)$.

Multiplying Eq. (6.14) by $R_\ell(r;k')$, and the corresponding equation with k' by $R_\ell(r;k)$, and subtracting the two equations, one finds

$$\begin{aligned} rR_\ell(r;k')\frac{d^2}{dr^2}[rR_\ell(r;k)] - rR_\ell(r;k)\frac{d^2}{dr^2}[rR_\ell(r;k')] \\ = (k'^2 - k^2)r^2 R_\ell(r;k')R_\ell(r;k). \end{aligned} \tag{6.21}$$

Integrating between 0 and R, one finds

$$\begin{aligned} \left[rR_\ell(r;k')\frac{d}{dr}[rR_\ell(r;k)] - rR_\ell(r;k)\frac{d}{dr}[rR_\ell(r;k')]\right]_0^R \\ = (k'-k)(k'+k)\int_0^R dr\, r^2 R_\ell(r;k')R_\ell(r;k). \end{aligned} \tag{6.22}$$

Expanding $R_\ell(r;k')$ in powers of $k'-k$ as $R_\ell(r;k') = R_\ell(r;k) + (k' - k)(d/dk)R_\ell(r;k) + \dots$, dividing by $k'-k$, and taking the limit $k' \to k$, one finds

$$\begin{aligned} \left[\frac{\partial}{\partial k}[rR_\ell(r;k)]\frac{\partial}{\partial r}[rR_\ell(r;k)] - rR_\ell(r;k)\frac{\partial^2}{\partial k\partial r}[rR_\ell(r;k)]\right]_0^R \\ = 2k\int_0^R dr\, r^2 R_\ell^2(r;k). \end{aligned} \tag{6.23}$$

The contribution from the lower limit of the left-hand side vanishes, while that coming from the upper limit can be evaluated by means of the asymptotic form, Eq. (6.16). One finds

$$\frac{1}{k}\left\{R + \frac{d\eta_\ell}{dk} - \frac{1}{2k}\sin\left[2\left(kR + \eta_\ell - \frac{\ell\pi}{2}\right)\right]\right\} = 2k\int_0^R dr\, r^2 R_\ell^2(r;k). \tag{6.24}$$

The corresponding result for the unperturbed radial solution $j_\ell(kr)$ is obtained by setting $\eta_\ell = d\eta_\ell/dk = 0$. Therefore, Eq. (6.20) becomes

$$\frac{1}{\mathcal{V}}\int^R d^3\mathbf{r}\left[|\psi(\mathbf{r})|^2 - |e^{i\mathbf{k}\cdot\mathbf{r}}|^2\right]$$
$$= \frac{2\pi}{\mathcal{V}k^2}\sum_{\ell=0}^{\infty}(2\ell+1)\left[\frac{d\eta_\ell}{dk} - \frac{1}{k}\sin\eta_\ell\cos(2kR + \eta_\ell - \ell\pi)\right]. \tag{6.25}$$

Multiplying the above equation by the density of states per spin in **k**-space, *viz.* $\mathcal{V}/4\pi^3$, and integrating with respect to k within the Fermi sphere, one may express the total number of particles displaced into the sphere or radius R as

$$\frac{1}{4\pi^3}\int_0^{k_\mathrm{F}} dk\, k^2 \int^R d^3\mathbf{r}\left[|\psi(\mathbf{r})|^2 - |e^{i\mathbf{k}\cdot\mathbf{r}}|^2\right]$$
$$= \frac{2}{\pi}\sum_{\ell=0}^{\infty}(2\ell+1)\eta_\ell(k_\mathrm{F}) - \frac{2}{\pi}\sum_{\ell=0}^{\infty}(2\ell+1)\int_0^{k_\mathrm{F}} \sin\eta_\ell\cos(2kR + \eta_\ell - \ell\pi). \tag{6.26}$$

With increasing R, the second term in the equation above oscillates, and must eventually integrate to zero. This is indeed consistent with the form of the density oscillations (Friedel oscillations), which are known to behave as $\sim \cos(2k_\mathrm{F}r)/r^3$. The remaining constant term must equal Z, *i.e.* the total number of electrons screening the impurity. One therefore obtains the *Friedel sum rule* as (Friedel, 1952)

$$Z = \frac{2}{\pi}\sum_{\ell=0}^{\infty}(2\ell+1)\eta_\ell(k_\mathrm{F}). \tag{6.27}$$

The Friedel sum rule has been generalized also to Bloch states for electrons interacting with a periodic external potential (Jones and March, 1986b). Langer and Ambegaokar (1961) also proved that it is robust with respect to the inclusion of electron-electron interaction, provided the resulting quasiparticle states do not decay in time, which is a tenet of Landau's theory of Fermi liquids, and is asymptotically valid close to the Fermi level at $T = 0$. Langer and Ambegaokar (1961) then derived a generalized version of Eq. (6.27) for interacting electrons in this limit, by use of an identity by Luttinger (1960); Luttinger and Ward (1960), which essentially guarantees the sharpness of the Fermi surface also in the presence of electron-electron interactions. The generalized Friedel sum rule for interacting electrons then reads (Langer and Ambegaokar, 1961)

$$Z = \frac{1}{2\pi i} \operatorname{Tr} \log S(\mu), \tag{6.28}$$

where $S(\zeta_\ell) = [\zeta_\ell - H - \Sigma(\zeta_\ell)]^{-1}$ is the scattering matrix, $\Sigma(\zeta_\ell)$ is the proper self-energy, $\zeta_\ell = \mu + (2\ell + 1)\pi i \beta^{-1}$ is related to the fermionic Matsubara frequencies at inverse temperature β, and μ is the chemical potential.

6.3 Second-order perturbation corrections for Dirac density matrix around an impurity in a metal

After setting out a perturbative series for the diagonal of Dirac's density matrix ρ in Sec. 6.1 (March and Murray, 1961), March and Murray (1962) derived a self-consistent expression for $\rho(\mathbf{r}, \mathbf{r}; k)$, valid at second order in the impurity potential $V(\mathbf{r})$. They start by reminding the series expansion (again, truncated at second order in the impurity potential)

$$\rho(\mathbf{r}, \mathbf{r}; k) = \rho_0(\mathbf{r}, \mathbf{r}; k) + \rho_1(\mathbf{r}, \mathbf{r}; k) + \rho_2(\mathbf{r}, \mathbf{r}; k), \tag{6.29a}$$

where

$$\rho_0(\mathbf{r}, \mathbf{r}; k) = \frac{k^3}{3\pi^2}, \tag{6.29b}$$

$$\rho_1(\mathbf{r}, \mathbf{r}; k) = -\frac{k^2}{2\pi^3} \int d^3\mathbf{r}_1 V(\mathbf{r}_1) \frac{j_1(2k|\mathbf{r} - \mathbf{r}_1|}{|\mathbf{r} - \mathbf{r}_1|^2}, \tag{6.29c}$$

$$\rho_2(\mathbf{r}, \mathbf{r}; k) = \frac{k^2}{4\pi^4} \int d^3\mathbf{r}_1 d^3\mathbf{r}_2 V(\mathbf{r}_1) V(\mathbf{r}_2) \times \frac{j_1(k|\mathbf{r} - \mathbf{r}_1| + k|\mathbf{r}_1 - \mathbf{r}_2| + k|\mathbf{r}_2 - \mathbf{r}|)}{|\mathbf{r} - \mathbf{r}_1| \cdot |\mathbf{r}_1 - \mathbf{r}_2| \cdot |\mathbf{r}_2 - \mathbf{r}|}. \tag{6.29d}$$

The excess charge density due to the presence of the impurity may be related to the potential via Poisson's equation as $\nabla^2 V = 4\pi(\rho_0 - \rho)$. Retaining only the second order corrections in V as in Eq. (6.29a), and separating the angular integrations, one obtains (March and Murray, 1962)

$$\frac{d^2}{dr^2}\left[rV + \int_0^\infty dt\, t\, V(t)\left(F[2k(r+t)] - F[2k|r-t|]\right)\right]$$
$$= -k^2 \int_0^\infty ds\, s\, V(s) \int_0^\infty dt\, t\, V(t) G(kr, ks, kt), \quad (6.30)$$

where (March and Murray, 1961, 1962)

$$F(u) = \frac{1}{4\pi}\left(4u - u\cos u - \sin u + (u^2+2)\int_u^\infty ds \frac{\sin s}{s}\right), \quad (6.31a)$$

$$G(kr, ks, kt) = \frac{1}{\pi^3}\int d\omega_1 d\omega_2\, r\, s\, t \times \frac{j_1(k|\mathbf{r}-\mathbf{s}| + k|\mathbf{s}-\mathbf{t}| + k|\mathbf{t}-\mathbf{r}|)}{|\mathbf{r}-\mathbf{s}|\cdot|\mathbf{s}-\mathbf{t}|\cdot|\mathbf{t}-\mathbf{r}|}. \quad (6.31b)$$

In the definition of $G(kr, ks, kt)$, the integrations are over the spheres $|\mathbf{s}| =$ const and $|\mathbf{t}| =$ const. Although further analytical progress is unlikely, March and Murray (1962) note the following closed results

$$\lim_{t\to 0} \frac{1}{t} G(r, s, t) = \frac{8}{\pi r s}\left[j_0(r + s + |r-s|) - j_0\left(2(r+s)\right)\right], \quad (6.32a)$$

$$\lim_{s\to 0}\lim_{t\to 0} \frac{1}{t} G(r, s, t) = \frac{16}{\pi r} j_1(2r), \quad (6.32b)$$

which may be used to check the numerical solution.

Equation (6.30) may also be viewed as a self-consistent equation for the impurity potential V, at second order in this potential. This can then be solved for given values of Z and k, the latter relating to density, both for attractive and repulsive potentials. However, for sufficiently large positive Z or sufficiently low electron density, bound states will be formed, which are beyond the grasp of this low-order perturbation theory.

6.4 Effect of impurities on the resistivity of a metal

While still in the context of effects due to an isolated impurity, another remarkable result is that of Huang (1948), who derived an exact expression for the excess resistivity of a dilute metallic alloy due to a spherical potential $V(r)$ (see March, 1990; Alonso and March, 1989, for a review). Huang (1948) starts by noting that the scattering cross-section of the electrons diffused by a single localized impurity into a solid angle $d\omega$ is given by

$$I(\theta) d\omega = |f(\theta)|^2 d\omega, \quad (6.33)$$

with $f(\theta)$ given by Eq. (6.17). Then, the resistivity due to impurity ions in a dilute solution, in which the ions may be regarded as independent scatterers of electrons, may be expressed as (Mott, 1936)

$$\rho = \frac{\hbar k_m}{e^2} x S \tag{6.34}$$

where x is the concentration of the impurity ions, and k_m is the wave number corresponding to the highest filled state, and

$$S = 2\pi \int_0^\pi d\theta \sin\theta (1 - \cos\theta) I(\theta). \tag{6.35}$$

Using Eq. (6.17) for $f(\theta)$, and exploiting well-known properties of the Legendre polynomials, Huang (1948) arrives at the exact result

$$\rho = \frac{2h}{e^2 k_m} \sum_{\ell=0}^{\infty} [(2\ell + 1)\sin^2\eta_\ell - 2\ell \sin\eta_\ell \sin\eta_{\ell-1} \cos(\eta_{\ell-1} - \eta_\ell)], \tag{6.36}$$

which can be cast into the equivalent, but more compact, form (March, 1990; Alonso and March, 1989)

$$\rho = \frac{2h}{e^2 k_m} \sum_{\ell=1}^{\infty} \ell \sin^2(\eta_{\ell-1} - \eta_\ell). \tag{6.37}$$

As noted by Huang (1948), one of the advantage of the above closed formula is that it does not depend on the details of the impurity potential, provided it is spherically symmetric.

As noted by March (1975a) (see March, 1990, for a review), Huang's result in the form of Eq. (6.37) can be combined with a result of Gerjuoy (1965) (and independently of Gaspari and Gyorffy, 1972), to show that it is actually related to a force-force correlation function. Gerjuoy (1965); Gaspari and Gyorffy (1972) preliminarly note that the radial functions $R_\ell(r)$ for scattering off a spherical potential $V(r)$ of finite range, but otherwise generic strength, satisfy the identity

$$\int_0^\infty dr\, r^2 R_{\ell-1}(r) \frac{\partial V(r)}{\partial r} R_\ell(r) = \sin(\eta_\ell - \eta_{\ell-1}). \tag{6.38}$$

Outside the range of the potential, $R_\ell(r)$ may be expressed as a 'rotation' of spherical Bessel and Neumann functions $j_\ell(r)$ and $n_\ell(r)$, respectively, the scattering angle η_ℓ being the 'angle' of such a 'rotation', *i.e.*

$$R_\ell(r) = j_\ell(r)\cos\eta_\ell - n_\ell(r)\sin\eta_\ell. \tag{6.39}$$

Thus, combining Eqs. (6.37) and (6.38), one finds

$$S = \sum_{\ell=1}^{\infty} \ell \int_0^{\infty} dr_1\, r_1^2\, R_{\ell-1}(r_1) \frac{\partial V(r_1)}{\partial r_1} R_\ell(r_1) \times \int_0^{\infty} dr_2\, r_2^2\, R_{\ell-1}(r_2) \frac{\partial V(r_2)}{\partial r_2} R_\ell(r_2). \quad (6.40)$$

According to an inverse transport theory proposed by Rousseau *et al.* (1972), the force-force correlation function F may be related to the energy derivative $\sigma(\mathbf{r}_1, \mathbf{r}_2)$ of the Dirac density matrix for potential V evaluated at the Fermi energy $E_{\mathrm{F}} = k_{\mathrm{F}}^2/2$ as

$$F = d^3\mathbf{r}_1 d^3\mathbf{r}_2 \frac{\partial V(\mathbf{r}_1)}{\partial \mathbf{r}_1} \cdot \frac{\partial V(\mathbf{r}_2)}{\partial \mathbf{r}_2} |\sigma(\mathbf{r}_1, \mathbf{r}_2)|^2, \quad (6.41)$$

where (March and Murray, 1960b)

$$\sigma(\mathbf{r}_1, \mathbf{r}_2) = \sum_\ell (2\ell + 1) \sigma_\ell(r_1, r_2) P_\ell(\cos\gamma), \quad (6.42)$$

where γ is the angle between $\mathbf{r}_1$ and $\mathbf{r}_2$. One may now observe that

$$\begin{aligned} f &= \frac{\partial V(\mathbf{r}_1)}{\partial \mathbf{r}_1} \cdot \frac{\partial V(\mathbf{r}_2)}{\partial \mathbf{r}_2} |\sigma(\mathbf{r}_1, \mathbf{r}_2)|^2 \\ &= \sum_\ell \sum_j (2\ell + 1) \sigma_\ell(r_1, r_2) P_\ell(\cos\gamma) \\ &\qquad \times (2j + 1) \sigma_j(r_1, r_2) P_j(\cos\gamma) \frac{\partial V}{\partial r_1} \frac{\partial V}{\partial r_2} \cos\gamma. \end{aligned} \quad (6.43)$$

Multiplying by $\sin\gamma d\gamma$ and integrating over γ between 0 and γ, one is led to consider integrals of the form

$$\int_0^{\pi} \cos\gamma P_\ell(\cos\gamma) P_j(\cos\gamma) \sin\gamma d\gamma,$$

which can be easily evaluated, using identities involving the Legendre polynomials. One eventually arrives at the result that

$$\int_0^{\pi} f \sin\gamma d\gamma = 4 \sum_{\ell=1}^{\infty} \ell \sigma_{\ell-1}(r_1, r_2) \frac{\partial V(r_1)}{\partial r_1} \frac{\partial V(r_2)}{\partial r_2}. \quad (6.44)$$

This result, together with the fact that $\sigma_\ell(r_1, r_2) \propto R_\ell(r_1) R_\ell(r_2)$, shows that the expression for the force-force correlation function F in Eq. (6.41) is essentially equivalent to Huang's formula, Eq. (6.37), apart from an inessential numerical factor, which is however exactly unity for a spherically symmetric potential (March, 1990).

Chapter 7

Pair potential and many-body force models for liquids

Most information concerning the structure of both solids (crystalline and amorphous) and liquids is encoded in the pair correlation function $g(r)$ (March, 1990; March and Tosi, 2002). In the case of liquids, where this function is usually spherically symmetric, the quantity $4\pi\rho r^2 g(r) d^3\mathbf{r}$ yields the number of atoms at a distance r from the origin, within a volume $d^3\mathbf{r}$, where ρ is the number density. Standard scattering theory then relates the pair correlation function directly to the (static) structure factor $S(k)$ through

$$S(k) = 1 + \rho \int (g(r) - 1)\, e^{i\mathbf{k}\cdot\mathbf{r}} d^3\mathbf{r}. \tag{7.1}$$

The static structure factor, in turn, is directly accessible through several scattering experiments, *e.g.* X-ray scattering. Introducing the total correlation function $h(r)$ as the excess correlation with respect to unity,

$$h(r) = g(r) - 1, \tag{7.2}$$

the theory of Ornstein and Zernike (1914) then divides it into a direct part $c(r)$ and an indirect part, the latter being implicitly defined by the relation (March, 1990; March and Tosi, 2002)

$$h(r) = c(r) + \rho \int h(|\mathbf{r} - \mathbf{r}'|) c(r') d^3\mathbf{r}'. \tag{7.3}$$

The indirect part is related to the correlation between atoms or molecules 1 and 2 in a liquid, say, through a third atom or molecule, 3 say. Standard Fourier analysis of Eqs. (7.1) and (7.3) then shows that

$$c(k) = \frac{S(k) - 1}{S(k)}, \tag{7.4}$$

where $c(k)$ denotes the Fourier transform of the Ornstein-Zernike (direct) correlation function, $c(r)$.

With this as a background, it is through the pair correlation functions that one can get important insights on the liquid structure, and in particular on the force field between atoms or molecules, as is briefly reviewed below, with specific examples for some liquid alkalis.

7.1 Force equation for pair potentials

In order to relate the structure factor to the interatomic force in a liquid, consider atoms 1, 2, and 3 at positions $\mathbf{r}_1$, $\mathbf{r}_2$, and $\mathbf{r}_3$, respectively. Given the probabilistic interpretation of the pair correlation function $g(r)$, it is natural to express such a quantity in terms of the average potential $U(r)$ as a Boltzmann factor

$$g(r) = \exp[-\beta U(r)]. \tag{7.5}$$

Here, $\beta = (k_{\mathrm{B}}T)^{-1}$, and we are neglecting an inessential overall normalization factor. Inversely, Eq. (7.5) may serve as a definition of the potential for the mean force experienced by an atom at position r, in those circumstances in which the pair correlation function is known.

The total force $-\partial U(|\mathbf{r}_1 - \mathbf{r}_2|)/\partial \mathbf{r}_1$ applied to atom 1 can then be decomposed into a direct part due to the interaction with atom 2, and into an indirect part due to the rest of the atoms as

$$-\frac{\partial U(|\mathbf{r}_1 - \mathbf{r}_2|)}{\partial \mathbf{r}_1} = -\frac{\phi(|\mathbf{r}_1 - \mathbf{r}_2|)}{\partial \mathbf{r}_1} - \rho \int \frac{\partial \phi(|\mathbf{r}_1 - \mathbf{r}_3|)}{\partial \mathbf{r}_1} \frac{g_3(\mathbf{r}_1, \mathbf{r}_2, \mathbf{r}_3)}{g(|\mathbf{r}_1 - \mathbf{r}_2|)} d^3\mathbf{r}_3. \tag{7.6}$$

Here, $\phi(|\mathbf{r}_1 - \mathbf{r}_2|)$ denotes the pair potential energy acting between two atoms 1 and 2, and $g_3(\mathbf{r}_1, \mathbf{r}_2, \mathbf{r}_3)$ is the probability density of finding atoms 1, 2, and 3 at positions $\mathbf{r}_1$, $\mathbf{r}_2$, and $\mathbf{r}_3$, respectively. Thus, the ratio $g_3(\mathbf{r}_1, \mathbf{r}_2, \mathbf{r}_3)g(|\mathbf{r}_1 - \mathbf{r}_2|)$ in the integral is the conditional probability that atom 3 is at $\mathbf{r}_3$, given that atoms 1 and 2 are at $\mathbf{r}_1$ and $\mathbf{r}_2$, respectively, and the integration takes into account for all possible positions of atom 3. The distinction of the total in a direct and an indirect part obviously reflects that in the definition of the Ornstein-Zernike correlation function, Eq. (7.3).

Although Eq. (7.6) provides the desired link between liquid structure and interatomic forces, the relation between the correlation function $g(r)$ and the pair potential $\phi(r)$ still requires explicit knowledge of the three-body correlation function $g_3(\mathbf{r}_1, \mathbf{r}_2, \mathbf{r}_3)$. While many-body theory would in principle allow to construct a hierarchy of equations involving in turn four-body correlations g_4 and so on (see March, 1990), several approximate structural theories exist, the simplest consisting in factoring out g_3 as $g_3(\mathbf{r}_1, \mathbf{r}_2, \mathbf{r}_3) \simeq g(r_{12})g(r_{23})g(r_{31})$. This results in the so-called Born-Green theory of liquid structure (Green, 1952; Rushbrooke, 1960), whose basic equation reads

$$\beta[U(r) - \phi(r)] = -\rho \int E(\mathbf{r} - \mathbf{r}')h(\mathbf{r}')d^3\mathbf{r}', \tag{7.7}$$

where

$$E(r) = \beta \int_r^\infty g(s)\frac{\partial \phi(s)}{\partial s}ds. \tag{7.8}$$

The hypernetted chain (HNC) approximation then corresponds to replacing $E(r)$ by $c(r)$ in Eq. (7.8), which is justified by the asymptotic behaviour of $c(r)$ at large r, far from any possible critical point (March, 1990). This yields, explicitly

$$\beta[U(r) - \phi(r)] = -\rho \int c(\mathbf{r} - \mathbf{r}')h(\mathbf{r}')d\mathbf{r}', \tag{7.9}$$

which, combined with the definition of $c(r)$ in Eq. (7.3), yields

$$\beta\phi_{\mathrm{HNC}}(r) = \beta U(r) + h(r) - c(r). \tag{7.10}$$

This enables, in principle, to determine $\phi_{\mathrm{HNC}}(r)$ through experimental measurements of $S(k)$ (Johnson and March, 1963).

7.2 Pair potentials for liquid Na near freezing

The possibility of inverting experimental measurements of the structure factor $S(k)$ in favour of the pair potential $\phi(r)$ (Johnson and March, 1963) has been first exploited for liquid sodium near freezing by Reatto *et al.* (1986), using data by Greenfield *et al.* (1971). Parallel efforts were devoted to obtain $\phi(r)$ from electron theory (Perrot and March, 1990). First, a Kohn-Sham density functional calculation was carried out for a Na nucleus in a jellium background, with average conduction electron density $\bar{n}$. The one-body potential was written as

$$V(\mathbf{r}) = V_{\mathrm{Hartree}}(\mathbf{r}) + V_{xc}(\mathbf{r}), \tag{7.11}$$

where $V_{xc}(\mathbf{r})$ denotes the exchange and correlation contribution. Two approximate forms of $V_{xc}(\mathbf{r})$ were considered, *viz.* that due to Ichimaru (1982), and that due to Hedin and Lundqvist (1970), but the difference of their effect on the displaced charge $\Delta n(r)$ was negligible. Perrot and March (1990)

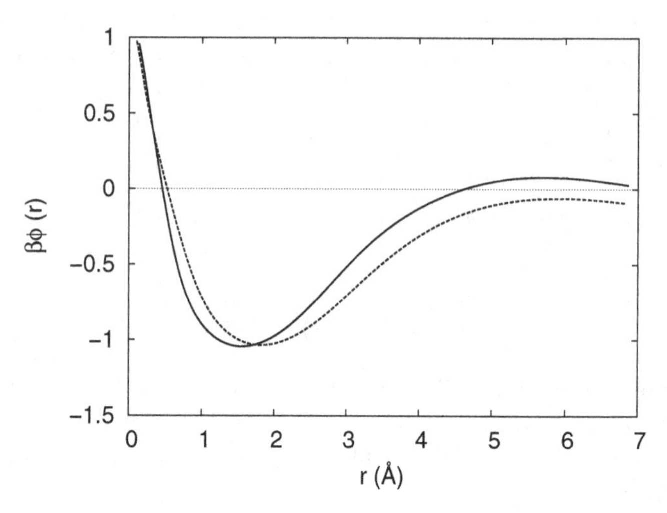

Fig. 7.1 Pair potentials $\beta\phi(r)$ for liquid Na near freezing ($T = 100°\mathrm{C}$, $\bar{n} = 0.929\ \mathrm{g\ cm}^{-3}$. Solid line is the result of Perrot and March (1990), from electron theory. Dashed line is the result of Reatto *et al.* (1986), from inversion of $S(k)$. Redrawn after Perrot and March (1990).

then expressed the pair interaction as in second-order perturbation theory as

$$\phi(r) = \frac{(Z^* e)^2}{r} + \int d^3\mathbf{q} e^{-i\mathbf{q}\cdot\mathbf{r}} w(q)\overline{\Delta n}(q), \tag{7.12}$$

where $\overline{\Delta n}(q)$ is constructed in such a way as to normalize the valence Z^* ($Z^* = 1$, for sodium), so as to compensate the Coulomb term [first term in Eq. (7.12)] as $r \to \infty$. The weak potential $w(q)$ in Eq. (7.12) is related to $\overline{\Delta n}(q)$ by means of linear response theory as

$$w(q) = \frac{\overline{\Delta n}(q)}{\chi(q)}, \tag{7.13}$$

where $\chi^{-1}(q) = \pi_0^{-1}(q) + (4\pi/q^2 + X)$ is the susceptibility, $\pi_0(q)$ is the random-phase approximation (RPA) density response function, and X the local field correction in the (q-independent) local density approximation for exchange and correlation,

$$X = \frac{dV_{xc}(\bar{n})}{d\bar{n}}. \tag{7.14}$$

Figure 7.1 shows the results of $\beta\phi(r)$ for Na near freezing by Perrot and March (1990) (solid line), compared with those by Reatto *et al.* (1986) from inversion of the liquid structure factor (dashed line). Table 7.1 also compares the main characteristics of such pair potential as obtained theoretically, with the corresponding quantities available from experiment.

Table 7.1 Characteristics of diffraction potential compared with electron theory form. After Perrot and March (1990).

Positions (Å)	1st node	1st minimum	1st maximum	2nd minimum
Diffraction	3.30	3.90	5.76	7.44
Electron theory	3.20	3.73	5.67	7.37

7.3 Possible empirical extrapolation of the valence-valence electron partial structure factor for liquid Mg near freezing

Egelstaff *et al.* (1974) proposed the extraction of electron correlation functions in liquids from scattering data. Here, we appeal to computer simulation by de Wijs *et al.* (1995) on the partial structure factor $S_{iv}(k)$ between ions (i) and valence electrons (v) for liquid Mg near freezing, to write the valence-valence partial structure factor $S_{vv}(k)$ in terms of $S_{iv}(k)$ and the neutron structure factor $S_{ii}(k)$, to high accuracy.

The simplest model to treat electron-electron correlations in metals goes back at least to Sommerfeld. His ideas are embodied in the so-called jellium model, in which the positive ion background, say for Na^+ ions on a body-centred cubic (bcc) lattice in metallic Na, are smeared out into a uniform background of positive charge to neutralize the interacting electrons which remain to be treated by many-body theory, as a function of the (uniform) density, ρ_0 say, defined usually in terms of an interelectronic separation r_s through

$$\rho_0 = \frac{3}{4\pi r_s^3}. \tag{7.15}$$

In simple s-p metals, r_s ranges from around $2a_0$ (where the hydrogen atomic radius $a_0 = \hbar^2/me^2$ is the atomic unit of length), which is appropriate roughly for Al metal, to the low density alkalis, where Cs has the lowest conduction electron density, with $r_s \sim 5.5$ a.u. (Ashcroft and Mermin, 1976).

Even for the jellium model, analytic theory remains, as yet, only possible for $r_s \ll a_0$, and therefore Ceperley and Alder (1980) carried out quantal simulations on this admittedly oversimplified model. Below, therefore, we shall use simple s-p liquid metals, like the alkalis, plus Mg and Al, to seek insight, via a combination of theory and experiment, on the electron-electron pair correlations. The specific example we take below, following

Leys and March (2003) is that of Mg, whose work was motivated by a pioneering proposal of Egelstaff *et al.* (1974, EMM below).

7.3.1 *Treating pure s-p liquid metals like Na near freezing as a two-component system*

Following EMM (Egelstaff *et al.*, 1974), we regard liquid Na near the melting temperature T_m as a two-component system (cf. NaCl, say), consisting of Na^+ ions (i) and valence electrons (v). Thus, by the analogy with the classical molten salt NaCl, three pair correlation functions are required to characterize the structure. Since experiments are normally in momentum ($\mathbf{k}$) space rather than coordinate ($\mathbf{r}$) space, the ion-ion pair correlation function $g_{ii}(r)$ has, via Fourier transform, the corresponding partial structure factor $S_{ii}(k)$, and this is well known to be directly accessible by neutron scattering. In principle, the remaining correlation functions $S_{iv}(k)$ and $S_{vv}(k)$ are also experimentally accessible by combining knowledge of $S_{ii}(k)$ with X-ray and electron diffraction studies. For Mg, on which most emphasis is placed below, $S_{iv}(k)$ has become available through the computer simulation studies of de Wijs *et al.* (1995). Motivated by this, Leys and March (2003) have presented calculations for the remaining partial structure factor $S_{vv}(k)$.

7.3.2 *Comparison of first-order theory for $S_{iv}(k)$ with computer simulation results*

Leys and March (2003, LM, below) noted that $S_{iv}(k)$ is given to first order by (Tosi and March, 1973)

$$S_{iv}(k) = \frac{1}{Z_v^{1/2}} \left(\pi(k) \frac{v_{iv}(k)}{\epsilon(k)} \right) S_{ii}(k), \tag{7.16}$$

where Z_v denotes the valency, $\pi(k)$ the proper jellium polarization function, $v_{iv}(k)$ the electron-ion pseudopotential, and $\epsilon(k) = 1 - v(k)\pi(k)$ the static dielectric function, with $v(k) = 4\pi/k^2$ the Fourier transform of the bare Coulombic interaction. Figure 7.2, redrawn from Leys and March (2003), compares the result for liquid Mg near freezing using Eq. (7.16) to the computer simulation work of de Wijs *et al.* (1995). For the dielectric function $\epsilon(k)$, a local field correction in the Hartree-Fock approximation was used. The pseudopotential $v_{iv}(k)$ was calculated by LM using the Ashcroft empty core approximation (Ashcroft, 1966) with the standard Mg

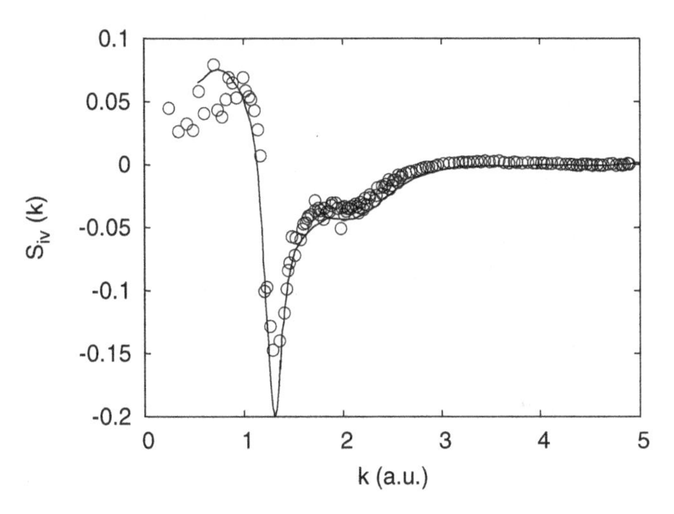

Fig. 7.2 Ion-valence electrons structure factor $S_{iv}(k)$ for Mg. Circles are computer simulation results from de Wijs *et al.* (1995). Solid line is linear response result, from Leys and March (2003). Redrawn after Leys and March (2003).

core radius $R_c = 1.39a_0$, and the ion-ion structure factor $S_{ii}(k)$ was approximated by X-ray data at $T = 680$ K of the SCM-LIQ database (Fig. 7.3). The above Fig. 7.2 shows that Eq. (7.16) then reproduces rather well the computer simulation results of de Wijs *et al.* (1995).

7.3.3 *Treatment of electron-electron structure factor $S_{vv}(k)$ by DFT*

If we denote the jellium model approximation to Mg at the appropriate interelectronic spacing $r_s \sim 2a_0$ for this divalent liquid metal by $S_{vv}^{\text{jellium}}(k)$, then one can follow Cusack *et al.* (1976) and correct for the ionic background by adding two terms labelled by 1 and 2 below

$$\delta S(k) = \delta S_1(k) + \delta S_2(k). \tag{7.17}$$

The first correction $\delta S_1(k)$ is local in the ionic structure factor, and is given by

$$\delta S_1(k) = \frac{1}{Z_v} \left| \pi(k) \frac{v_{iv}(k)}{\epsilon(k)} \right|^2 S_{ii}(k). \tag{7.18}$$

Here, as above, Z_v denotes the valency ($Z_v = 2$, for Mg), $\pi(k)$ is the proper jellium polarization, and $\epsilon(k)$ the static dielectric function, given by

$$\epsilon(k) = 1 - v(k)\pi(k), \tag{7.19}$$

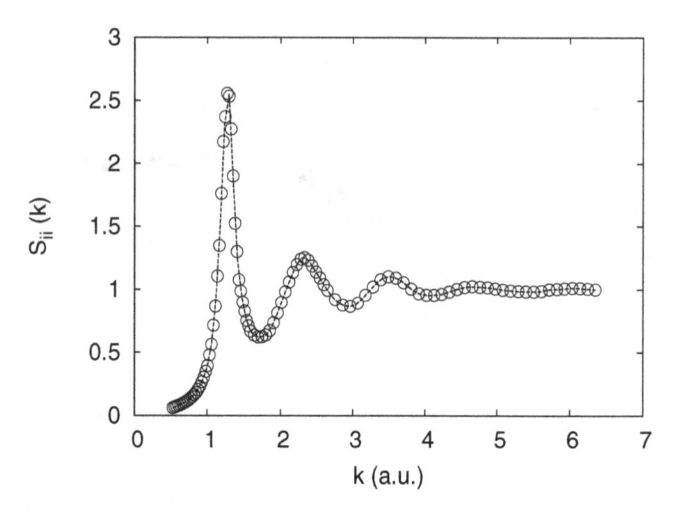

Fig. 7.3 Ion-ion structure factor $S_{ii}(k)$ from X-ray data at $T = 680$ K for Mg. Data from SCM-LIQ. Dashed line is a guide to the eye.

with $v(k) = 4\pi/k^2$ representing the bare Coulomb interaction.

As was shown by March and Murray (1960b), the resulting electron density profile when one perturbs an initially uniform electron gas of density ρ_v^0 with a weak spherically symmetric pseudopotential is given in linear response by

$$\rho_v(\mathbf{r}) = \rho_v^0(\mathbf{r}) + \int F(|\mathbf{r} - \mathbf{r}'|)V(\mathbf{r}')d^3\mathbf{r}'. \tag{7.20}$$

Here, the response function $F(\mathbf{r})$ is the Fourier transform of the proper polarization $\pi(\mathbf{k})$, while $V(\mathbf{r})$ is the screened pseudopotential. Then the correction $\delta S_1(k)$ in Eq. (7.17) takes the form

$$\delta S_1(k) = \frac{1}{Z_v}|\rho_v(k)|^2 S_{ii}(k), \tag{7.21}$$

with $\rho_v(k)$ being the linear response approximation to the Fourier transform of the displaced density profile around every ion: frequently termed the 'form factor'. The correction $\delta S_1(k)$ can then be understood as the correlation that results between electrons since they form neutralizing shells around the correlated ions.

The second term $\delta S_2(k)$ entering Eq. (7.17) is more complicated, but fortunately for liquid Mg is much smaller. It is given by Cusack *et al.* (1976) as

$$\delta S_2(k) = \frac{1}{2Z_v}\int \frac{d^3\mathbf{q}}{(2\pi)^3} S_{ii}(q)|v_{iv}(q)|^2 \chi_4(\mathbf{k}, -\mathbf{k}; -\mathbf{q}, \mathbf{q}), \tag{7.22}$$

where $\chi_4(\mathbf{k}, -\mathbf{k}; -\mathbf{q}, \mathbf{q})$ is the interacting four-body jellium response function which Leys and March Leys and March (2003) calculated for liquid Mg using the expression derived by Cusack *et al.* (1976) by means of density functional theory (DFT).

7.3.4 *Analytic relation between $\delta S_1(k)$ and $S_{iv}(k)$*

Next we here combine $S_{iv}(k)$, the partial structure factor of ion-valence electron correlations, with the dominant correction $\delta S_1(k)$ to the jellium approximation for the partial structure factor $S_{vv}(k)$.

Squaring Eq. (7.16) it follows that

$$\frac{S_{iv}^2(k)}{S_{ii}(k)} = \frac{1}{Z_v}\left(\pi(k)\frac{v_{iv}(k)}{\epsilon(k)}\right)^2 S_{ii}(k). \tag{7.23}$$

But from Eq. (7.18) this yields simply

$$\delta S_1(k) = \frac{S_{iv}^2(k)}{S_{ii}(k)}. \tag{7.24}$$

In the long wavelength limit $k \to 0$, we have the conditions for charge neutrality (Watabe and Hasegawa, 1973; Chihara, 1973; March and Tosi, 1973)

$$S_{iv}(0) = Z_v S_{ii}(0), \tag{7.25a}$$

$$S_{vv}(0) = Z_v^2 S_{ii}(0). \tag{7.25b}$$

Using Eqs. (7.25) in the right-hand side of Eq. (7.24), it follows that

$$\delta S_1(k)|_{k\to 0} \to Z_v^2 S_{ii}(0). \tag{7.26}$$

Thus, since $S_{ii}(0)$ is known from fluctuation theory to be related to thermodynamics by (see *e.g.* March and Tosi, 2002)

$$S_{ii}(0) = \rho k_{\mathrm{B}} T \kappa_T, \tag{7.27}$$

where ρ is the atomic number density, $k_{\mathrm{B}}T$ the thermal energy, and κ_T the isothermal compressibility, $S_{ii}(0)$ near freezing for *s*-*p* metals is always small (often ~ 0.02), the polyvalent metals have normally the largest corrections to the jellium structure factor $S_{vv}^{\mathrm{jellium}}$ as $k \to 0$.

Summarizing, the main achievement of the present Section is to relate the most difficult partial structure $S_{vv}(k)$ in an *s*-*p* liquid metal like Mg or Al to the sum of the jellium structure factor for the appropriate r_s, plus as its main correction $\delta S_1(k)$. But this latter quantity is usefully approximated by Eq. (7.24), the numerator of which is accessible via computer simulation, following de Wijs *et al.* (1995), while the denominator can be measured directly by neutron scattering. Hence, for *s*-*p* metals, a combination of theory plus a neutron scattering experiment, is enough to bypass the use of electron diffraction measurements, advocated by Egelstaff *et al.* (1974, EMM), which are beset in liquid metals by considerably greater difficulties than their X-ray or neutron counterparts.

As to future directions, the presence of the factor $Z_v^2 S_{ii}(0)$ in the long wavelength limit of $S_{vv}(k)$ quoted in Eq. (7.25) suggests the obvious interest of focussing next on some polyvalent metals (*e.g.* Al and Bi: in the latter case, Table 1 of EMM (Egelstaff *et al.*, 1974) gives references already to data from neutrons, X-rays and also electrons at the principal peaks of the scattering factors).

7.4 Magnetic susceptibility of expanded fluid alkali metals

While treating correlations in the alkali metals close to their critical points, it seems appropriate, at this point, to shortly refer to the magnetic properties of the expanded alkali metals. In particular, their magnetic susceptibility χ along the liquid-vapor coexistence curve shows signatures of electron correlations.

Chapman and March (1988) (see also March *et al.*, 1979) considered a correlation-induced metal-insulator transition at $T = 0$, and wrote the ground-state energy per atom of a half-filled band in the metallic state as a power expansion in the magnetization m per atom and in the quasiparticle renormalization factor q, as

$$E(m,q) = E_0 + am^2 + \cdots + bq + cq^2 + \cdots + eqm^2 + \cdots . \tag{7.28}$$

Such quasiparticle renormalization factor q is related to the discontinuity in the single-particle occupation number at the Fermi level [cf. Eq. (7.32) below], and decreases continuously to zero at the transition. The volume magnetic susceptibility associated with energy E in Eq. (7.28) is thus

$$\chi = \frac{n_0 \mu_0 \mu_B^2}{2(a+eq)}, \tag{7.29}$$

where n_0 is the number density. In the Brinkman and Rice (1970) model, where $a = 0$, one finds moreover $b = (U_c - U)/8$, $c = U/32$, $e = [1 + \frac{3}{2}\bar{\epsilon}N(\epsilon_F)/2N(\epsilon_F)$, where U is the Hubbard onsite interaction, U_c is the value thereof at the metal-insulator transition, $\bar{\epsilon}$ is the band energy in the absence of correlations, and $N(\epsilon_F)$ is the electronic density of states per atom at the Fermi level. With these values for the coefficients, one obviously finds $\chi \to q^{-1}$.

Consider next the situation at $T \neq 0$. Then, the energy E in Eq. (7.28) naturally extends into the free energy per atom, whose expansion reads

$$F(m,q,T) = E_0 + a(T)m^2 + \cdots + b(T)q + c(T)q^2 + \cdots + e(T)qm^2 + \cdots, \tag{7.30}$$

where the coefficients naturally acquire an explicit T-dependence. The renormalization factor q is here interpreted as the average number of doubly occupied sites (Brinkman and Rice, 1970). Furthermore, as $T \to \infty$, one can anticipate that the susceptibility asymptotically develops a Curie-like behaviour, say when T exceeds the Fermi temperature. This suggests that $a = \alpha T$, while $e \approx$ const, so that $\chi \approx n_0\mu_0\mu_B^2/2\alpha T$ as $T \to \infty$. This reduces to the Curie law for electrons, in the limit $\alpha = \frac{1}{2}k_{\mathrm{B}}$. In order to regain the Pauli susceptibility, since $\chi \to n_0\mu_0\mu_B^2/2eq$ as $T \to 0$, one requires that $1/2e \to N(\epsilon_F)$ as $q \to 1$ (high density).

Motivated by the phenomenology of the heavy-fermion materials, Rice *et al.* (1985) extended the model of Brinkman and Rice (1970) in order to embrace finite temperature. This enables to express the free energy per atom as

$$F = \sum_{\mathbf{k}\sigma} q_\sigma \epsilon_{\mathbf{k}} n_{\mathbf{k}\sigma} + Ud + k_{\mathrm{B}}T \sum_{\mathbf{k}\sigma} w_{\mathbf{k}} \left(n_{\mathbf{k}\sigma} \log n_{\mathbf{k}\sigma} + (1 - n_{\mathbf{k}\sigma}) \log(1 - n_{\mathbf{k}\sigma}) \right), \tag{7.31}$$

where

$$n_{\mathbf{k}\sigma} = \frac{1}{1 + \exp[\beta q(\epsilon_{\mathbf{k}} - \mu)/w_{\mathbf{k}}]}, \tag{7.32}$$

is the average occupation number in state $(\mathbf{k}, \sigma)$, $w_{\mathbf{k}}$ is a renormalization factor in $\mathbf{k}$ space, introduced to account for the nonorthogonality of the quasiparticle states, and d is the average number of doubly occupied sites. The factor $w_{\mathbf{k}}$ is generally unknown, but must fulfill the constraints $\bar{w} = \sum_{\mathbf{k}} w_{\mathbf{k}} = \left((\frac{1}{2} - d) \log(\frac{1}{2} - d) + d \log 2 \right) / \log 2$, $\bar{w}_{-1} = \sum_{\mathbf{k}} w_{\mathbf{k}}^{-1} = 2$, and $w_{\mathbf{k}} \to 1$ as $\mathbf{k} \to \mathbf{k}_F$. Taking $q = \left(\sqrt{d(n_\uparrow - d)} + \sqrt{d(n_\downarrow - d)} \right)^2 / n_\uparrow n_\downarrow$, and expanding the free energy to second order in the magnetization, Chapman and March (1988) find for the susceptibility in the $T \to 0$ limit

$$\chi = \frac{n_0\mu_0\mu_B^2 N(\epsilon_F)}{q} \left(1 + 2\bar{\epsilon}N(\epsilon_F) \left(1 - \frac{1}{4(1-2d)^2} \right) \right)^{-1}, \tag{7.33}$$

while for $k_{\mathrm{B}}T \gg q\epsilon_F$ they obtain $\chi = n_0\mu_0\mu_B^2/k_{\mathrm{B}}T$. Hence, there is a crossover between enhanced Pauli paramagnetism and Curie behaviour, when the temperature is of the order of the Fermi temperature. One may also note that, in the low-temperature, high-density regime, Eq. (7.33) provides a mechanism for the observed enhancement of the susceptibility over the Pauli value. Assuming that the reduction of d from its uncorrelated value $\frac{1}{4}$ is small, and writing $d = \frac{1}{4} - \delta$, one finds $q = 1 + O(\delta)$, and, from Eq. (7.33),

$$\chi = \frac{n_0 \mu_0 \mu_B^2 N(\epsilon_F)}{1 + 16\bar{\epsilon} N(\epsilon_F)\delta}. \tag{7.34}$$

Since $\bar{\epsilon} = \sum_{\mathbf{k}\sigma} \epsilon_{\mathbf{k}} n_{\mathbf{k}\sigma} < 0$, χ shows a Stoner-like enhancement, of reduced importance, as density decreases, while at the same time the factor $1/q$ becomes more important.

In summary, a phenomenological treatment of a correlation-induced metal-insulator transition allows to interpret the behaviour of the magnetic susceptibility of expanded fluid alkalis. In particular, as a consequence of correlations, the momentum distribution at the Fermi level is quantitatively very different from that in jellium at the same density.

Chapter 8

Anderson localization in disordered systems

In 1958, Anderson advanced the hypothesis that transport in a lattice (*e.g.* charge transport in a semiconductor) could be hindered by disorder not in real space, but in the Hilbert space, so to speak. In other words, while scattering centres are still located at regular positions in a lattice, it is their scattering strengths that are taken to be randomly distributed, according to a given probability distribution, thereby providing electron localization, *e.g.* in a semiconductor. Rather than continuously decreasing with increasing disorder, the conductivity of a disordered system will reach a minimum, at a certain critical value of the impurity scattering strength. This is an instance of a metal-to-insulator transition, another instance (of different, and often competing, origin) being that studied by Mott (1949, 1968); Ioffe and Regel (1960), where the driving force of localization is the strong Coulomb repulsion between electrons, thereby reducing their mean free path below their de Broglie wavelength. It is now established that Anderson's model (1958) is a generic effect of wave diffusion, that applies equally well to any kind of classic or quantum waves, ranging from electromagnetic, to acoustic, to spin waves (Van Tiggelen, 1999).

We briefly review Anderson's model for strong localization in Sec. 8.1, whereas its precursor effect, *viz.* weak localization, will be addressed, also briefly, in Sec. 8.2 (see Thouless, 1974; Lee and Ramakrishnan, 1985; Giamarchi and Schulz, 1988; Fisher *et al.*, 1989, for a more comprehensive review).

The case of isolated magnetic impurities, or of a periodic array thereof, is of interest by itself, and will be briefly dealt with in Appendix D, dedicated to the Kondo problem and to Kondo lattice model, with particular reference to the heavy fermion systems.

8.1 Anderson's model for localization in random lattices

Originally, Anderson's model (1958) was developed for an assembly of spins on a lattice (see *e.g.* Jones and March, 1986b, for an introduction). Let us suppose that the spin residing on lattice site j has an energy ϵ_j, whose value is randomly distributed according to a given probability distribution $P(\epsilon)d\epsilon$, characterized by a width w. Moreover, let t_{jk} denote the hopping matrix element between sites j and k. The probability amplitude a_j for a spin to reside on site j is then governed by the equation (Jones and March, 1986b)

$$i\frac{\partial a_j}{\partial t} = \epsilon_j a_j + \sum_{k\neq j} t_{jk} a_k. \tag{8.1}$$

Rather than studying the above diffusion problem, Anderson (1958) addresses the corresponding stationary problem. Thus, let E_α and ψ_α denote the allowed energies and eigenstates, respectively. One can then expand the wave functions $\psi_\alpha(\mathbf{r})$ in terms of the site wave functions $\psi_i(\mathbf{r})$ as

$$\psi_\alpha(\mathbf{r}) = \sum_i (\alpha|i)\psi_i(\mathbf{r}). \tag{8.2}$$

Similarly, the Green function

$$G(\mathbf{r},\mathbf{r}';E) = \sum_\alpha \frac{\psi_\alpha^*(\mathbf{r})\psi_\alpha(\mathbf{r}')}{E - E_\alpha} \tag{8.3}$$

also admits a representation in terms of the site wave functions as

$$G(\mathbf{r},\mathbf{r}';E) = \sum_{ij} G_{ij}(E)\psi_i^*(\mathbf{r})\psi_j(\mathbf{r}'). \tag{8.4}$$

Comparing the above equations, one finds immediately

$$G_{ij}(E) = \sum_\alpha \frac{(i|\alpha)(\alpha|j)}{E - E_\alpha}. \tag{8.5}$$

In particular, the diagonal element reads

$$G_{ii}(E) = \sum_\alpha \frac{|(\alpha|i)|^2}{E - E_\alpha}. \tag{8.6}$$

It turns out that the asymptotic behaviour of $|(\alpha|i)|^2$ with the number of sites N can be derived from the knowledge of the asymptotic behaviour of the amplitude $a_j(t)$ as $t \to \infty$. Taking the Laplace transform $a_j(s) = \int_0^\infty e^{-st} a_j(t)dt$ of Eq. (8.1), one finds

$$i[sa_j(s) - a_j(t=0)] = \epsilon_j a_j(s) + \sum_{k\neq j} t_{jk} a_k(s). \tag{8.7}$$

This takes the form

$$a_j(s) = \frac{i\delta_{0j}}{is - \epsilon_j} + \sum_{k \neq j} \frac{1}{i\delta - \epsilon_j} t_{jk} a_k(s), \tag{8.8}$$

which implicitly takes into account for the boundary condition $a_0(t = 0) = 1$. This can be solved iteratively, to yield

$$a_j(s) = \frac{i}{is - \epsilon_0} t_{j0} a_0(s) + \sum_k \frac{1}{is - \epsilon_k} t_{k0} a_0(s) + \cdots . \tag{8.9}$$

In particular, assuming that a spin occupies site $j = 0$ at $t = 0$, one finds

$$a_0(s) = \frac{i}{is - \epsilon_0} + \sum_k \left(\frac{t_{0k} t_{k0}}{is - \epsilon_k} + \sum_\ell t_{0k} \frac{1}{is - \epsilon_k} t_{kl} \frac{1}{is - \epsilon_\ell} t_{\ell 0} + \cdots \right). \tag{8.10}$$

Such a procedure is analogous to deriving the self-energy, in many-body parlance.

Denoting $V_c(s)$ the quantity in brackets in Eq. (8.10), and retaining only its first term, one finds

$$V_c(s) = \sum_k t_{0k}^2 \left(\frac{-\epsilon_k}{s^2 + \epsilon_k^2} - \frac{is}{s^2 + \epsilon_k^2} \right). \tag{8.11}$$

Now, the long-time limit of the amplitude at site j is related to the $s \to 0^+$ limit of its Laplace transform by

$$a_j(t \to \infty) = \lim_{s \to 0^+} s a_j(s). \tag{8.12}$$

A finite value of $a_0(t \to \infty)$ implies a finite value also $a_j(t \to \infty)$ on neighbouring sites, and this evidently corresponds to localization. To this aim, one studies the limit as $s \to 0$ of Eq. (8.11), to find

$$\lim_{s \to 0} V_c(s) = -i \sum_k t_{0k}^2 \delta(\epsilon_k) - \lim_{s \to 0} \sum k, E_k \neq 0 \frac{t_{0k}^2}{E_k^2} \equiv -\frac{i}{\tau} - i \lim_{s \to 0} sK. \tag{8.13}$$

The solution for $a_0(s)$ can then be cast into the form

$$a_0(s) = \frac{i}{is(1 + K) + (i/\tau) - (\epsilon_0 - \Delta E^{(2)})}, \tag{8.14}$$

where $\Delta E^{(2)}$ is the second-order perturbation energy. If τ is finite, it can be interpreted as the lifetime of a state having energy $\epsilon_0 - \Delta E^{(2)}$. But if $\tau \to \infty$, this yields

$$a_0(s) = \frac{1}{s(1 + K) + i(\epsilon_0 - \Delta E^{(2)})}, \tag{8.15}$$

which implies a finite value of $a_0(t)$ as $t \to \infty$, with simply a reduction of $1/(1+K)$ in the probability amplitude on to neighbouring sites. One is therefore led to study $V_c(s)$ more in detail, and in particular its imaginary part (Jones and March, 1986b). The probability distribution of the imaginary part of $V_c(s)$ can be related to that of the energies, having width w. The main predictions of Anderson's model are that one gets localization in one- and two-dimensional lattices (Abrahams *et al.*, 1979), regardless of the energy probability distribution width w, whereas in three-dimensional lattices one has localization when w exceeds a critical value.

8.2 Beyond simple scaling

In early work on disorder within the framework of weak localization theory (see *e.g.* Lee and Ramakrishnan, 1985; Altshuler and Aronov, 1985, for a review), it proved possible to display the dependence of the conductance G on (i) sample size L, (ii) temperature T, and (iii) frequency ω. The scaling hypothesis based on the assumption that the dimensionless conductance

$$g = hG/e^2 \tag{8.16}$$

was the only scaling parameter allowed (Abrahams *et al.*, 1979) to support the absence of extended states in $d = 2$ dimensions (see also Dancz *et al.*, 1973) and to develop at least a qualitative description of the metal-insulator transition for $d > 2$.

However, it is now established that there are limitations on the validity of such results. The point emphasized by Altshuler *et al.* (1988) is that there are serious limitations on the results mentioned above, which apply to the conductance averaged over the ensembles of samples with macroscopically identical characteristics which differ from each other only by realization of the random impurity potential. But conductance fluctuations from one sample to another (which have been termed *mesoscopic fluctuations*) turn out to be of the same order of magnitude as quantum corrections to g in Eq. (8.16). What is of major moment is that these remain finite for $L \to \infty$ (Altshuler, 1985; Lee and Stone, 1985). These authors obtained the dispersion of the mesoscopic fluctuations as follows

$$\langle(\delta g)^2\rangle = \left(\frac{h}{e^2}\right)^2 (\langle\epsilon^2\rangle - \langle\epsilon\rangle^2) \approx 1, \tag{8.17}$$

where $\langle\cdots\rangle$ denotes ensemble averaging.

Such mesoscopic fluctuations lead to unusual properties of disordered conductors. These can show up, for instance, as aperiodic reproducible oscillations of G as a function of magnetic field or Fermi energy (see Imry, 1986; Washburn and Webb, 1986, for a review).

Altshuler *et al.* (1988) considered statistical properties of these mesoscopic fluctuations. These authors, within the context of the so-called broadened σ-model, studied the irreducible fluctuation moments (cumulants) $\langle\langle(\delta g)^k\rangle\rangle$, where double brackets here denote irreducible averages, in order to determine the distribution function $f(\delta g)$ of mesoscopic fluctuations. They show that this function is Gaussian (and universal) only in the regime of weak localization with δg being not too large. Even in this region, the probability of large fluctuations ($\delta g > g_0^{1/2}$) drastically exceeds the Gaussian form: the asymptotics of the distribution function proving to be logarithmically normal. These long tails are (i) not described in the one-parameter scaling, and (ii) turn out to be nonuniversal, depending on both the renormalized value of the dimensionless conductance $g = g(L)$ and its bare value $g_0 = g(L = \ell)$, where ℓ denotes the mean free path.

Altshuler *et al.* (1988) add that similar non-exponential long tails manifest themselves in the current relaxation processes in disordered conductors. They note, in particular, that the long-time asymptotic behaviour of relaxation current after a step change in voltage proves to be logarithmically normal, $\sim \exp(-\log^2 t)$, rather than exponential. Also noted is that the analogy between the asymptotics of the fluctuations distribution and electric current relaxation is no accident. It is stressed that the long-time character of relaxation current is intimately connected with mesoscopic fluctuations of phase relaxation times.

8.2.1 *Current relaxation in disordered conductors*

Altshuler *et al.* (1988) note further that the electric current density $\mathbf{j}(t)$ is governed for arbitrary time dependence of the electric field $\boldsymbol{\mathcal{E}}(t)$ by the frequency dependent conductivity $\sigma(\omega) = e^2 g(\omega) L^{2-d}/h$ as

$$\langle \mathbf{j}(t)\rangle = \int \boldsymbol{\mathcal{E}}(t-t')\langle\sigma(t')\rangle dt'. \tag{8.18}$$

Here, $\sigma(t)$ denotes the Fourier transform of $\sigma(\omega)$ and is customarily referred to as the response function, since it corresponds to the current response to the $\delta(t)$ shape of the electric field.

The classical equation for the average of the response function

$$\langle\sigma_0(t)\rangle = \frac{\sigma_0}{\tau}\exp(-t/\tau) \tag{8.19}$$

results from the Drude formula, $\sigma(\omega) = \sigma_0/(1 - i\omega\tau)$. Here, σ_0 is the classical result for the conductivity related to the electron density of states per spin ν and diffusion constant $D = e^2/\ell\tau$ from the Einstein relation $\sigma_0 = 2e^2 D\nu$, with ℓ the mean free path and τ the time of free path.

The response function is valid when Eq. (8.19) is used only for $t < \tau$. For $t > \tau$, the quantity $\sigma(t)$ is due totally to quantum effects. The customary correction to $\langle\sigma(\omega)\rangle$ is given according to Altshuler *et al.* (1988) by

$$\langle\sigma_1(\omega)\rangle = -\frac{4e^2}{h}\int\frac{d^dq}{(2\pi)^d}\frac{1}{q^2 - i\omega D + L_{\phi_0}^{-2}}. \tag{8.20}$$

Here, $L_{\phi_0} = D\tau_{\phi_0}$, where τ_{ϕ_0} is the phase breaking time due to inelastic processes. The contribution to the response function follows as

$$\langle\sigma_1(t)\rangle = -\frac{e^2}{\pi h}t^{-d/2}(4\pi D)^{1-d/2}\exp(-t/t_0). \tag{8.21}$$

Here, t_0 is equal to τ_{ϕ_0} for the isolated sample, while

$$t_0 = \min\{\tau_{\phi_0}, L^2/D\} \tag{8.22}$$

for a cube of volume L^d with massive contacts. Therefore, one finds from Eq. (8.21) the attenuation of the averaged relaxation current change from a power law at $\tau < t < t_0$ to an exponential law for $t > t_0$. With decreasing temperature T, the phase breaking time τ_{ϕ_0} increases in such a manner that the attenuation time t_0 becomes equal to the time L^2/D of electron diffusion through the sample.

However, the quantum correction Eq. (8.20) is not sufficient to obtain the response function for very long times t. Allowing for the higher order quantum corrections to $\sigma(\omega)$ leads to attenuation which is not really as rapid as the exponential form, although it exceeds the power-law attenuation (Altshuler *et al.*, 1988):

$$\langle\sigma(t)\rangle = -(\sigma/t)\exp[-(\log(t/\tau) - 2u)^2/8u]. \tag{8.23}$$

In the above equation, $\sigma = e^2 g(L)L^{2-d}/h$ is the observable value of the static conductivity, while the quantity u is defined as

$$u = \log(\sigma_0/\sigma), \tag{8.24}$$

where $\sigma_0 = e^2 g_0\ell^{2-d}/h$ denotes the bare (classical) conductivity, and $g_0 = g(L = \ell)$.

Altshuler *et al.* (1988) then consider qualitatively the relaxation current $\mathbf{j}(t)$ after a step-wise switching on of the electric field, $\boldsymbol{\mathcal{E}}(t) = \boldsymbol{\mathcal{E}}\theta(t)$. After a short time $t \approx \tau$, the current rises to its classical value, $\mathbf{j}_0 = \sigma_0\boldsymbol{\mathcal{E}}$. Following that, it decreases at first as a power law (or logarithmically, at $d = 2$), then by the exponential law in Eq. (8.23) above, and finally by the logarithmically normal law down to its steady-state value $\mathbf{j} = \sigma\boldsymbol{\mathcal{E}}$. With increasing u, the relaxation slows down so that the steady-state regime is reached only after quite a long time.

Following Altshuler *et al.* (1988), we next discuss the reason for the time-delayed relaxation in Eq. (8.23). The non-exponential decay of $\sigma(t)$ is due to the faint singularity in the frequency dependence of the ac conductivity $\sigma(\omega)$ as $\omega \to 0$. Such a singularity was exposed by Kravtsov and Lerner (1984). To reveal the singularity, it is useful to expand $\sigma(\omega)$ as

$$\sigma(\omega) = \sigma_0 \sum_{n=0}^{\infty} C_n(i\omega\tau)^n + \frac{e^2}{h}\sum_{n=0}^{\infty} S_n(i\omega t_0)^n. \tag{8.25}$$

Neglecting quantum corrections, one obtains $C_n = 1$ and $S_n = 0$, which then recovers the Drude formula. Taking into account of quantum corrections, the coefficients C_n result from expanding the electron Green's function in ω, while the S_n result from expanding the 'diffuson' in ω.

It turns out that it is the first term in Eq. (8.25) which governs the asymptotics expressed in Eq. (8.23). This is caused by the rapid increase in the coefficients C_n with n, *viz.*

$$C_n \sim \exp\left(2u(n^2 + n - \frac{1}{2})\right). \tag{8.26}$$

To sum the first asymptotic series appearing in Eq. (8.25), Altshuler *et al.* (1988) use an analogous procedure to the Borel summation, which invokes the identity

$$\exp\left(2u(n^2+n)\right) = \frac{1}{(8\pi u)^{1/2}} \int_0^{\infty} (t_\phi/t)^n \exp\left(-[\log(t_\phi/t) - 2u]^2/8u\right) \frac{dt_\phi}{t_\phi}. \tag{8.27}$$

Substituting Eqs. (8.26) and (8.27) into Eq. (8.25), changing the order of summation and integration, and carrying out the Fourier transform involved, one reaches the asymptotics of the response function as

$$\langle\sigma(t)\rangle = \sigma \int \exp(-t/t_\phi) f(t_\phi) dt_\phi/t_\phi, \tag{8.28}$$

where $f(t_\phi)$ has the form

$$f(t_\phi) \sim t_\phi^{-1} \exp\left(-[\log(t_\phi/t) - 2u]^2/8u\right). \tag{8.29}$$

The result in Eq. (8.23) is then recovered by performing the integration in Eq. (8.28).

As to the form of the distribution function $f(t_\phi)$, it may be noted that this type also results for the mesoscopic fluctuations of the static conductance and density of states. But we must refer the interested reader to Altshuler *et al.* (1988).

8.3 Hopping conductivity in disordered systems: reduction to a percolation problem

Although a sufficient degree of positional and spectral disorder implies localized electronic states around impurities, and therefore minimal conductivity, Mott (1969b) anticipated as early as in 1969 that phonon-induced tunneling of electrons between such localized states should nonetheless allow for a finite conductivity, which, at a finite temperature T, he predicted to behave characteristically as

$$\sigma(T) \propto \exp[-(T_0/T)^{1/4}], \tag{8.30}$$

where $k_\mathrm{B}T_0 = \lambda\alpha^3/n_\mathrm{F}$, with α the coefficient of the exponential decay of the localized states, n_F the density of states at the Fermi level, and λ a dimensionless constant (Mott, 1969b,a). Such a temperature dependence has been found to be consistent with the dc conductivity of several amorphous semiconductors, within temperature ranges up to room temperature. Mott (1969b) assumed that the transition rate p for phonon-induced quantum hopping of a single particle between two localized scattering centres could be expressed as

$$p = p_0 e^{-2\alpha r} e^{-\beta\Delta E}, \tag{8.31}$$

where r is the distance between the scattering centres, ΔE the activation energy for the hopping process, $\beta = (k_\mathrm{B}T)^{-1}$ is the inverse temperature, and p_0 a constant depending on the properties of the material, but not on the aforementioned variables. He then argued that $\Delta E \propto (n_\mathrm{F}r^3)^{-1}$. By maximizing p with respect to r, Mott (1969b) could thus derive Eq. (8.30), which also implies

$$\log(\sigma/\sigma_0) \propto -\beta^{1/4}. \tag{8.32}$$

Soon afterwards, Ambegaokar *et al.* (1971) re-derived Mott's prediction, Eq. (8.30), within a microscopic model based on the idea of percolation. More specifically, Ambegaokar *et al.* (1971) considered a system of impurities with finite density, whose positions and energies are distributed randomly. Then, they assumed that electron hopping were allowed only between sites such that

$$|\Delta E| < U e^{-\alpha r}, \tag{8.33}$$

where ΔE is the energy separation between their energy levels, r is again their mutual distance, and U is of the order of the binding energy for an isolated impurity. This situation may be equivalently described as of a network, whose nodes are connected by bonds when the condition of Eq. (8.33) is fulfilled. Usually, and in particular if the impurity eigenstates are well localized, these bonds will appear in clusters. However, there will be a finite probability for an electron to delocalize throughout the whole sample if such a network of bonds extends across the sample volume. More formally, this happens when the number of bonds per site exceeds the critical percolation number η^*, say, which can be roughly estimated as (Ambegaokar *et al.*, 1971)

$$\eta^* = n_\mathrm{F} \int d^3\mathbf{r} \int d(\Delta E) = \frac{16\pi n_\mathrm{F} U}{\alpha^3}, \tag{8.34}$$

where the integration was extended to the domain defined by the hopping condition, Eq. (8.33). One expects η^* to be of the order of a few bonds.

Denoting the hopping probability between site i and j by p_{ij}, and assuming that p_{ij} depends only on the distance $r = |\mathbf{r}_i - \mathbf{r}_j|$ between the sites, and on the difference $E_i - E_j$ between the relevant energies E_i, E_j on the two sites, respectively, Ambegaokar *et al.* (1971) assumed a low-temperature expression for p_{ij} of the form[1]

$$p_{ij} = \begin{cases} p_0 e^{-2\alpha r - \beta(E_j - E_i)}, & \text{for } E_j > E_i, \\ p_0 e^{-2\alpha r}, & \text{for } E_j < E_i, \end{cases} \tag{8.35}$$

to be compared with Eq. (8.31) by Mott (1969b). The average transition rate from site i to site j is then

$$P_{ij} = \langle p_{ij} n_i (1 - n_j) \rangle, \tag{8.36}$$

[1] An even more drastic assumption has been studied later by Bernasconi (1973), who considered a nonvanishing probability $p_{ij} = p_0 \exp(-\beta \Delta E)$, only when i and j are nearest-neighbour sites, and zero otherwise. Such a situation describes more realistically systems composed of small particles, when hopping is possible only if two such particles are in contact with each other. Then, a different behaviour in temperature is found for the conductivity, which however could still be cast into the form $\log(\sigma/\sigma_0) = -\beta^\kappa$, which is compatible with Eq. (8.32) for a dilute system, for which Mott's value $\kappa = 1/4$ applies.

where $\langle \cdots \rangle$ denotes a time average, and n_i is the occupation number at site i. At equilibrium, $\langle n_i \rangle \equiv n_i^0 = [1 + \exp(\beta E_i)]^{-1}$, while $\langle n_i n_j \rangle = \langle n_i \rangle \langle n_j \rangle$, for statistically independent sites. Combining Eqs. (8.35) and (8.36), Ambegaokar *et al.* (1971) they predict an average transition rate in the form

$$P_{ij}^{(0)} = p_0 \exp[-2\alpha r - \beta(|E_i| + |E_j| + |E_i - E_j|)/2] \tag{8.37}$$

at thermal equilibrium.

In order to obtain the conductivity of such a network, an external electric field $\boldsymbol{\mathcal{E}}$ perturbs the hopping probability Eq. (8.35) via the substitution $E_i - E_j \mapsto E_i - E_j + e\boldsymbol{\mathcal{E}} \cdot \mathbf{r}$. Also the occupation numbers n_i will differ from their equilibrium values n_i^0 through a change in the chemical potential, $\delta\mu_i$. Assuming that such an electric field does not induce correlation between the occupation numbers at different sites, so that they can still be assumed independent, at first order in the electric field one finds for the average net flow between sites i and j the expression (Ambegaokar *et al.*, 1971)

$$P_{ij}(\boldsymbol{\mathcal{E}}) - P_{ji}(\boldsymbol{\mathcal{E}}) = \beta P_{ij}^{(0)}(e\boldsymbol{\mathcal{E}} \cdot \mathbf{r} + \delta\mu_i - \delta\mu_j). \tag{8.38}$$

Ambegaokar *et al.* (1971) are able to express the conductance between sites i and j as

$$G_{ij} = \frac{e^2}{k_B T} P_{ij}^{(0)}(r, E_i, E_j). \tag{8.39}$$

They then define a critical conductance G_c as the largest value of the conductance, such that the subset of resistors with $G_{ij} > G_c$ still contains a connected network which spans the whole system, and claim that the overall conductivity of the system be given by the ratio $\sigma = G_c/L$, where L is a characteristic length scale for the system. Eq. (8.37) then implies that

$$2\alpha r + \frac{\beta}{2}(|E_i| + |E_j| + |E_i - E_j|) < \log(p_0/P_0), \tag{8.40}$$

where $P_0 = k_B T G_c/e^2$, or, in an equivalent but dimensionless form,

$$\frac{r}{R_{\max}} + \frac{|E_i| + |E_j| + |E_i - E_j|}{2E_{\max}} < 1, \tag{8.41}$$

where

$$R_{\max} = \frac{1}{2\alpha} \log(p_0/P_0), \tag{8.42a}$$

$$E_{\max} = k_B T \log(p_0/P_0). \tag{8.42b}$$

Any link with $r > R_{\max}$ will violate Eq. (8.41), or its equivalent form Eq. (8.40), regardless of E_i and E_j. Similarly, any node of the network with $|E_i| > E_{\max}$ will violate those inequalities, regardless how small be the distance r, and thus will be disconnected from those clusters in the network which do contribute to the conductivity.

Assuming a random distribution in space of the impurity sites, and a uniform distribution of the impurity energies in the interval $-E_{\max} \leq E_i \geq E_{\max}$, an estimate of the total number of sites with $|E_i| \geq E_{\max}$ is then $n = 2n_{\mathrm{F}}E_{\max}$. On the other hand, one expects $nR_{\max}^3 = \nu_c$, where ν_c is a dimensionless number, of order unity. Ambegaokar *et al.* (1971) estimate $\nu_c \simeq 4$. Combining Eqs. (8.42) with these estimates, one obtains (Ambegaokar *et al.*, 1971)

$$\log(p_0/P_0) = \left(\frac{4\nu_c\alpha^3}{n_{\mathrm{F}}k_{\mathrm{B}}T}\right)^{1/4}, \tag{8.43}$$

which ultimately implies Mott's formula, Eq. (8.30), with $k_{\mathrm{B}}T_0 = \lambda\alpha^3/n_{\mathrm{F}}$, and $\lambda = 4\nu_c \approx 16$.

8.4 Icosahedra in amorphous Ni-Al alloys

To end this Chapter on disorder, it is noteworthy that Wang *et al.* (2002, WBY, below) have reported structural simulations of clusters in a liquid Ni-Al alloy. In particular, the evolution of atomic clusters in the liquid Ni_3Al alloy during rapid solidification were performed by the above group using the molecular dynamics simulation technique. A tight-binding potential function was adopted to characterize the interaction between atoms. Structural parameters such as pair correlation functions, bonded pairs and polyhedral clusters were used by WBY to characterize how the type and number of clusters depend on temperature during a rapid cooling process.

Notable features of the WBY finding were that the number of Ni_3Al clusters at 2000 K was small, and were made up of defect icosahedra. However, this number was found to increase rapidly at the cooling rate of 4×10^{13} K/s, and finally icosahedral clusters were found to occur at low temperature. It was concluded that a network consisting of the icosahedra and defect icosahedral clusters can be formed in the amorphous systems.

Chapter 9

Statistical field theory: especially models of critical exponents

Critical behaviour is characteristic of many-body systems, and the concept of universality allows to connect quite diverse physical systems or models, ranging from condensed matter (Cardy, 1996; Kardar, 2007; Reichl, 2009), to astrophysics, to biological or social complex systems. Within condensed matter, six critical exponents at or near the critical point of a second-order phase transition display universal behaviour in various materials. Here, some insight will be provided, by relating the critical exponents in several models, ranging from the Gaussian model (Cardy, 1996), a model of string theory (Iqbal *et al.*, 2011), the Ising model (Ising, 1925; Onsager, 1944; Zhang, 2007, 2013), the XY and the Heisenberg models (Cardy, 1996).

The Gaussian model is a classical statistics model with Gaussian-type distribution constant. On the other hand, the anti-de Sitter/conformal field (CFT/AdS) theory correspondence is a conjectured relationship between two kinds of physical theories: on one side of the correspondence are conformal field theories (CFT) which are quantum field theories describing elementary particles. On the other side are anti-de Sitter spaces (AdS) which are used in theories of quantum gravity, formulated in terms of string theory (see Sachdev, 2011a, 2012; Zaanen *et al.*, 2015, for recent reviews and an introduction to the overall field). Since there is some duality between quantum field theory and quantum statistical mechanics, the AdS/CFT model may provide some connection between the phase transitions in string theory and condensed matter theory. Furthermore, the Ising model, the XY model, and the Heisenberg model are $O(n)$ models with $n = 1$, 2, and 3, respectively. Besides, it is also very interesting to discuss the large n limit of the $O(n)$ model, before one establishes a final theory for the $O(n)$ model of critical exponents.

To gain orientation, it is first shown that the critical exponents of the

Gaussian model (GM) are embraced by the AdS/CFT model of Iqbal *et al.* (2011, ILM, below). While the GM exponents involve solely the dimensionality d, we then demonstrate how the universality class n can also be included via the ILM model in the case of sufficiently large n. These latter results can be viewed as approximate summations of the ϵ-expansion to all orders for sufficiently large n. For the three exponents β, ν and η, the range of validity may be wider in the variable n. However, the exponent γ is always unity in the ILM model, independently of the choice of n and d. Generalizations of the ILM string theory model should therefore be sought. Furthermore, Section 9.4 discusses the critical exponents expressed in terms of both the dimensionality d and the universality class n, and especially, some formulas for large n limit have been given.

After providing a general overview of the Ising model and its solution in 2D due to Onsager (1944) in Sec. 9.1, we will propose some quite general relations among various critical exponents, involving dimensionality d and universality class n in $O(n)$ models in Sec. 9.4. Section 9.5 then accounts for the mapping of a generalization of the Heisenberg model in 2D, *viz.* the Baxter or eight-vertex model, to the Luttinger model in 1D. Appendix E provides a self-contained account of the Luttinger model.

9.1 Ising model

The Ising model was originally introduced by Lenz in 1920 and solved by Ising (1925) in dimensions $d = 1$ to describe the transition from a para- to a ferromagnetic phase in a magnetic lattice. Later, it has become a paradigm for several different systems, including antiferromagnets, lattice gases, and large biological molecules. The reason of its relevance and popularity resides perhaps on the fact that it accounts for an order-disorder transition on a lattice by dealing with a minimum number of variables and external parameters and, of course, that exact solutions are available also close to a phase transition in one dimension (Ising, 1925, where, however, it does not present a phase transition; cf. Sec. 9.1.1 below), and in two dimensions, first in a vanishing external field [Onsager (1944); see also Kaufman (1949); Kaufman and Onsager (1949)], and then in a nonzero external field (Yang, 1952, Sec. 9.1.2 below). This allows to extract all details of the model, including its critical exponents, which can then be compared with approximate or numerical estimates of similar models. In three and more dimensions, a mean-field approximation is still capable of grasping most of the features of the model.

Consider a d-dimensional lattice with N sites, and assume that the state of each lattice site, labelled by i, with $i = 1, \ldots N$, can be characterized by the value of a single variable, s_i say, taking only the possible values $s_i = \pm 1$. For the sake of definiteness, we might think of magnetic spins residing on such sites, with $s_i = +1$ corresponding to a spin up, and $s_i = -1$ corresponding to a spin down (Reichl, 2009).[1] The Ising model is then a minimal model allowing for interaction between spins residing at nearest-neighbour sites in the lattice, and for spins with an external magnetic field, H say. The 'Hamiltonian' (*i.e.* the classical energy) of the model is then given by

$$\mathcal{H}\{s_i\} = \sum_{\langle ij \rangle} \epsilon_{ij} s_i s_j - H \sum_i s_i. \tag{9.1}$$

Here, ϵ_{ij} is the interaction energy between sites i and j, and $\langle ij \rangle$ restricts that to nearest-neighbouring sites only. In the following, for the sake of simplicity, we shall assume $\epsilon_{ij} = -\epsilon < 0$. The number of nearest neighbours, or coordination number, z, is determined by the geometry of the lattice, being $z = 2d$ for a cubic lattice in d-dimensions.

The partition function of the model can be written as

$$Z(T, N, H) = \exp\left(-\beta G(T, N, H)\right) = \sum_{s_1 = \pm 1} \cdots \sum_{s_N = \pm 1} e^{-\beta \mathcal{H}}, \tag{9.2}$$

where $G(T, N, H)$ is the Gibbs free energy. The thermal average of the magnetization $M = \sum_{i=1}^{N} s_i$ can then be extracted from the partition function as

$$\langle M \rangle = k_{\mathrm{B}} T \frac{\partial}{\partial H} \log Z(T, N, H), \tag{9.3}$$

and can be treated as an order parameter (cf. Chapter 5), with $\langle M \rangle \neq 0$ in the ordered phase.

9.1.1 *Exact solution in one dimension*

For $d = 1$ dimensions, the Ising Hamiltonian, Eq. (9.1) reduces to

$$\mathcal{H}\{s_i\} = -\epsilon \sum_{i=1}^{N} s_i s_{i+1} - H \sum_{i=1}^{N} s_i, \tag{9.4}$$

[1]The lattice gas model corresponds to having $s_i = +1$ when a site is occupied by an atom, and $s_i = -1$ when it is empty.

where one usually imposes periodic boundary conditions, $s_{N+1} = s_1$. The partition function can be written as

$$Z(T,N) = \sum_{s_1=\pm 1} \cdots \sum_{s_N=\pm 1} \exp\left(\beta \sum_{i=1}^{N} \left(\epsilon s_i s_{i+1} + \frac{1}{2} H(s_i + s_{i+1})\right)\right), \tag{9.5}$$

where we have used the fact that $S = \sum_i s_i = \frac{1}{2}\sum_i (s_i + s_{i+1})$, on account also of the periodical boundary conditions. It is now convenient to introduce the matrix P given by

$$P = \begin{pmatrix} e^{\beta(\epsilon+H)} & e^{-\beta\epsilon} \\ e^{-\beta\epsilon} & e^{\beta(\epsilon-H)} \end{pmatrix}, \tag{9.6}$$

or equivalently having elements (in Dirac notation)

$$\langle s_i|P|s_j\rangle = e^{\beta[\epsilon s_i s_j + \frac{1}{2}H(s_i+s_j)]}. \tag{9.7}$$

Thus, the partition function may be written as

$$\begin{aligned} Z(T,N) &= \sum_{s_1=\pm 1} \cdots \sum_{s_N=\pm 1} \langle s_1|P|s_2\rangle\langle s_2|P|s_3\rangle \cdots \langle s_N|P|s_1\rangle \\ &= \sum_{s_1=\pm 1} \langle s_1|P^N|s_1\rangle = \operatorname{Tr} P^N = \lambda_+^N + \lambda_-^N, \end{aligned} \tag{9.8}$$

where

$$\lambda_\pm = e^{\beta\epsilon}\left(\cosh(\beta H) \pm \sqrt{\cosh^2(\beta H) - 2e^{-2\beta\epsilon}\sinh(2\beta\epsilon)}\right) \tag{9.9}$$

are the two eigenvalues of the matrix P, with $\lambda_+ > \lambda_-$.

In the thermodynamic limit, $N \to \infty$, only the largest eigenvalue, λ_+ contributes to the partition function, and all thermodynamic function, *e.g.* the Gibbs free energy per site

$$\begin{aligned} g(T,H) &= \lim_{N\to\infty} G(T,N,H) \\ &= -k_{\mathrm{B}}T \lim_{N\to\infty} \frac{1}{N} \log Z(T,N,H) \\ &= -k_{\mathrm{B}}T \log \lambda_+ \\ &= -\epsilon - k_{\mathrm{B}}T \log\Big(\cosh(\beta H) \\ &\qquad + \sqrt{\cosh^2(\beta H) - 2e^{-2\beta\epsilon}\sinh(2\beta\epsilon)}\Big). \end{aligned} \tag{9.10}$$

The average magnetization per site is then given by

$$\langle m\rangle = -\left.\frac{\partial g}{\partial H}\right|_H = \frac{\sinh \beta H}{\sqrt{\cosh^2(\beta H) - 2e^{-2\beta\epsilon}\sinh(2\beta\epsilon)}}. \tag{9.11}$$

Since $\langle m \rangle \to 0$ as $H \to 0$, the order parameter does not form spontaneously, and thus the Ising model is not characterized by any phase transition in $d = 1$ dimensions.[2]

9.1.2 *Exact solution in two dimensions*

The Ising was first solved for a square lattice in $d = 2$ dimensions by Onsager (1944) for zero external field, $H = 0$ (see also Kaufman, 1949; Kaufman and Onsager, 1949), and later by Yang (1952) for a nonzero external field.[3]

Some results are listed in the following (cf. Reichl, 2009). The Helmholtz free energy per site is $(H = 0)$

$$a(T, H = 0) = -k_{\mathrm{B}}T \log(2\cosh 2\beta\epsilon) - \frac{k_{\mathrm{B}}T}{2\pi} \int_0^{\pi} d\phi \, \log \frac{1}{2}(1+\sqrt{1 - \delta^2 \sin^2 \phi}), \tag{9.12}$$

where

$$\delta^2 = \frac{2\sinh 2\beta\epsilon}{\cosh^2 2\beta\epsilon}. \tag{9.13}$$

The internal energy per site is

$$u(T, H = 0) = \epsilon \coth(2\beta\epsilon) \left(1 + \frac{2}{\pi}(2\tanh^2 2\beta\epsilon - 1)K(\delta)\right), \tag{9.14}$$

where $K(\delta) = \int_0^{\pi/2} (1 - \delta^2 \sin^2 \phi)^{-1/2} d\phi$ is the complete elliptic integral of the first kind (Gradshteyn and Ryzhik, 1994). Differentiating with respect to T, one obtains the heat capacity per site as

$$\begin{aligned} c(T, H = 0) = \frac{4k_{\mathrm{B}}}{\pi}(\beta\epsilon \coth 2\beta\epsilon)^2 \Big\{ & K(\delta) - E(\delta) \\ & - (1 - \tanh^2 2\beta\epsilon) \left[\frac{\pi}{2} + (2\tanh^2 2\beta\epsilon - 1)K(\delta)\right]\Big\}, \end{aligned} \tag{9.15}$$

where $E(\delta) = \int_0^{\pi/2} d\phi \, \sqrt{1 - \delta^2 \sin^2 \phi}$ is the complete elliptic integral of the second kind (Gradshteyn and Ryzhik, 1994). Since the complete elliptic integral of the first kind $E(\delta)$ has a logarithmic singularity as $\delta \to 1$, the Ising model in $d = 2$ dimensions is characterized by a phase transition. The condition $\delta = 1$ implicitly defines an inverse critical temperature β_c given by

$$(\sinh 2\beta_c\epsilon - 1)^2 = 0, \tag{9.16}$$

[2]It should be emphasized that this result has nothing to do with Mermin-Wagner theorem (cf. Chapter 5), as the symmetry possibly broken in the Ising model is a discrete, rather than a continuous, one.

[3]The analytical solution of the Ising model in $d \geq 3$ dimensions is still an open problem.

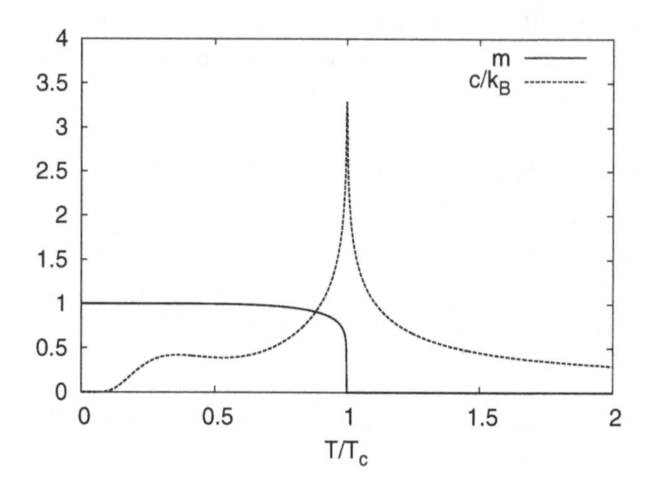

Fig. 9.1 Spontaneous order parameter per site M (Eq. (9.19), solid line) and heat capacity $c/k_{\rm B}$ per site (Eq. (9.15), dashed line) of the Ising model in dimensions $d = 2$.

or

$$k_{\rm B}T_c = \frac{2}{\log(1+\sqrt{2})}\epsilon \simeq 2.269\,\epsilon. \tag{9.17}$$

Close to $T = T_c$ the specific heat diverges logarithmically as

$$C(T) \sim \frac{2k_{\rm B}}{\pi}(2\epsilon\beta_c)^2 \left[-\log\left|1 - \frac{T}{T_c}\right| + \log\left(\frac{1}{2\epsilon\beta_c}\right) - \left(1 + \frac{\pi}{4}\right)\right], \tag{9.18}$$

while the spontaneous order parameter per site is ($H = 0$; Yang, 1952)

$$\langle m \rangle = \begin{cases} 0, & T > T_c \\ (1 - \sinh^4(2\beta\epsilon))^{1/8}, & T \leq T_c. \end{cases} \tag{9.19}$$

Figure 9.1 depicts the temperature behaviour of the specific heat and of the magnetization for the Ising model in $d = 2$ dimensions.

9.1.2.1 *Critical exponents*

Equation (9.18) shows that for the Ising model in $d = 2$ dimensions one has $\alpha = 0$, while Eq. (9.19) yields $\beta = 1/8$. A third critical exponent of interest is that related to the susceptibility $\chi(T) = (d\langle m \rangle/dT)_{H=0} \sim A_\pm |T - T_c|^{-\gamma}$. One finds (Wu *et al.*, 1976) $\gamma = 7/4$, with $A_+/A_- \simeq 12\pi$ (see also Kardar, 2007, and references therein, for a review).

In dimension $d \geq 5$ it is known (Aizenman, 1981) that the Ising model on the cubic lattice exhibits mean-field critical exponents at the critical temperature. In particular and moreover, the specific heat obeys the mean-field exponent $\alpha = 0$ for $d \geq 4$.

Current numerical and some analytical results for the critical exponents of the Ising model mainly in $d = 3$ dimensions have been reviewed by Pelissetto and Vicari (2002).

9.1.2.2 *Lattice gas model*

An equivalent version of the Ising model is the lattice gas model, which is used to describe the critical properties of the liquid-gas transition in a fluid. Such a model consists in dividing volume V into $\mathcal{N}$ cells of smaller volume $v_0 \ll V$, each of which can accommodate only one particle. One then introduces occupation number x_i for the ith cell, such that $x_i = 1$ if the cell is occupied, or $x_i = 0$ if the cell is empty. One immediately has $\sum_{i=1}^{\mathcal{N}} x_i = N$, where N is the total number of particles. The interaction potential energy can then be written as $U = -\frac{1}{2}\sum_{ij} \phi(|\mathbf{r}_i - \mathbf{r}_j|) x_i x_j$, where $\phi(r)$ is the pair potential, which is consistently approximated by a hard-core potential. The lattice gas model Hamiltonian can then mapped onto the Ising model Hamiltonian via the identification $s_i = 2x_i - 1$.

Besides being applied to the study of the liquid-vapour transition in fluids, or to the study of binary alloys, the lattice gas model has been used to describe the adsorption *e.g.* of hydrogen on metal surfaces, and, more recently, of graphene. The critical behaviour of the heavy alkali metals, such as Rb and Cs, in their fluid phases, and of the d-dimensional Ising model has been reviewed by March (2014b) (see also Leys *et al.*, 2002). Denoting by ρ_ℓ and ρ_v their liquid and vapour densities, respectively, let $\Delta = (\rho_\ell - \rho_v)/\rho_c$ and $\zeta = (\rho_\ell + \rho_v)/2\rho_c$ their density difference and average, respectively, both scaled with respect to the critical density, ρ_c. Then, in particular, $\zeta = 1$ at the critical point. Generalizing the work of March *et al.* (1988), Leys (2003) derived an empirical relationship between these two quantities as

$$\zeta = 1 + \kappa \Delta^\Gamma, \tag{9.20}$$

with $\Gamma = 2$ for Rb and Cs, and $\Gamma = 3$ for a number of insulating liquids (March *et al.*, 1988; March, 2014b). These results can be subsumed into the differential equation (March, 2014b)

$$\mu \frac{d^2\Delta}{d\zeta^2} + (\Gamma - 1)\left(\frac{d\Delta}{d\zeta}\right)^2 = 0 \tag{9.21}$$

with the appropriate choice of Γ as discussed above.

Leys (2003) then examined the critical behaviour of Δ, and found empirically that, for Cs, $\Delta \to 0$ as

$$\Delta = \text{const} \times \left(1 - \frac{T}{T_c}\right)^{1/3}, \tag{9.22}$$

where the critical exponent compares well with the value (0.35 ± 0.02) extracted from experimental data (see March, 2014b, and references therein). This would then imply a departure from the law of rectilinear diameters (Rowlinson, 1986) as

$$\zeta = 1 + \text{const} \times \left(1 - \frac{T}{T_c}\right)^{2/3}. \tag{9.23}$$

9.1.3 *Ising model in one- and four-dimensions in a magnetic field of arbitrary strength*

Most of the exact results listed thus far in the given dimensionality d are available for the Ising model without an applied magnetic field, $H = 0$ (see, however Yang, 1952). In the case of the Ising chain ($d = 1$), however, Kowalski (1972) was able to derive a partial differential equation for the partition function of such a model in a magnetic field $H \neq 0$.

Kowalski (1972) starts by relating the partition function $Z_{N+1} \equiv Z(T, N+1, H)$ of an Ising chain with $N+1$ nodes to that $Z_N \equiv Z(T, N, H)$ of a chain with N nodes, Eq. (9.2). Assuming nearest-neighbour interactions and periodic boundary conditions (*i.e.* a translationally invariant Ising ring), Kowalski (1972) finds

$$\begin{aligned}\frac{Z_{N+1}}{Z_N} &= 2(\cosh\beta\epsilon - \sinh\beta\epsilon)\Big[\cosh\beta H(\cosh 2\beta\epsilon + \sinh\beta\epsilon\cosh\beta\epsilon) \\ &\quad + \langle s_i\rangle_N \sinh\beta H \sinh 2\beta\epsilon - \frac{1}{2}\langle s_i s_{i+1}\rangle_N \cosh\beta H \sinh 2\beta\epsilon\Big],\end{aligned} \tag{9.24}$$

where $\langle\cdots\rangle_N$ denotes a statistical average with respect to the Hamiltonian with N sites. In particular, one has $\langle s_i\rangle_N = (N\beta)^{-1}\partial\log Z_n/\partial H$, and $\langle s_i s_{i+1}\rangle_N = (N\beta)^{-1}\partial\log Z_n/\partial\epsilon$. An exact result could also be obtained for the Ising chain with free ends, for which we refer the interested reader to the original paper of Kowalski (1972). Setting $z = \lim_{N\to\infty}\frac{Z_{N+1}}{Z_N}$, and assuming the existence of its partial derivatives with respect to the parameters of the model, in the limit $N \to \infty$, Kowalski (1972) then obtains

$$\left(\frac{2}{\beta}\frac{\partial z}{\partial H}\sinh\beta H - \frac{1}{\beta}\frac{\partial z}{\partial \epsilon}\cosh\beta H\right)\sinh 2\beta\epsilon$$
$$= z^2 e^{\beta\epsilon} - 2\cosh\beta H\left(\cosh 2\beta\epsilon + \frac{1}{2}\sinh 2\beta\epsilon\right)z. \quad (9.25)$$

Equation (9.25) is a quasilinear partial differential equation satisfied by the thermodynamic limit of the reduced partition function z for the Ising chain (periodic boundary conditions). In particular, Eq. (9.25) shows that $z = z(\beta\epsilon, \beta H)$.

For $H = 0$, Eq. (9.25) reduces to an ordinary differential equation (the Bernoulli equation), which may be solved by standard means under the initial condition $z(0,0) = 2$ as $z = \cosh 2\beta\epsilon$.

For $H \neq 0$, the solution of the Cauchy problem associated with Eq. (9.25) under the initial condition $z(\beta\epsilon, 0) = 2\cosh\beta\epsilon$ is (Kowalski, 1972)

$$z(\beta\epsilon, \beta H) = e^{\beta\epsilon}\cosh\beta H + \left(e^{2\beta\epsilon}\cosh^2\beta H - 2\sinh 2\beta\epsilon\right)^{1/2}. \quad (9.26)$$

Squaring Eq. (9.26), March (2014a) then expressed z implicitly as the solution of the second order algebraic equation

$$z^2 - 2ze^{\beta\epsilon}\cosh\beta H + 2\sinh 2\beta\epsilon = 0. \quad (9.27)$$

In view of the fact, as mentioned above, that $z(\beta\epsilon, 0) = 2\cosh\beta\epsilon$, one may express $\beta\epsilon$ in favour of $z(\beta\epsilon, 0)$ and conclude therefore that (March, 2014a)

$$z^2(\beta\epsilon, \beta H) = \mathcal{F}_1[z(\beta\epsilon, 0)] + z(\beta\epsilon, \beta H)\mathcal{F}_2[z(\beta\epsilon, 0)] \times \cosh\beta H, \quad (9.28)$$

where $\mathcal{F}_1[\cdot]$ and $\mathcal{F}_2[\cdot]$ are known functionals of the given argument. Thus, $z(\beta\epsilon, \beta H)$ is completely determined by its zero-field limit $z(\beta\epsilon, 0)$, and by the factor $\cosh\beta H$, as

$$z(\beta\epsilon, \beta H) = \mathcal{G}[z(\beta\epsilon, 0); \cosh\beta H]. \quad (9.29)$$

Such a result was then generalized by March (2014a) in the case of $d = 4$ dimensions. One starts by writing (Reichl, 2009)

$$Z_N(\beta\epsilon, \beta H) = [2\cosh\beta E]^N, \quad (9.30)$$

where $E \equiv E(\beta\epsilon, \beta H) = \frac{1}{2}z_c\beta\epsilon\langle s\rangle + \beta H$, and z_c is the coordination number (*i.e.* the number of nearest-neighbours in the lattice under consideration) in $d = 4$ dimensions. The reduced partition function $z_4 = \lim_{N\to\infty} Z_{N+1}/Z_N$ in $d = 4$ dimensions is thus

$$z_4(\beta\epsilon, \beta H) = 2\cosh\left(\frac{1}{2}z_c\beta\epsilon\langle s\rangle + \beta H\right). \quad (9.31)$$

Within mean-field theory, the average spin per site in non-zero magnetic field, $\langle s \rangle$, satisfies the implicit equation (Reichl, 2009)

$$\langle s \rangle = \tanh\left(\frac{1}{2}\beta\epsilon\langle s \rangle + \beta H\right). \tag{9.32}$$

Inverting Eq. (9.31), one finds

$$\cosh^{-1}\left(\frac{1}{2}z_4(\beta\epsilon, \beta H)\right) = \frac{1}{2}z_c\beta\epsilon\langle s \rangle + \beta H, \tag{9.33}$$

which for $H = 0$ reduces to

$$\cosh^{-1}\left(\frac{1}{2}z_4(\beta\epsilon, 0)\right) = \frac{1}{2}z_c\beta\epsilon\langle s \rangle_0. \tag{9.34}$$

But from Eq. (9.32) for $H = 0$ one has $\tanh^{-1}\langle s \rangle_0 = \frac{1}{2}z_c\beta\epsilon\langle s \rangle_0$, and therefore

$$\cosh^{-1}\left(\frac{1}{2}z_4(\beta\epsilon, \beta H)\right) = \langle s \rangle \left(\tanh^{-1}\langle s \rangle_0\right) + \beta H. \tag{9.35}$$

Eliminating further z_c using Eq. (9.32) results in

$$\tanh^{-1}\langle s \rangle = \frac{\langle s \rangle \tanh^{-1}\langle s \rangle_0}{\langle s \rangle_0} + \beta H. \tag{9.36}$$

The above equation shows again that $\langle s \rangle$ is functionally related to its zero-field value, $\langle s \rangle_0$, and to βH. One can similarly conclude that $z_4(\beta\epsilon, \beta H)$ is determined entirely by its zero field limit $z_4(\beta\epsilon, 0)$ and magnetic field βH, plus the number of nearest neighbours z_c (March, 2014a), thereby generalizing to $d = 4$ the result in Eq. (9.29) valid for $d = 1$.

9.2 Yang-Lee theory of phase transitions

In a series of two seminal papers (Yang and Lee, 1952; Lee and Yang, 1952), Yang and Lee provided some exact results which enable to characterize the equilibrium phase transitions of a system, by looking at the (non)analytical properties of its grand partition function as a function of complex fugacity (see Reichl, 2009, for an introduction) (see Bena *et al.*, 2005, for a review). Although Yang and Lee's results formally apply to infinite systems (*i.e.* in the thermodynamic limit), any macroscopic system is composed of a sufficiently large number of particles (of the order of the Avogadro number, say), so that it behaves like an infinite system, except perhaps in a narrow (and experimentally inaccessible) domain in the vicinity of the transition point, where finite-size effects may become relevant. Fisher (1965, 1967) then generalized Yang and Lee's results to temperature-driven phase transitions. Recent advances concern their extension to non-equilibrium phase transitions, *i.e.* transitions between non-equilibrium steady states of a system (see again Bena *et al.*, 2005, for a review). Specifically, Yang and Lee's theory was originally applied to determine the main features, including the critical exponents, of the Ising model (Sec. 9.1).

9.2.1 *Yang-Lee zeroes*

Quite generally, Yang and Lee (1952) considered systems of particles characterized by two-body central interactions, described by a potential $u(r)$, with r the interparticle distance. In order to allow for a correct thermodynamic limit (see Bena *et al.*, 2005, for a discussion), such a potential was assumed *(i)* to vanish in a sufficiently rapid way at infinity, *i.e.* $|u(r)| < C_1/r^{d+\epsilon}$ as $r \to \infty$, where d is the dimensionality of the system, and $C_1, \epsilon > 0$; *(ii)* the potential $u(r)$ should have a repulsive behaviour at sufficiently small distances, in order to avoid the system's collapse at high particle number densities; *(iii)* the potential should be bounded from below, $u(r) \geq -u_0$, with $u_0 > 0$. One such potential is for instance the hard-core potential, with $u(r) = \infty$, for $r \leq r_0$, $u(r) = -u_0$, for $r_0 < r \leq r_1$, and $u(r) = 0$, for $r > r_1$. In particular, if b denotes the co-volume of each particle, then $r_0 \sim b^{1/d}$. Moreover, the number of particles which could be fitted within a box of given volume V is limited from above by the finite number $M = \lfloor V/b \rfloor$, where $\lfloor x \rfloor$ is the greatest integer contained in x.

The classical Hamiltonian of such a system with N particles may then be written as

$$H \equiv H(\mathbf{r}_i, \mathbf{p}_i) = \sum_{i=1}^{N} \frac{p_i^2}{2m} + \frac{1}{2} \sum_{i \neq j} u(|\mathbf{r}_i - \mathbf{r}_j|). \tag{9.37}$$

Thus, the grand partition function takes the form

$$\begin{aligned} \mathcal{Z}(T, V, \mu) &= \sum_{N=0}^{M} \frac{e^{\beta\mu N}}{N!} \frac{1}{h^{dN}} \int\int d^N\mathbf{r} d^N\mathbf{p}\, e^{-\beta H} \\ &= \sum_{N=0}^{M} \frac{z^N}{N!} Q_N(V, T), \end{aligned} \tag{9.38}$$

where the (Gaussian) integral over momenta has been performed analytically,

$$Q_N(V, T) = \int d^N\mathbf{r} \exp\left(-\frac{\beta}{2} \sum_{i \neq j} u(|\mathbf{r}_i - \mathbf{r}_j|) \right) \tag{9.39}$$

is the so-called configuration integral, and

$$z = \frac{e^{\beta\mu}}{\lambda^d} \tag{9.40}$$

is a generalized fugacity. In the above expressions, $\beta = (k_{\rm B}T)^{-1}$ is the inverse temperature, μ denotes the chemical potential, h is the action constant, and $\lambda = h/\sqrt{2\pi m k_{\rm B}T}$ is the thermal wavelength (it coincides with de Broglie wavelength, when h is identified with Planck's constant).

In Eq. (9.38), the configuration integral, Eq. (9.39), is such that $Q_N(V,T) \equiv 0$, for $N > M$, because of the hard-core nature of the interparticle interaction. Therefore, for a finite volume V, the grand canonical partition function, Eq. (9.38) is a polynomial of order M in the variable z. For the fundamental theorem of algebra, it then possesses exactly N zeroes $z_i \equiv z_i(T) \in \mathbb{C}$ $(i = 1, \ldots N)$ in the complex plane. Moreover, since the coefficients of the polynomial are all real and positive, $Q_N > 0$, all its zeroes are complex conjugate, away from the real, positive semiaxis. It is then possible to write Eq. (9.38) as

$$\mathcal{Z}(T,V,z) = \prod_{i=1}^{M} \left(1 - \frac{z}{z_i}\right), \tag{9.41}$$

apart from an inessential multiplicative constant. One can then express the (finite-volume) grand canonical potential in terms of $\mathcal{Z}(T,V,z)$ as

$$\Omega \equiv -pV = -k_{\rm B}T \log \mathcal{Z}(T,V,z), \tag{9.42}$$

from which the (finite-volume) pressure p can be extracted as

$$p(z) = \frac{k_{\rm B}T}{V} \sum_{i=1}^{M} \log\left(1 - \frac{z}{z_i}\right), \tag{9.43}$$

which can also be viewed as an equation of state, in parametric form, while the reciprocal of the specific volume $v = V/N$ can be defined as

$$\frac{1}{v} = \frac{z}{V}\frac{d}{dz} \sum_{i=1}^{M} \log\left(1 - \frac{z}{z_i}\right). \tag{9.44}$$

Similarly, all other thermodynamic potentials can be derived from the grand canonical partition function, thus showing that they are entire functions of z, as long as M is finite.

In particular, by analytical continuation, Eq. (9.43) defines an entire function of the complex variable z, which is analytical in any region of the complex z-plane that is free of zeroes of the partition function. In other words, any nonanalyticity of $p(z)$ would occur at a zero of $\mathcal{Z}(T,V,z)$, or at an accumulation point of zeroes thereof. Since, for a finite system, these zeroes occur away from the real, positive semiaxis, no phase transition may take place. On the other hand, in the thermodynamic limit ($V \to \infty$, $N \to \infty$, with $N/V = \text{const}$), the number M of such zeroes will diverge, and their locations will proliferate and roam in the complex plane. In the event that a point $z = z_0$, with $\operatorname{Im} z_0 = 0$ and $\operatorname{Re} z_0 > 0$, may develop, such that it is either a zero of $\mathcal{Z}(T,V,z)$, or an accumulation point of zeroes thereof, then a phase transition would occur. In general, while the pressure $p(z)$ might be continuous through z_0, the density or other higher derivatives of $p(z)$ may develop discontinuities.

Let $\mathcal{C} \subset \mathbb{C}$ denote the region of the complex plane where the zeroes of the partition function accumulate in the thermodynamic limit. One can then introduce the local density of zeroes $\rho(z)$. This is a real-valued distribution, having support over $\mathcal{C}$, such that $\rho(z) = \rho(z^*)$, and normalized in such a way as to make

$$\int_{\mathcal{C}} \rho(z)\, d\sigma = \frac{M}{V}, \tag{9.45}$$

where $d\sigma = dz \wedge dz^*$ is the area element over the complex plane. Observing that $\rho(z) = 0$ for $z \notin \mathcal{C}$, all the integrations involving $\rho(z)$ can be extended to the whole complex plane. In particular, in the thermodynamic limit, one may define the complex pressure as

$$p(z) = k_{\mathrm{B}} T \int d\sigma'\, \rho(z') \log\left(1 - \frac{z}{z'}\right). \tag{9.46}$$

Denoting

$$\varphi(z) = \frac{1}{k_{\mathrm{B}} T} \operatorname{Re} p(z) = \int d\sigma'\, \rho(z') \log\left|1 - \frac{z}{z'}\right| \tag{9.47}$$

the real part of $p/k_{\mathrm{B}}T$, and recalling that $\log|z|$ is essentially the Green's function of the Laplacian $\Delta = \partial_x^2 + \partial_y^2$ in two dimensions, one may also write

$$\rho(z) = \frac{1}{2\pi} \Delta\varphi(z). \tag{9.48}$$

This enables to interpret $\varphi(z)$ as the electrostatic potential associated to a distribution of charges having density $\rho(z)$. The equipotential lines are then given by $\varphi(z) = \text{const}$, while the lines defined by $\psi(z) = \text{const}$, with

$$\psi(z) = \frac{1}{k_{\mathrm{B}} T} \operatorname{Im} p(z) \mod 2\pi, \tag{9.49}$$

can be interpreted as the lines of force of such a 'charge' distribution, with a field given by $\nabla\varphi(z)$. One has, furthermore, $\varphi(z) = \varphi(z^*)$ and $\psi(z) = -\psi(z^*)$. It is also possible to show that $\varphi(z)$ is continuous in $\mathcal{C}$, and in particular at the transition point, *i.e.*

$$\phi(z_0^-) = \phi(z_0^+), \tag{9.50}$$

where $z_0^\pm$ denote points along the real positive semiaxis infinitesimally on the right and on the left of z_0, respectively. Other similar relations may be derived for the higher derivatives of φ and ψ, which enable to characterize the order of the transition according to Ehrenfest. In particular, if $\mathcal{C}$ is a smooth curve in the complex fugacity plane, intersecting the real positive semiaxis in z_0, and parametrized by the real parameter s, such that $s = 0$ corresponds to z_0, let $\lambda(s)$ denote the line density of zeroes along $\mathcal{C}$. Then, it is possible to show (cf. Bena *et al.*, 2005, and references therein) that

$$\frac{d}{ds}\psi(z_0^+) - \frac{d}{ds}\psi(z_0^-) = 2\pi\lambda(0). \tag{9.51}$$

In the thermodynamic limit, and under the general hypotheses concerning the mutual interacting potential already introduced, Yang and Lee (1952) demonstrated the following two theorems (Reichl, 2009; Bena *et al.*, 2005).

Theorem 9.1 (Yang and Lee, 1952). *For all $z > 0$, $V^{-1}\log\mathcal{Z}(V,T,z)$ approaches, as $V \to \infty$, a limit which is independent of the shape of the volume V. Furthermore, the limit is a continuous, monotonically increasing function of z.*

Theorem 9.2 (Yang and Lee, 1952). *In a region $\mathcal{R}$ of the complex fugacity plane, containing a segment of the positive real axis, and always free of roots of the grand canonical partition function, then in this region, as $V \to \infty$, all the functions*

$$\left(z\frac{d}{dz}\right)^n \frac{1}{V}\log\mathcal{Z}(V,T,z) \quad (n = 0,1,2,\ldots) \tag{9.52}$$

approach limits which are analytic with respect to z. Furthermore, the operators $z(d/dz)$ and $\lim_{V\to\infty}$ commute in $\mathcal{R}$, so that in particular

$$\lim_{V\to\infty} z\frac{d}{dz}\frac{1}{V}\log\mathcal{Z}(V,T,z) = z\frac{d}{dz}\lim_{V\to\infty}\frac{1}{V}\log\mathcal{Z}(V,T,z). \tag{9.53}$$

From Eq. (9.44), therefore, one has

$$\frac{1}{v} = z\frac{d}{dz}\left(\frac{p}{k_{\mathrm{B}}T}\right). \tag{9.54}$$

As already mentioned, this implies that the pressure of an infinite system must be continuous at all values of the fugacity, but its derivatives can exhibit discontinuities where $\mathcal{C}$ touches (or 'pinches') the real positive semiaxis. This corresponds to a phase transition, and its order (according to Ehrenfest classification) can be related to the order n of the derivative $d^n p/dz^n$ which first develops a discontinuity at z_0.

9.2.2 Yang-Lee theory of the Ising model

In their second seminal paper, Lee and Yang (1952) studied the zeroes of the grand canonical partition function associated to a quite general class of lattice models, including the Ising model in $d = 2$ dimensions (Sec. 9.1). Specifically, they proved the following

Theorem 9.3 (Lee and Yang, 1952). *For a gas on a lattice, interacting through a potential u such that $u = +\infty$ whenever two particles are on the same site, and $u \leq 0$ otherwise, the locus $\mathcal{C}$ of the zeroes of the grand partition function is a subset of a circumference centred in the origin in the complex fugacity plane.*

In the particular case of the Ising model in $d = 2$ dimensions, this circumference turns out to be the unit circle, $|\zeta| = 1$, where the so-called 'activity' $\zeta = \exp(-2\beta H)$ is proportional to the fugacity, and H is the modulus of the external magnetic field. However, the circle theorem applies regardless of the specific form of the interaction potential, of dimensionality d, and of the nature of the lattice (whether periodic or not, for example).

Since the zeroes of the grand partition function belong to the unit circle, it is possible to parametrize them by an angle θ, say $\zeta_i = e^{i\theta_i}$. Passing to the continuum limit, let $g(\theta) \equiv g(\theta, T)$ denote the (temperature-dependent) density of the zeroes along the unit circle. One finds that $g(\theta) = g(-\theta)$, and that $g(\theta)$ is nonzero only along two arcs of the unit circle defined by $\theta_c \leq |\theta| \leq \pi$, with $\theta_c > 0$ a temperature-dependent angle. The zero distribution function is normalized as

$$2 \int_{\theta_c}^{\pi} g(\theta, T)\, d\theta = b, \tag{9.55a}$$

and the complex pressure and complex particle-number density correspondingly read

$$p = k_{\mathrm{B}}T \int_{\theta_c}^{\pi} g(\theta, T) \log(\zeta^2 - 2\zeta \cos\theta + 1)\, d\theta, \tag{9.55b}$$

$$n = 2\zeta \int_{\theta_c}^{\pi} g(\theta, T) \frac{\zeta - \cos\theta}{\zeta^2 - 2\zeta\cos\theta + 1}\, d\theta. \tag{9.55c}$$

Lee and Yang (1952) provide the explicit expression of the zero density $g(\theta)$ for the Ising model in $d = 1$ dimension with nearest-neighbour interactions, whereas in the mean field approximation it was evaluated by Katsura (1954, 1955). Numerical approximations for the zero density $g(\theta)$ were obtained by Kortman and Griffiths (1971) for the Ising model on a square lattice ($d = 2$), and for the Ising model on a diamond lattice ($d = 3$).

Similarly, the free energy, the magnetization

$$M = 1 - 4\zeta \int_{\theta_c}^{\pi} g(\theta, T) \frac{\zeta\cos\theta}{\zeta^2 - 2\zeta\cos\theta + 1}\, d\theta, \tag{9.56a}$$

the spontaneous magnetization,

$$M_0 = \lim_{H\to 0} M = 2\pi g(0, T), \tag{9.56b}$$

the zero-field isothermal magnetic susceptibility

$$\chi \simeq 8\beta \int_{\theta_c}^{\pi} \frac{g(\theta, T) - g(0, T)}{\theta^2}\, d\theta, \tag{9.56c}$$

and the specific heat at constant volume

$$C_V \simeq \frac{\partial^2}{\partial T^2} \int_{\theta_c}^{\pi} g(\theta, T) \log\theta\, d\theta \tag{9.56d}$$

can be related to the zero density $g(\theta, T)$ (Suzuki, 1967a,b).

9.2.2.1 *Zeroes distribution function, $d = 1$*

For a one dimensional ($d = 1$) spin lattice with N sites and nearest-neighbour interaction, the grand canonical partition function of the Ising model can be cast into the closed form recorded in Eq. (9.8), *i.e.* as

$$\mathcal{Z}_N = \lambda_+^N + \lambda_-^N, \tag{9.57}$$

with $\lambda_\pm$ given explicitly by Eq. (9.9), and here rewritten in terms of ζ as

$$\lambda_\pm = \left[\frac{1}{2}(1+\zeta) \pm \left(\frac{1}{4}(1-\zeta)^2 + e^{-4\beta J}\zeta\right)^{1/2}\right] e^{\beta(J+H)}. \tag{9.58}$$

The zeroes of the grand partition function, Eq. (9.8), are then given implicitly by the condition

$$\lambda_- = \lambda_+ e^{ip\pi/N}, \tag{9.59}$$

where p is an odd integer such that $|p| < N$. This reduces the problem of finding the zeroes of an N-degree polynomial to that of finding the zeroes of $N/2$ second-order polynomials, which is of course analytically solvable. One finds

$$\zeta_p = \cos\theta_p + i\sin\theta_p, \tag{9.60a}$$

$$\cos\theta_p = \left(1 - e^{-4\beta J}\right)\cos\left(\frac{p\pi}{N}\right) - e^{-4\beta J}. \tag{9.60b}$$

It is then possible to construct the zeroes distribution function, which in the thermodynamic limit reads

$$g(\theta, T) = \begin{cases} \dfrac{1}{2\pi}\dfrac{|\sin(\theta/2)|}{\sin^2(\theta/2) - e^{-4\beta J}}, & \theta_c < |\theta| \leq 0, \\ 0, & |\theta| \leq \theta_c, \end{cases} \tag{9.61}$$

where, as anticipated, θ_c is a temperature-dependent critical angle, given by

$$\cos\theta_c = 1 - e^{-4\beta J}. \tag{9.62}$$

In particular, one finds that $\theta_c \neq 0$ for all $T > 0$, so that the grand canonical partition function zeroes never touch the real ζ axis, at any finite temperature. This implies that the spontaneous magnetization of the Ising model in $d = 1$ is always zero at any finite temperature, thereby implying the absence of any magnetic phase transition at a finite temperature in $d = 1$. On the other hand, one has

$$\lim_{T\to 0} g(\theta, T) = \frac{1}{2\pi}. \tag{9.63}$$

Hence, at $T = 0$, all spins in the one-dimensional Ising model are collinear: the absolute value of the spontaneous magnetization becomes abruptly maximum, and the ground state of the Ising model in $d = 1$ is in fact ferromagnetic.

At the other limit, one may also observe that

$$\lim_{T\to\infty} g(\theta, T) = \delta(\theta - \pi). \tag{9.64}$$

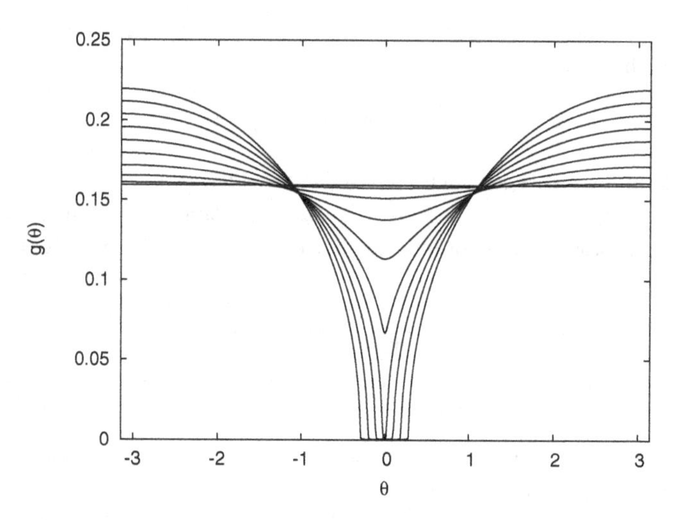

Fig. 9.2 Mean field approximation of the Yang-Lee zeroes density distribution function, $g(\theta, T)$, for various values of the ratio $T/T_c = 0.1 - 1.5$.

9.2.2.2 *Zeroes distribution function, mean field*

By comparing the series expansion of the magnetization M, given by Eq. (9.56a), in terms of powers of ζ, and the Fourier series expansion of the zeroes density, it is possible to relate these two quantities as (Binek, 1998)

$$g(\theta, T) = \lim_{|\zeta|\to 1^-} \operatorname{Re} M(\zeta). \tag{9.65}$$

Although this relation is exact, it can be used to evaluate $g(\theta, T)$ either *(i)* approximately, when only approximate values of M are known, or *(ii)* empirically, when experimental values of M are available.

Within the mean field approximation, the spontaneous magnetization M is implicitly given by

$$\zeta = \frac{1-M}{1+M} \exp\left(2M\frac{T_c}{T}\right). \tag{9.66}$$

Inverting the above relation numerically in favour of M at a given temperature T, and using Eq. (9.65), one may then construct the zeroes density ditribution function $g(\theta, T)$ within this mean field approximation. The numerical results are shown in Fig. 9.2 for various values of the ratio T/T_c (cf. Suzuki, 1967b). Again, one finds $g(\theta, T) = 0$ for $|\theta| < \theta_c$ and $T > T_c$, whereas $g(\theta, T) \neq 0$ at $\theta = 0$ for $T \leq T_c$, thereby signalling a phase transition.

On the other hand, by employing a Padé approximation for the available experimental values of the magnetization of $FeCl_2$ (a layered antiferromagnet) as a function of temperature, Binek (1998) (see also Binek *et al.*, 2001) was able to empirically reconstruct the shape of the zeroes density distribution function for that material.

9.2.3 *Fisher zeroes*

An extension of Yang and Lee theory of phase transitions is due to Fisher (1965, 1967) (see also Grossmann and Rosenhauer, 1967, 1969; Grossmann and Lehmann, 1969). When looking at the partition function for a canonical ensemble, one may characterize the canonical partition function $Z_N(\beta, V)$ through its zeroes in the complex plane of inverse temperature, say $\beta \to \mathcal{B} = \beta + i\tau/\hbar$ (with τ having then the explicit dimensions of a time). Such a generalization turns out to be useful also to describe finite systems (Mülken *et al.*, 2001) and nonequilibrium critical phenomena (Bena *et al.*, 2005).

Then, the free energy of a finite system takes the form

$$F = -\frac{k_{\rm B}T}{N} \log Z_N(\beta, V) = -\frac{k_{\rm B}T}{N} \sum_i \log\left(1 - \frac{\mathcal{B}}{\mathcal{B}_i}\right). \tag{9.67}$$

As in Yang-Lee theory, a phase will be represented by a region of analyticity in the complex $\mathcal{B}$ plane, while a phase transition will be characterized by the singular behaviour of F (or of its derivatives) in the proximity of some $\mathcal{B}_i$, the latter denoting a zero of the canonical partition function Z_N. Such zeroes will again occur in complex conjugate pairs, and will in general be isolated points for finite systems, with nonzero imaginary parts. However, in the thermodynamic limit, one expects that they will proliferate into lines (or even two-dimensional patches), with non-empty intersections with the real $\mathcal{B}$ axis. The point $\beta_c \equiv \mathcal{B}_c$ where a limiting line of zeroes cuts the real axis will locate the critical temperature. In this limit, the summation in Eq. (9.67) will become an integration, possibly weighted by a conveniently defined zero distribution function (see Eq. (9.71) below).

9.2.3.1 *Ising model in two dimensions*

Given Onsager (1944) solution of the Ising model in $d = 2$ dimensions (cf. Sec. 9.1.2), for which model the free energy per spin, $f = F/N$ can be written as

$$-\frac{f}{k_{\mathrm{B}}T}=\frac{1}{2}\int_{-\pi}^{\pi}\frac{d\varphi_1}{2\pi}\int_{-\pi}^{\pi}\frac{d\varphi_2}{2\pi}\log\left(\cosh 2K\cosh 2K'\right.$$
$$\left.-\sinh 2K\cos\varphi_1-\sinh 2K'\cos\varphi_2\right), \quad (9.68)$$

where $K=J/2k_{\mathrm{B}}T$, $K'=J'/2k_{\mathrm{B}}T$ are the scaled nearest-neighbours coupling constants along the two directions associated with a square lattice, and using the explicit expression for the canonical partition function (Kaufman, 1949; Kaufman and Onsager, 1949)

$$Z_N=\prod_{r=1}^{m}\prod_{s=1}^{n}\left(\frac{1+v^2}{1-v^2}\frac{1+v'^2}{1-v'^2}-\frac{2v}{1-v^2}\cos\frac{2\pi r}{m}-\frac{2v'}{1-v'^2}\cos\frac{2\pi s}{n}\right), \quad (9.69)$$

where $v=\tanh K$ and $v'=\tanh K'$, Fisher (1965) was able to determine the zeroes of Z_N in the complex v plane in the symmetric case ($K=K'$, $v=v'$) as

$$v_\theta=\pm 1+\sqrt{2}e^{i\theta}, \quad (9.70)$$

where $0\leq\theta\leq 2\pi$. Eq. (9.70) defines two circles centred at $v=\pm 1$, having radius $\sqrt{2}$ (Fig. 9.3). These intersect the real v axis at $v=\pm(1+\sqrt{2})$ and $\pm(\sqrt{2}-1)$. The former correspond to complex (and thus nonphysical) temperatures, while the latter determine the ferromagnetic and antiferromagnetic transition temperatures, as expected.

The zero density distribution function can be expressed as

$$g(\theta)=|\sin\theta|F(\theta), \quad (9.71)$$

where $F(\theta)$ is analytic and periodic in θ, and $g(\theta)\sim|\theta|$ near the real v axis (cf. Lu and Wu, 2001). This implies a logarithmically singular behaviour of the specific heat close to the transition,

$$C_V(T)\sim\log\left|1-\frac{T}{T_c}\right|, \quad (9.72)$$

corresponding to a value of the critical exponent $\alpha=0$, in agreement with Onsager (1944) exact solution (see Section 9.1 above).

The same result can be equivalently recovered by inspection of Eq. (9.68). Performing one integration, one obtains (Fisher, 1965)

$$-\frac{f}{k_{\mathrm{B}}T}=\log\cosh 2K+\frac{1}{2\pi}\int_0^{\pi}d\psi\,\log\left[\frac{1}{2}\left(1+\sqrt{1-\kappa^2\sin^2\psi}\right)\right], \quad (9.73)$$

where

$$\kappa^2=\frac{2\sinh 2K}{\cosh^2 2K}, \quad (9.74)$$

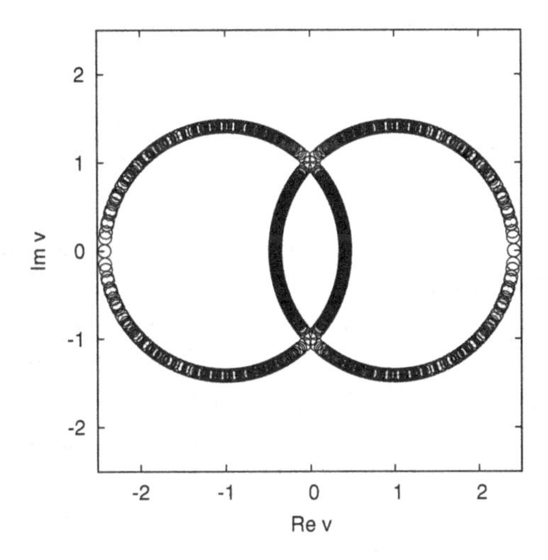

Fig. 9.3 Fisher zeroes for the symmetric $d = 2$ dimensional Ising model in the complex $v = \tanh(J/2k_{\rm B}T)$ plane, Eqs. (9.69) and (9.70), with $n = m = 100$.

with $\kappa = 1$ at $T = T_c$. Since $U = -\partial \log Z/\partial\beta$, differentiating under the integral sign, one may then express the internal energy per spin as

$$U = -\frac{J}{2}\coth 2K\left(1 + \kappa'\frac{2}{\pi}K(\kappa)\right), \tag{9.75}$$

where $K(\kappa)$ denotes the complete elliptic integral of the first kind, and

$$\kappa' = \pm\sqrt{1-\kappa^2} = 2\tanh^2(2K) - 1. \tag{9.76}$$

From this, it follows that (Fisher, 1965; Gradshteyn and Ryzhik, 1994)

$$U(T) = U(T_c) + a(T - T_c)\log|T - T_c| + \cdots \tag{9.77}$$

so that the internal energy is continuous at the critical point, but with an infinite-slope vertical tangent there, corresponding to a logarithmic singularity in the specific heat.

9.2.3.2 *Transition order and critical exponents*

More generally, the critical exponents (Itzykson *et al.*, 1983), as well as the order and strength of the phase transition (Janke and Kenna, 2001), can be related to the behaviour of the line distribution and density of the Fisher zeroes, in the proximity of the critical zero, by means of a finite-size scaling (FSS) analysis. For a d-dimensional system of linear size L, the jth partition function zero behaves asymptotically as (Itzykson *et al.*, 1983)

$$t_j(L) \sim (j/L^d)^{1/\nu d}, \qquad (h=0), \tag{9.78a}$$
$$h_j(L) \sim (j/L^d)^{(d+2-\eta)/2d}, \qquad (t=0), \tag{9.78b}$$

where $t \equiv T/T_c - 1$ is the reduced temperature, ν is the critical exponent associated with the correlation length, and η is the anomalous dimension. Eqs. (9.78) above are related to Fisher and Yang-Lee zeroes, respectively.

Let z collectively denote the field- or temperature-dependent complex variable, with respect to which the appropriate partition function has been characterized in terms of its zeroes as $Z_L(z) \propto \prod_j (z - z_j(L))$. Such a variable is to be identified with the fugacity or with the inverse temperature, in the case of an analysis based on the Yang-Lee or Fisher zeroes, respectively, Eqs. (9.78). Assuming that the zeroes z_j touch the real axis (as $L \to \infty$, *i.e.* in the thermodynamic limit) at the critical point z_c along the line

$$z_j = z_c + r_j e^{i\varphi}, \tag{9.79}$$

where φ is the impact angle, one may then define the density of zeroes as (Janke and Kenna, 2001)

$$g_L(r) = \frac{1}{L^d} \sum_j \delta\left(r - r_j(L)\right). \tag{9.80}$$

Their cumulative distribution turns then out to be

$$\begin{aligned} G_L(r) &= \int_0^r dr'\, g_L(r') \\ &= j/L^d, \qquad \text{for } r_j < r < r_{j+1}, \\ &= (j - \tfrac{1}{2})/L^d, \qquad \text{for } r = r_j. \end{aligned} \tag{9.81}$$

In the thermodynamic limit ($L \to \infty$) and for a first-order phase transition, the integrated density of zeroes tends to

$$G_\infty(r) = g_\infty(0) r, \qquad \text{(1st order)}, \tag{9.82a}$$

where the slope at the origin, $g_\infty(0)$, is proportional to the latent heat and to the magnetization, in the Fisher and Yang-Lee case, respectively. In addition, for a first-order phase transition, one finds (Yang and Lee, 1952; Lee and Yang, 1952)

$$g_\infty(r) = g_\infty(0) + ar^b + \dots, \tag{9.82b}$$

where $g_\infty(0) \neq 0$, and the impact angle of the line of zeroes onto the real z axis has to be $\varphi = \pi/2$ (Grossmann and Rosenhauer, 1967, 1969; Grossmann and Lehmann, 1969). For a second-order phase transition, the corresponding expressions are

$$G_\infty(r) \propto r^{2-\alpha} \qquad \text{(2nd-order, Fisher zeroes)},$$
$$G_\infty(r) \propto r^{2d/(d+2-\eta)} \qquad \text{(2nd-order, Yang-Lee zeroes)}. \tag{9.82c}$$

More generally, for a higher-order phase transition, one finds (Janke *et al.*, 2005)

$$\alpha = 1 - b, \tag{9.83a}$$

$$\gamma = \tan\left(\varphi - \frac{\pi}{2}\right), \tag{9.83b}$$

$$\operatorname{Im} z_j \sim L^{-1/\nu}, \tag{9.83c}$$

where b is defined in the expansion of the zeroes density, Eq. (9.82b).

Using the results above, which relate the main critical exponents to the FSS of the zeroes density, it is possible to derive the usual scaling relations among them, as well as additional scaling relations for the logarithmic corrections thereto (Kenna *et al.*, 2006a,b).

9.2.3.3 *Complex temperature*

The analytical continuation of the partition function to complex temperatures might at first seem a mathematical artifact. Several exact results have been obtained indeed for the Ising model in $d = 2$ dimensions exploiting complex temperatures by Matveev and Shrock (1995, 1996).

More recently, Blythe and Evans (2002) have studied nonequilibrium steady states, *i.e.* states close to equilibrium, that allow for the flow of mass or energy (see also Bena *et al.*, 2005, for a review). After defining a suitable generalization of the partition function for these nonequilibrium states, Blythe and Evans (2002) can characterize phase transitions when the zeroes thereof accumulate close to the real axis. Again, a study of the zeroes distribution enables to extract the order and critical exponents of the transition.

Quantum systems exhibiting a nonequilibrium behaviour are also ultracold atomic gases close to a Bose-Einstein condensation (Greiner *et al.*, 2002a,b). These systems can be described as isolated quantum systems, in which one parameter of the Hamiltonian changes abruptly so that equilibrium is destroyed. This process is described as a 'quantum quench', where the initial state $|\Psi_i\rangle$ of the system is usually described by a non-thermal superposition of eigenstates of the Hamiltonian H (cf. Sec. 5.2.7). One is often interested in cases where a quantum quench results in a quantum phase transition (Sachdev, 2011b). This has been studied in the one-dimensional Bose-Hubbard model (Läuchli and Kollath, 2008), which describes a Mott-insulator-to-superfluid quantum phase transition in ultracold atoms trapped in an optical lattice (Chen *et al.*, 2011). Experimental observations of the non-equilibrium dynamics of a density wave of ultracold bosonic atoms in an optical lattice in the regime of strong correlations have been indeed reported quite recently (Trotzky *et al.*, 2012).

Following Heyl *et al.* (2013), the overlap between the initial state $|\Psi_i\rangle$ at $t = 0$ and the time-evolved state at a time $t = \tau > 0$ will then be given by

$$G(\tau) = \langle\Psi_i|e^{-iH\tau/\hbar}|\Psi_i\rangle. \tag{9.84}$$

This bears an obvious similarity with the canonical partition function $Z(\beta) = \operatorname{Tr} e^{-\beta H}$ of the same system at equilibrium, which prompted Heyl *et al.* (2013) to introduce the so-called boundary partition function, defined as

$$Z(\mathcal{B}) = \langle\Psi_i|e^{-\mathcal{B}H}|\Psi_i\rangle, \tag{9.85}$$

for complex $\mathcal{B} \in \mathbb{C}$. For a purely imaginary $\mathcal{B} \equiv i\tau/\hbar$, this reduces to the overlap amplitude $G(\tau)$, Eq. (9.84), while for real $\mathcal{B} \equiv \beta$, it can be interpreted as the partition function of an appropriate field theory associated with the Hamiltonian H. This is consistent with the definition of the quantum mechanical partition function $Z(\mathcal{B})$ of Mülken *et al.* (2001), with $\mathcal{B} \equiv \beta + i\tau/\hbar$, which reads

$$Z(\mathcal{B}) = \operatorname{Tr}\left(e^{-\mathcal{B}H}\right) = \langle\Psi_{\text{can}}|e^{-i\tau H/\hbar}|\Psi_{\text{can}}\rangle = \langle\Psi_{\text{can}}(t=0)|\Psi_{\text{can}}(t=\tau)\rangle, \tag{9.86}$$

where the 'canonical state' $|\Psi_{\text{can}}\rangle = \sum_k e^{-\beta\epsilon_k/2}|\phi_k\rangle$ is defined as a Boltzmann-weighted superposition of eigenstates. Denoting by $\mathcal{B}_k = \beta_k + i\tau_k/\hbar$ the Fisher zeroes of the quantum mechanical partition function, Eq. (9.86), their imaginary parts τ_k can be interpreted as those times for which the overlap of the initial canonical state with the time-evolved state vanishes. Heyl *et al.* (2013) specifically analyze the transverse field Ising model on a one dimensional chain, with periodic boundary conditions (see also Calabrese *et al.*, 2011),

$$H(g) = -\frac{1}{2}\sum_{i=1}^{N-1} \sigma_i^z \sigma_{i+1}^z + \frac{g}{2}\sum_{i=1}^{N} \sigma_i^x, \tag{9.87}$$

where σ_i^x, σ_i^z are spin matrices on site i, and g the coupling parameter to a transverse field. At zero temperature, the model undergoes a quantum phase transition at $g = g_c = 1$ from a ferromagnetic ground state ($g > 1$), to a paramagnet ($g < 1$). (See also Dziarmaga, 2005, for exact solutions of the quantum Ising model.) The system is initially prepared in the ground state of the Hamiltonian $H(g_0)$ for some value, $g = g_0$ say, of the coupling parameter, and is then allowed to evolve according to the quenched Hamiltonian $H(g_1)$, with $g_1 \neq g_0$. Heyl *et al.* (2013) then determine exactly the zeroes of the partition function, which are shown to arrange according to a family of lines in the complex plane, their shape depending on whether the system is quenched within the same phase or across the quantum critical point. In the latter case, the partition function zeroes cross the real time axis, giving rise to a nonanalytic behaviour.

9.2.4 *Concluding remarks*

The Yang-Lee or Fisher zeroes have been studied for more general Ising-like models, including the Potts model (Alves *et al.*, 1991; Kim and Creswick, 1998; Biskup *et al.*, 2004), the Blume-Capel model (Ghulghazaryan *et al.*, 2007), the N-vector model (Kurtze, 1983), and the ϕ^3 field theory (Fisher, 1978). The analysis of the zeroes of the partition function has not been limited to models defined on regular lattices only. The Fisher zeroes of the Ising model on the Sierpinski gasket (thereby interpolating between $d = 1$ and $d = 2$, in the sense of fractals) have been studied by Burioni *et al.* (1999), as well as on aperiodic lattices, such as the Fibonacci chain and other quasicrystals (Baake *et al.*, 1995; Simon *et al.*, 1995; Simon and Baake, 1997; Yessen, 2013).

The exact partition function zeroes for a model describing protein folding have been computed by Lee (2013a,b), using both analytic and numerical methods. Earlier, the helix-coil transition in a polypteptide had been studied by Alves and Hansmann (2000) by means of a finite-size scaling analysis of the partition function zeroes.

Besides the theoretical motivation discussed in Sec. 9.2.3.3 above, recent experimental findings allowed to access the complex temperature domain of quantum systems (Peng *et al.*, 2015). These concern the experimental determination of the quantum coherence of a probe spin coupled to an Ising-type spin bath. The quantum evolution of the probe spin introduces a complex phase factor and therefore effectively realizes an imaginary magnetic field (Matveev and Shrock, 1995, 1996). Peng *et al.* (2015) could then determine the Yang-Lee zeroes, thereby reconstructing the free energy of the spin bath and thus determine its phase transition temperature.

The development of long-range order in Ising-like models has been related to the occurrence of Fisher zeroes in the partition function via the study of correlation functions at a complex temperature (Beichert *et al.*, 2013).

9.3 Tonks 1D model and its generalization

The Tonks 1D gas (Tonks, 1936) is a special case of the so-called Takahashi nearest-neighbour gas. The latter model is characterized by a total potential which is the sum of two-body potentials of the form

$$v(x) = \begin{cases} \infty, & \text{for } |x| < a \text{ (hard core)}, \\ \phi(x-a), & \text{for } a < |x| < 2a, \\ 0, & \text{for } |x| \geq 2a. \end{cases} \tag{9.88}$$

Consider N particles in a box of length L, with density $\rho = N/L$. The configurational partition function Z and the free energy F are given by

$$Z \equiv \exp(-\beta F) = \frac{1}{N!}\int_0^L dx_1 \cdots \int_0^L dx_N \exp\left(-\beta \sum_{i>j} v(x_i - x_j)\right). \tag{9.89}$$

Now observe that the integral in the above equation is $N!$ times the integral over the simplex S defined by $0 \leq x_1 \leq x_2 \leq \ldots x_N \ldots L$. The introduction of such a region S is in fact often a useful procedure in 1D many-body theory. Unfortunately, it has no useful analogue in higher dimensions.

First, let us take the original Tonks model in which $\phi = 0$ in Eq. (9.88) above. Then, since only the hard core remains, the integral over S can be readily carried out by using as variables $y_j = x_j - (j-1)a$, $\ell = L - (N-1)a$.

Hence one can readily obtain

$$\begin{aligned} Z &= \int_0^\ell dy_N \cdots \int_0^{y_3} dy_2 \int_0^{y_2} dy_1 \\ &= \frac{\ell^N}{N!} \to \exp N(-\log\rho + \log(1-\rho a) + 1), \end{aligned} \tag{9.90}$$

the last relation holding in the thermodynamic limit. The exact partition function in Eq. (9.90) in this limit can next be generalized to the case when $\phi \neq 0$. Then, one notes that, due to the third line in Eq. (9.88), the only terms in the potential energy that need to be retained are $v(x_i - x_j)$, where $i = j+1$. Thus the partition function generalizes to

$$Z = \int_0^\ell dy_N \cdots \int_0^{y_3} dy_2 \exp[-\beta\phi(y_3 - y_2)] \int_0^{y_2} dy_1 \exp[-\beta\phi(y_2 - y_1)]. \tag{9.91}$$

This has the form of an iterated Laplace convolution, and by taking the Laplace transform of Z as a function of ℓ one finds

$$\bar{Z}(s) = \int_0^\infty Z(\ell)e^{-s\ell}d\ell = s^{-2}[K(s)]^{N-1}, \tag{9.92}$$

where $K(s)$ is defined as

$$K(s) = \int_0^\infty \exp[-sx - \beta\phi(x)]dx. \tag{9.93}$$

To find $Z(\ell)$, the steepest descent method can be utilized to invert the Laplace transform, Eq. (9.92), which procedure becomes exact for sufficiently large N. The desired result is

$$\bar{Z}(s) = -s\ell + \log Z(\ell), \tag{9.94}$$

where

$$s = \left.\frac{\partial \log Z(\ell)}{\partial \ell}\right|_{\bar{\ell}} = \beta P(\bar{\ell}), \tag{9.95}$$

with P denoting the pressure of the gas. Gürsey (1950) subsequently produced graphs of typical isotherms, to which the interested reader is referred.

To conclude this section, it is important to note that the two-particle distribution $g(x_1, x_2)$ can also be obtained for the Takahashi model. We merely quote below the bulk limit in which $g(x_1, x_2)$ becomes a function of $x = x_1 - x_2$ only. Then its Laplace transform $\bar{g}(k)$ has the form

$$\bar{g}(k) = \rho\{\exp\beta[\mu(\beta P + k) - \mu(\beta P)]\} - 1. \tag{9.96}$$

P is the pressure entering Eq. (9.95), while μ is the chemical potential given by

$$\beta\mu = \log\frac{Z_N}{Z_{N+1}}. \tag{9.97}$$

9.4 Critical exponents in terms of d and η

In this Section, following closely March (2015), a connection is provided among the Gaussian model, the Ising model and the anti-de Sitter/conformal field theory duality (AdS/CFT) via analysis of the critical exponents thereof. It proves that the AdS/CFT model embraces the Gaussian case. The second aim is to have a novel view on the $O(n)$ model of critical exponents. This is done by utilizing the results as approximate summations of the ϵ expansion for Ising, XY and Heisenberg models. Furthermore, this Section discusses the critical exponents expressed in terms of both the dimensionality d and the universality class n. Some formulas in the large n limit have been derived.

9.4.1 *String theory model of critical exponents*

As a preliminary background, we briefly review the so-called gauge/gravity duality (or anti-de Sitter/Conformal Field Theory, *i.e.* AdS/CFT, or holographic duality) (Maldacena, 1999; Gubser *et al.*, 1998; Witten, 1998), and its relevance for the statistical mechanics of condensed phases of matter (see Iqbal *et al.*, 2011, ILM, below). This relates a classical gravity theory in a weakly curved $(d+1)$-dimensional anti-de Sitter (AdS_{d+1}) spacetime to a strongly coupled d-dimensional quantum field theory defined on its boundary (CFT_d).

Such a duality has suggested that complex issues arising in strongly interacting many-body systems can be mapped onto simpler, and possibly exactly solvable, problems in classical gravity, where a geometrical formulation is usually available (see also Sachdev, 2011a, 2012; Zaanen *et al.*, 2015). In particular, the analysis of a black hole, with its spherical symmetry and rotational invariance, being characterized by simple and universal properties such as its mass and charge, has provided remarkable insights within the AdS/CFT duality for the statistical properties of non-Fermi liquids, such as the 'strange' metallic phase in the underdoped cuprate superconductors, and other many-electron systems at finite density, such as the heavy Fermion compounds (Iqbal *et al.*, 2011). Current understanding holds that both the aforementioned material systems lie in the proximity of a quantum critical point (QCP), *i.e.* a point in parameter space (electron density, for the cuprates, and pressure and/or applied magnetic field, for the heavy Fermion compounds) separating two different phases at $T=0$. Therefore, a transition between such phases would not be driven by thermal, but rather by quantum fluctuations.

One of the oddities of such a 'strange' metallic phase (*e.g.* in the normal phase of the cuprate superconductors) is the presence of a 'blurred' Fermi surface, as is signalled, *e.g.*, by a broad peak in angle-resolved photoemission spectroscopy (ARPES; see *e.g.* Randeria and Campuzano, 1999, and references therein). This has been interpreted as a breakdown of the conventional Landau theory of Fermi liquids, and has been described in terms of a nontrivial self-energy $\Sigma(k,\omega)$ in the inverse quasiparticle propagator, $G^{-1}(k,\omega) \sim \omega - v_{\mathrm{F}}(k-k_{\mathrm{F}}) + \Sigma(k,\omega)$. In particular, within the so-called marginal Fermi liquid (MFL) theory, this is modeled by $\mathrm{Re}\,\Sigma(k,\omega) \approx \omega \log \omega$ and $\mathrm{Im}\,\Sigma(k,\omega) \approx \omega$, rather than $\mathrm{Im}\,\Sigma(k,\omega) \approx \omega^2$, as in conventional Fermi liquid theory (Varma *et al.*, 1989). One could synthetically describe such a state of affairs as a 'Fermi surface without quasiparticles' (in the sense of Landau). In the parlance of renormalization group (RG), it must be that the system is flowing towards a nontrivial fixed point distinct from that of a Fermi liquid.

One of the aspects of the holographic duality from AdS_{d+1} to a CFT_d on the boundary of the former is the possibility of identifying the additional or 'holographic' coordinate of the AdS spacetime, z say, as the scaling parameter in the renormalization flow of the CFT. In other words, the radial direction z can be identified as the RG scale of the boundary theory.

Specifically, Iqbal *et al.* (2011) consider a spherically symmetric black hole in $d+1$ dimensions, whose action corresponds to the anti-de Sitter metric. In order to account for finite density in the holographic field theory, they also include charge, and therefore a coupling with a gauge field A_μ, the value of whose time component at the boundary is to be identified with the chemical potential of the CFT, $A_t(z \to 0) = \mu$. Then, the action becomes

$$S = \frac{1}{2\kappa^2} \int d^{d+1}x \sqrt{-g} \left[\mathcal{R} + \frac{d(d-1)}{R^2} + \frac{R^2}{g_F^2} F_{MN} F^{MN} \right], \qquad (9.98)$$

where $x \equiv (\vec{x}, t, z)$, R is the AdS curvature radius, $\mathcal{R}$ is the Ricci scalar, F_{MN} is the gauge field tensor, g_F a suitable coupling to the gauge field, κ^2 is Newton's gravitational constant, and g is the metric tensor, so that

$$ds^2 = \frac{R^2}{z^2}(-f dt^2 + d\vec{x}^2) + \frac{R^2}{z^2} \frac{dz^2}{f}, \qquad (9.99)$$

is the Reissner-Nordström metric, with $f = 1 + Q^2 z^{2d-2} - M z^d$, $A_t = \mu[1 - (z/z_0)^{d-2}]$, where M and Q are the black hole's mass and charge, respectively, and z_0 is a constant, with $z = z_0$ defining the horizon through the condition $f(z_0) = 0$. Iqbal *et al.* (2011) stress that different systems may differ in their matter field spectrum on the gravity side, depending on the above parameters, but the underlying geometry would be the same.

The field theory quantities dual to this black hole are given in terms of the parameters as

$$\mu = g_F z_0^{d-2} \sqrt{\frac{d-1}{2(d-2)}} Q, \qquad \text{chemical potential,} \tag{9.100a}$$

$$\rho = \sqrt{2(d-1)(d-2)} \frac{R^{d-1}}{\kappa^2} \frac{Q}{g_F}, \qquad \text{charge density,} \tag{9.100b}$$

$$s = 2\pi \frac{R^{d-1}}{\kappa^2} \frac{1}{z_0^{d-1}}, \qquad \text{entropy density,} \tag{9.100c}$$

$$\epsilon = \frac{d-1}{2} \frac{R^{d-1}}{\kappa^2} M, \qquad \text{energy density,} \tag{9.100d}$$

$$T = \frac{d}{4\pi z_0} \left(1 - \frac{d-2}{d} Q^2 z_0^{2d-2} \right), \qquad \text{(Hawking) temperature.} \tag{9.100e}$$

All of these can be further expressed in terms of a single length scale z_*, which is completely determined by the black hole charge as

$$Q = \sqrt{\frac{d}{d-2}} \frac{1}{z_*^{d-1}}. \tag{9.101}$$

As the system flows towards the horizon during the RG process specified above, the boundary flows towards a point which is dual to $\mathrm{AdS}_2 \times \mathbb{R}^{d-1}$. This is what Iqbal *et al.* (2011) term a *semi-local quantum liquid* (SLQL), whose correlation properties can be brought into contact with the fermion field theory under consideration.

Specifically, Iqbal *et al.* (2011) are then able to relate a quantum field theory defined in the 'bulk' AdS_{d+1} space, to a quantum field theory on its d-dimensional boundary. In particular, the scaling dimension, δ_k say, of a Dirac field $\phi(t, \vec{k}, z)$ in the SLQL can be expressed as $\delta_k = \frac{1}{2} + \nu_k$, where ν_k is related to k ($\vec{k}$ being conjugate to $\vec{x}$ under Fourier transformation), and to the curvature R of the AdS_2 spacetime. At the Fermi wavevector, $k = k_\mathrm{F}$, Iqbal *et al.* (2011) find that the self-energy in the retarded Green's function behaves as $\Sigma(\omega) \sim \omega^{2\nu}$, where $\nu \equiv \nu_\mathrm{F}$. One therefore recovers a Fermi liquid-like behaviour for $\nu > \frac{1}{2}$, a non-Fermi liquid behaviour for $\nu < \frac{1}{2}$, with no well-formed quasiparticles but with a Fermi surface, and a marginal Fermi liquid at $\nu = \frac{1}{2}$.

9.4.2 *Proof that ILM model embraces Gaussian case*

We first emphasize the merit of the Gaussian model of critical exponents (see Cardy, 1996), the greatest asset being its expression of the six critical exponents α, β, γ, δ, η, ν in terms solely of dimensionality d. Secondly we prove that this model is embraced by the string theory model derived by Iqbal *et al.* (2011, ILM, below). Unfortunately, however, both these models have the severe limitation that $\gamma = 1$, whereas for the $2D$ Ising model ($n = 1$) we have the Onsager exact value as $7/4$ (Onsager, 1944, see Section 9.1 above). Zhang's value $\gamma = 5/4$ is obtained via two conjectures in the $3D$ Ising case, and it is now established to take that value to 'experimental' accuracy. For discrete values of $d = 4$, 3 and 2, we can therefore usefully write that

$$\gamma(d, n = 1) = (a_1 + b_1)/d, \tag{9.102}$$

where a_1 and b_1 are proposed in Eq. (9.110) below.

As a starting point, Table 9.1 summarizes the Gaussian exponents. As we already mentioned above, that $\gamma = 1$ in the ILM model seems first to have been pointed out by Zhang and March (2013). Relating to the result of the ILM theory in that (Iqbal *et al.*, 2011; Zhang and March, 2013)

$$\nu = \frac{1}{2} d\nu_{\mathrm{ILM}}, \tag{9.103}$$

where ν_{ILM} is the model parameter, it is readily seen that to embrace $\nu = \frac{1}{2}$ in Table 9.1, this parameter must be chosen such that

$$\nu_{\mathrm{ILM}} = \frac{1}{d} \tag{9.104}$$

Furthermore Zhang and March (2013) show that the exponent η is explicitly

$$\eta = 2 - 2d\nu_{\mathrm{ILM}} \tag{9.105}$$

and hence for the parameter choice Eq. (9.104) we immediately recover $\eta = 0$ as in Table 9.1. But Hubbard and Schofield (1972) (see also Zhang and March, 2013) have shown that

$$\delta = \frac{d + (2 - \eta)}{d - (2 - \eta)} \tag{9.106}$$

This leads immediately to the result for Gaussian model in Table 9.1 when η is set equal to zero in Eq. (9.106). Next, the exponent β can be found using the well-known result that

$$\gamma = \beta(\delta - 1). \tag{9.107}$$

Since, using δ in Table 9.1,

$$\delta - 1 = \frac{4}{d-2} \tag{9.108}$$

we regain β from Eqs. (9.107) and (9.108) to be as in Table 9.1. Finally from the Rushbrooke relation that

$$\alpha + 2\beta + \gamma = 2 \tag{9.109}$$

we find α as in Table 9.1 by inserting β and γ as there recorded into Eq. (9.109). We must stress though that while it is an elegant feature of the Gaussian model to have α, β, δ as explicit functions of dimensionality d, it is embraced by the ILM model only because $\gamma = 1$ in Table 9.1, in common with ILM (Iqbal *et al.*, 2011). Any final theory of the $O(n)$ model of critical exponents might be anticipated to have the above critical exponents expressed in terms of both dimensionality d (compare the examples in Table 9.1) but now plus the universality class n, the latter being 1, 2 and 3 for Ising, XY and Heisenberg models respectively. In the above context, it is then relevant to note that for the Ising model corresponding to $n = 1$, $\gamma = \frac{7}{4}$ for $d = 2$ (Onsager, 1944), $\gamma = 1$ for $d = 4$ (see, *e.g.* Kardar, 2007). For $d = 3$, it is known by now that, from the ϵ-expansion plus the work of Zhang (2007) based on two conjectures, $\gamma = \frac{5}{4}$ to within conceivable experimental accuracy (see Kardar, 2007) for $\gamma = 1.238$ from high order ϵ-expansion (Cardy, 1996); also Wilson (1972) already gave $\gamma = 1.244$ at low order). Thus, we can write a simple closed formula for γ, valid only at discrete $d = 2$, 3 and 4, for the universality class $n = 1$. This is readily verified to read

$$\gamma(d, n = 1) = \frac{1}{4} + \frac{3}{d}, \quad \text{(integral } d = 2,\ 3,\ 4 \text{ only)}. \tag{9.110}$$

For the XY model corresponding to $n = 2$, Kardar's book (2007) for $d = 3$ gives $\gamma = 1.32$ from high temperature series for an fcc lattice. Therefore, we assume $\gamma = \frac{4}{3}$ for $d = 3$, and March (2015) proposed to write

$$\gamma(d, n = 2) = a_2 + \frac{b_2}{d}, \quad \text{with } 3a_2 + b_2 = 4. \tag{9.111}$$

Similarly, for $n = 3$ (as in Heisenberg's model) write, from Kardar's tables (2007), that $\gamma \simeq 1.4$ which is $\frac{7}{5}$ for $d = 3$, as

$$\gamma(d, n = 3) = a_3 + \frac{b_3}{d}, \quad \text{with } 3a_3 + b_3 = \tfrac{21}{5}. \tag{9.112}$$

Thus, we have assumed above that at integral values of d equal to 2, 3 and 4,

$$\gamma(d, n) = a_n + \frac{b_n}{d}, \tag{9.113}$$

where a_1 and b_1 are given explicitly in Eq. (9.110). Relations between a_n and b_n are proposed already in Eqs. (9.111) and (9.112). Next, we discuss the string theory 'parameter' ν_{ILM} modeled as a function of universality class and dimensionality, in a form appropriate at large n for arbitrary d. From Kardar's book [2007; p. 93 and Eq. (5.76)] we model the string theory 'parameter' ν_{ILM} by the approximate form at sufficiently large n:

$$(n+8)\nu_{\mathrm{ILM}} = \frac{n+2}{2} + \frac{4-n}{d}. \tag{9.114}$$

As shown by Zhang and March (2013), the critical exponent ν is related to ν_{ILM} by

$$\nu(n,d) = \frac{1}{2} d\nu_{\mathrm{ILM}}. \tag{9.115}$$

Equation (9.114) then has the considerable merit that it yields the exact form $\nu = (d-2)^{-1}$ as $n \to \infty$. Using the Josephson relation $d\nu = 2 - \alpha$, we have

$$\alpha(n,d) = \frac{dn - 4n - 4d + 16}{dn - 2n + 2d + 8}, \tag{9.116}$$

which we record along with ν in Table 9.2. Note from Table 9.2 that in the limit $n \to \infty$, $\beta = \nu$, where ν is exactly $(d-2)^{-1}$ (Kardar, 2007). Next we discuss dimensionality dependence as the universality class n tends to infinity. Kardar (2007) gives

$$\nu = \frac{1}{d} - 2 \tag{9.117}$$

for the critical exponent ν, as $n \to \infty$. Evidently, from the Josephson relation

$$d\nu = 2 - \alpha \tag{9.118}$$

we immediately find that α in this limit $n \to \infty$ is given by

$$\alpha = \frac{d-4}{d-2}. \tag{9.119}$$

Applying next the well known relation

$$\gamma = (2-\eta)\nu, \tag{9.120}$$

we obtain by employing Eq. (9.117)

$$\eta = 2 - \gamma(d-2). \tag{9.121}$$

From the Hubbard-Schofield relation (Hubbard and Schofield, 1972) (see also Zhang and March, 2013)

$$\delta = \frac{d + (2-\eta)}{d - (2-\eta)}, \tag{9.122}$$

we find by substitution of Eq. (9.121) that

$$\delta = \frac{d + \gamma(d-2)}{d - \gamma(d-2)}. \tag{9.123}$$

Finally, utilizing the Rushbrooke relation $\alpha + 2\beta + \gamma = 2$ and invoking Eq. (9.119) the result for β is

$$\beta = \frac{d}{2}(d-2) - \frac{1}{2}\gamma. \tag{9.124}$$

Appealing once again to the string theory model of ILM, we have, in terms of their single parameter ν_{ILM} that (Zhang and March, 2013)

$$\nu = \frac{1}{2} d\nu_{\mathrm{ILM}}. \tag{9.125}$$

Using Eq. (9.117) it immediately follows that

$$\nu_{\mathrm{ILM}} = \frac{d-2}{2d}. \tag{9.126}$$

But ILM give, from their string theory model (Iqbal *et al.*, 2011),

$$\alpha = 2 - \frac{1}{2}\nu_{\mathrm{ILM}} = 2 - \frac{d}{d-2} \tag{9.127}$$

which readily leads back to Eq. (9.119). A further result of the ILM model (Iqbal *et al.*, 2011) [see also Eq. (3) in Zhang and March (2013)] reads

$$\delta = \frac{\frac{1}{2} + \nu_{\mathrm{ILM}}}{\frac{1}{2} - \nu_{\mathrm{ILM}}}. \tag{9.128}$$

Substituting Eq. (9.126) into Eq. (9.128) yields

$$\delta = d\left(1 - \frac{1}{d}\right). \tag{9.129}$$

Combining the general Eq. (9.123) and Eq. (9.122) with this 'prediction', they agree only if $\gamma = 1$, which as Zhang and March (2013) stressed, is true for the ILM model, independently of the choice of their parameter ν_{ILM}. Returning then to Eq. (9.121) this yields

$$\eta = 4 - d. \tag{9.130}$$

We collect these results of combining the ILM string theory model with the large n limit for ν in Eq. (9.117) in Table 9.2. Choosing the most important dimension $d = 3$, the critical exponents obtained as $n \to \infty$ are recorded in the last column of Table 9.2. The results for ν and hence α are indeed exact. Table 9.3 finally shows in the second column how, for $\gamma = 3 - \frac{d}{2}$,

Table 9.1 The Gaussian exponents. After March (2015).

α	β	γ	δ	η	ν
$2-\frac{d}{2}$	$\frac{1}{4}(d-2)$	1	$\dfrac{d+2}{d-2}$	0	$\frac{1}{2}$

Table 9.2 The approximate form for the critical exponents at sufficiently large n, in the limit $n \to \infty$, and in the case $d = 3$, $n \to \infty$. After March (2015).

α	$\dfrac{dn-4d-4n+16}{dn+2d-2n+8}$	$\dfrac{d-4}{d-2}$	-1
β	$\dfrac{3d+n-4}{dn+2d-2n+8}$	$\dfrac{1}{d-2}$	1
γ	1	1	
δ	$\dfrac{(d-1)n+5d+4}{3d+n-4}$	$d-1$	2
η	$\dfrac{(4-d)n-2d+8}{n+8}$	$4-d$	1
ν	$\dfrac{n+8}{(d-2)n+2d+8}$	$\dfrac{1}{d-2}$	1

Rushbrooke's relation becomes $\alpha+2\beta = 1$, when $\gamma = 1$ as in the ILM model.

the other critical exponents depend on d at sufficiently large n. The third column gives results for general d and γ in the same large n limit.

In summary, we have proved in this Section that the known critical exponents for the Gaussian model are embraced by ILM AdS/CFT string theory with the choice of their parameter $\nu_{\rm ILM} = d - 1$. Using admittedly approximate data, mainly from high temperature series expansion, we have proposed already the approximate form Eq. (9.117) for $\gamma(d, n)$ for $n = 1$, 2 and 3, with a_1 and b_1 known. Finally, we discuss string theory 'parameter' $\nu_{\rm ILM}$ modeled as a function of universality class and dimensionality, in a form appropriate at large n for arbitrary d.

9.5 Baxter model

9.5.1 *The Baxter or eight-vertex model*

The Baxter model in two dimensions (1971; 1972b) [see also (Baxter, 1972a, 1973a,b,c)], also called the eight-vertex model, which is a generalization of

Table 9.3 Critical exponents in the limit as the universality class, characterized by n, tends to infinity. Second column corresponds to $\gamma = 3 - \frac{d}{2}$ as in Kardar (2007). After March (2015).

	Case $\gamma = 3 - \frac{d}{2}$	General d and γ
α	$\frac{d-4}{d-2}$	$\frac{d-4}{d-2}$
β	$\frac{d^2-6d+12}{4(d-2)}$	$\frac{1}{2}\left(\frac{d}{d-2}-\gamma\right)$
γ	$3-\frac{d}{2}$	γ
δ	$\frac{-d^2+10d-12}{d^2-6d+12}$	$\frac{d+\gamma(d-2)}{d-\gamma(d-2)}$
η	$8-4d+\frac{1}{2}d^2$	$2-\gamma(d-2)$
ν	$\frac{1}{d-2}$	$\frac{1}{d-2}$

the Ising model, prompted (Luther and Peschel, 1974, 1975) to construct a relationship between the Baxter model in 2D and the Luttinger model in 1D. They then used it to calculate critical exponents for the Baxter model from appropriate Luttinger model correlation functions (see Appendix E for a self-contained account of the Luttinger model).

Since the work of (Schultz *et al.*, 1964), the intimate relation between the 2D Ising model and the noninteracting Fermion gas in 1D has been widely recognized. The solution of the Baxter model in 2D prompted Luther and Peschel (LP) to raise the question concerning a possible 1D Fermion equivalent of this model also. LP point out that the solutions of the Thirring (1958) and Luttinger (1963) models also exhibit nonuniversal behaviour similar to that of the Baxter model.

This led LP to discuss first the spin $\frac{1}{2}$ Heisenberg-Ising (HI) model in 1D. Already Sutherland (1970) had shown that it could be used to obtain orrelation functions at the critical point in the Baxter model. A new feature of the study of LP which we stress below is the use of a continuum generalization of the HI model and their construction of the continuum spin-$\frac{1}{2}$ algebra for this model. After these transformations, the HI Hamiltonian turns out to be the same as the Luttinger Hamiltonian, and the spin-$\frac{1}{2}$ operators, used previously in the calculation of the Luttinger model correlation functions (Luther and Peschel, 1974). Below, their calculation of these spin correlation functions is summarized, and related to the corresponding Baxter functions. This determines the critical exponents η and δ: a focal point in what follows.

When the temperature in the Baxter model is not equal to the critical temperature, the above equivalence leads to the appearance of a gap in the Luttinger model excitation spectrum, which in relation to the Thirring model is equivalent to a mass term. Assuming homogeneity, LP then calculate the exponents which involve this mass term, thereby completing the known critical exponents of the Baxter model.

For the special case of the Ising model, KP reproduce the usual exponents already discussed earlier in the present Chapter. In addition, the LP method (plus a Wick rotation) allows calcilation of the asymptotic behaviour of all higher spin functions as well. LP note the similarity to the results for the Thirring and Luttinger models. This is because LP find that the spin-density operator is essentially the square root of the free Fermion operator. With the Boson representation of these operators, the critical exponent calculations are reduced to a harmonic oscillator problem.

LP next note that the value they find for the exponent γ differs somewhat from previous calculations for the Baxter model on a lattice (Johnson *et al.*, 1973). LP comment that this is to be expected, since they use the continuum Fermi model, rather than the lattice Fermion model, for the transfer matrix. Comparing their two results thereby provides some measure of the accuracy of the continuum Fermion approximation for models of the interacting Fermion gas on a lattice.

One of the consequences of having calculated all (electric) critical exponents is that the coupling constant may be eliminated between any two, which led them to the additional scaling law

$$4\beta = 2\nu - 1. \tag{9.131}$$

LP argue that this result, Eq. (9.131), is universal, independent of the particular parametrization of the 2D model, and is consistent with a conjecture for β by Baxter and Kelland (1974) for the lattice model. This new law reduces the parameters required to specify all exponents to one, and this remaining parameter is model dependent. The three different models referred to above, *viz.* Thirring, Luttinger, and Baxter, differ concerning this additional parameter.

LP next note that an intermediate step in their discussion involves solving the HI model now generalized to the continuum case. They discuss both results of correlation functions and their exponents. In addition, LP derive the exponents associated with basal-plane anisotropy. They find that the exponents depend on the ration of longitudinal to transverse exchange. Their result at the XY point is that the exponent of the transverse spin correlation function is $\frac{1}{2}$, confirming previous results (see McCoy, 1968), increasing to 1.06 at the Heisenberg point. The XY result turns out to be the same in both lattice and continuum approaches.

The LP discussion, we should stress here, uses the Ising representation of the Baxter model (Kadanoff and Wegner, 1971; Wu, 1971), in which one regards the system as consistent of two interpenetrating Ising lattices, with Ising variables say μ_i and μ_i' associated with each site. These are related to the original Baxter variables α_i by $\alpha_i = \mu_i \mu_{i+1}$, where $i+1$ is nearest-neighbour to i. Additional to the usual nearest-neighbour two-spin interactions $-J\mu_i\mu_{i+1}'$ and $-J'\mu_i'\mu_{i+1}$, there is a four-spin interaction $J_4\mu_i\mu_{i+1}\mu_i'\mu_{i+1}'$ among four nearest-neighbour spins. This constitutes the coupling between the two Ising lattices.

Correlation functions can be calculated by diagonalization of the transfer matrix T. LP use a simpler approach based on finding an associated Hamiltonian operator $\mathcal{H}$ which commutes with T. Since T and $\mathcal{H}$ then share the same eigenvectors, correlation functions are then accessible by appropriate choice of the eigenstate.

For the eight-vertex model under discussion, Sutherland (1970) has proved that the 1D spin-$\frac{1}{2}$ XYZ Hamiltonian

$$\mathcal{H}_{XYZ} = -\sum_i (J_x S_i^x S_{i+1}^x + J_y S_i^y S_{i+1}^y + J_z S_i^z S_{i+1}^z) \tag{9.132}$$

commutes with T, where the coupling constants, for simplicity, have been taken such that

$$J_x = 1, \tag{9.133a}$$

$$J_y = \sinh 2K \sinh 2K' + \cosh 2K \cosh 2K' \tanh 2L, \tag{9.133b}$$

$$J_z = \tanh 2L. \tag{9.133c}$$

Here, $K = J/k_{\mathrm{B}}T$, $K' = J'/k_{\mathrm{B}}T$, $L = J_4/k_{\mathrm{B}}T$. The XY model (corresponding to universality class $n = 2$) is recovered in the case that $J_4 = 0$, with corresponding Hamiltonian

$$\mathcal{H}_{XY} = -\sum_i (S_i^x S_{i+1}^x + \sinh 2K \sinh 2K' S_i^y S_{i+1}^y). \tag{9.134}$$

According to Onsager's solution, set out in Sec. 9.1, the phase transition takes place when $\sinh 2K \sinh 2K' = 1$. This is the point where $\mathcal{H}_{XY}$ becomes isotropic. As LP then note, correlation functions for $\mathcal{H}_{XY}$ then reduce to those corresponding to two non-interacting Ising models. For $J_z \neq$

0, the critical temperature is still determined by requiring the equality of the two coupling constants in $\mathcal{H}_{XYZ}$. As LP then record, the Hamiltonian at the critical temperature therefore becomes

$$\mathcal{H}_{XYZ} = -\sum_i (S_i^x S_{i+1}^x + S_i^y S_{i+1}^y + J_z S_i^z S_{i+1}^z). \tag{9.135}$$

Taking account of the $y \leftrightarrow z$ interchange, the correlation function along a line of the square lattice in the Baxter model becomes

$$\langle \alpha_i \alpha_{i+n} \rangle = 4\langle \theta | S_i^y S_{i+n}^y | \theta \rangle, \tag{9.136}$$

where $|\theta\rangle$ is the ground state of $\mathcal{H}_{XYZ}$. Analogously, multioperator correlation functions are given by

$$\langle \alpha_{i_1} \alpha_{i_2} \cdots \alpha_{i_n} \rangle = 2^n \langle \theta | S_{i_1}^y S_{i_2}^y \cdots S_{i_n}^y | \theta \rangle. \tag{9.137}$$

Calculation of these correlation functions then enables one to determine the correlation functions also in the Baxter model, and to evaluate the corresponding electric critical exponents.

9.5.2 *Correlation functions for the XYZ model having spin $\frac{1}{2}$*

The calculation of the correlation functions for the spin-$\frac{1}{2}$ XYZ model in 1D was next carried out by LP by relating the XYZ model to that of Luttinger for an interacting one-dimensional spinless Fermion gas. Then the solutions of this model can be utilized to obtain the desired asymptotic properties. Normally, the spin-$\frac{1}{2}$ XYZ model is defined using a lattice and LP show how its continuum generalization can be used to allow application of the continuum Fermion field theory.

LP then write the Hamiltonian for the lattice model as

$$\mathcal{H}_{XYZ} = -\sum_i (S_i^x S_{i+1}^x + S_i^y S_{i+1}^y + J_z S_i^z S_{i+1}^z) + m_0 (S_i^x S_{i+1}^x - S_i^y S_{i+1}^y). \tag{9.138}$$

Here, $J_x = 1$ is the transverse exchange, J_z the corresponding longitudinal quantity, m_0 denotes the basal-plane anisotropy, $\mathbf{S}_i = (S_i^x, S_i^y, S_i^z)$ is a spin-$\frac{1}{2}$ operator, and i runs over the N sites of a ring. Using the Jordan-Wigner transformation to Fermion operators a_i yields (Jordan and Wigner, 1928)

$$S_i^+ = S_i^x + iS_i^y = a_i^\dagger \exp\left(i\pi \sum_{j=1}^{i-1} a_j^\dagger a_j \right), \tag{9.139a}$$

$$S_i^z = a_i^\dagger a_i - \frac{1}{2}. \tag{9.139b}$$

Fourier transforming, with

$$a_k = \frac{1}{\sqrt{N}} \sum_j e^{ikx_j} a_j, \tag{9.140}$$

the Hamiltonian takes then the form

$$\mathcal{H}_{XYZ} = -\sum_k \cos(ks)[a_k^\dagger a_k + J_z N^{-1} \rho(k)\rho(-k)] + m_0 \sum_k e^{iks}(a_k^\dagger a_{-k}^\dagger + \text{H.c.}). \tag{9.141}$$

Here, s denotes the lattice constant, the sum over k being restricted to the first Brillouin zone, and

$$\rho(k) = \frac{1}{N} \sum_j a_j^\dagger a_j e^{ikx_j} \tag{9.142}$$

is the density operator.

In zero external magnetic field ($H = 0$), there is no net magnetization, the mean Fermion number being then equal to $\frac{1}{2}N$, corresponding to a half-filled band. The pairing term containing m_0 in Eqs. (9.138) and (9.141) is seen to be proportional to $T - T_c$, which is responsible of a 'mass gap' in the excitation spectrum. In the limit of no anisotropy ($J_z = 0$) and vanishing interaction ($m_0 = 0$), the single-particle excitation spectrum is simply given by $-\cos(ks)$, where the states with $-\frac{\pi}{2s} < k < \frac{\pi}{2s}$ are occupied. The Fermi points are thus given by $k_F s = \pm\frac{\pi}{2}$, and one expects that the asymptotic properties of the correlation functions are governed by states close to those points. One is thus led to consider a linearized spectrum around k_F, with group velocities ± 1, exactly as in the Luttinger model. One then introduces the operators $a_{1,k}$ ($a_{2,k}$) to describe low-lying excitations close to the Fermi point with positive (negative) group velocity, and the associated fields $\psi_\alpha(x) = L^{-1/2} \sum_k a_{\alpha,k} e^{ikx}$ ($\alpha = 1, 2$), where L is the length of the chain.

It then remains to associate the density-density coupling constant J_z in Eq. (9.141) to an appropriate interaction within the Luttinger model. This is achieved by requiring that the matrix elements of the density-density interaction term in Eq. (9.141) for states near the Fermi points be reproduced in the continuum model. LP first decouple $\rho(k) = \rho_1(k) + \rho_2(k)$, where $\rho_\alpha(k) = \sum_p a_{\alpha,k+p}^\dagger a_{\alpha,k}$ is the continuum Fermion density operator around the Fermi point labeled α. While this procedure correctly accounts for the small-momentum transfer scattering processes, it excludes backward scattering, characterized by a momentum transfer $k \simeq 2k_F = \frac{\pi}{s}$. For the spinless Fermion case, backward scattering can be included by allowing terms such as $-\rho_1(k)\rho_2(-k)$, with $k \simeq 0$. At $T = T_c$ (*i.e.* $m_0 = 0$), LP then arrive at the expression

$$\mathcal{H}_{XYZ} = \sum_k k(a_{1,k}^\dagger a_{1,k} - a_{2,k}^\dagger a_{2,k}) - \frac{4J_z}{N} \sum_k \rho_1(k)\rho_2(-k). \qquad (9.143)$$

Close to T_c ($m_0 \neq 0$), the asymptotic behaviour of the correlation functions in the case $J_z \neq 0$ can be obtained by mapping the interaction term in Eq. (9.141) as

$$m_0 \sum_k e^{iks}(a_k^\dagger a_{-k}^\dagger + \text{H.c.}) \mapsto 2m_0 i \sum_{k>0} a_{1,k}^\dagger a_{2,-k}^\dagger + \text{H.c.}, \qquad (9.144)$$

which correctly accounts for the mixing of the states around the two Fermi points. Eqs. (9.143) and (9.144) then complete the program of finding the continuum Fermion Hamiltonian, equivalent to the XYZ model.

9.5.3 *Generalization to the continuum*

To effect the generalization to the continuum, LP next find the spin operators, the resulting representation for the continuum case being

$$S^-(x) = \frac{1}{\sqrt{2s}}[\psi_1(x) + \psi_2(x)]e^{-N(x)}, \qquad (9.145\text{a})$$

$$S^+(x) = [S^-(x)]^\dagger, \qquad (9.145\text{b})$$

$$2S_z(x) = \rho_1(x) + \rho_2(x) + \psi_1^\dagger(x)\psi_2(x) + \psi_2^\dagger(x)\psi_1(x). \qquad (9.145\text{c})$$

Then, LP express the spin correlation functions as correlation functions in the continuum Fermion problem. Their result is

$$-4s^2\langle S^y(x,t)S^y(0)\rangle = \langle[\psi(x,t)e^{-N(x,t)} - \psi^\dagger(x,t)e^{+N(x,t)}] \times[\psi e^{-N} - \psi^\dagger e^{+N}]\rangle, \qquad (9.146\text{a})$$

$$4\langle S^z(x,t)S^z(0)\rangle = \langle\rho_1(x,t)\rho_1\rangle + \langle\rho_2(x,t)\rho_2\rangle + \langle\psi_1^\dagger(x,t)\psi_2(x,t)\psi_2^\dagger\psi_1\rangle + \langle\psi_2^\dagger(x,t)\psi_1(x,t)\psi_1^\dagger\psi_2\rangle. \qquad (9.146\text{b})$$

Here, the time evolution and averaging is with respect to the Hamiltonian, Eq. (9.143), plus the mass term, Eq. (9.144), and $\psi(x) = \psi_1(x) + \psi_2(x)$.

These functions, following LP, can be evaluated since the density operators satisfy Boson algebra. Using the Boson representation (Luther and Peschel, 1974; Mattis, 1974) for the operators $\psi_1(x)$ and $\psi_2(x)$, LP obtain

$$\psi(x)e^{-N(x)} = \frac{1}{\sqrt{2\pi\alpha}}(e^{\frac{3}{2}\phi_2(x) - \frac{1}{2}\phi_1(x) + 2ik_F x + \tau} + e^{\frac{1}{2}\phi_1(x) + \phi_2(x) + \tau}), \qquad (9.147)$$

where

$$\phi_{1,2}(x) = \mp\frac{2\pi}{L}\sum_k \rho_{1,2}(k)\frac{e^{-ikx}}{k}e^{-\frac{1}{2}\alpha|k|}, \tag{9.148}$$

the upper (lower) sign going with the ρ_1 (ρ_2) operator, α being a cutoff parameter of the order of the bandwidth $\sim s$, and $2\tau = \phi_2(0) - \phi_1(0)$. LP point out that it has been assumed $J_x > 0$. If $J_x < 0$, the '1' and '2' single-particle branches are interchanged and one must take $k_{\mathrm{F}} \mapsto -k_{\mathrm{F}}$ in the definition of the field operators. As a consequence, the $e^{2ik_{\mathrm{F}}x}$ phase factor annulls the first term of Eq. (9.147), but appears in the second term. LP use also a result by Lieb and Mattis that Eq. (9.143) can be expressed entirely in terms of Boson density operators

$$\mathcal{H}_{XYZ} = \frac{2\pi}{L}\sum_k[\rho_1(-k)\rho_1(k) + \rho_2(-k)\rho_2(k)] - \frac{4J_z}{L}\sum_k \rho_1(k)\rho_2(-k), \tag{9.149}$$

where the density variables obey 'bosonic' commutation relations, *i.e.*

$$[\rho_1(-k), \rho_1(k')] = [\rho_2(k'), \rho_2(-k)] = \frac{L}{2\pi}k\delta_{kk'}. \tag{9.150}$$

The Hamiltonian, Eq. (9.149), is seen to be bilinear in 'bosonic' operators, whereas the operators in the correlation functions are either 'bosonic' or exponentials thereof, which allows their explicit evaluation.

It follows then that the leading asymptotic behaviour of the transverse equal-time correlation function takes the form

$$4\pi s^2\langle S^+(x)S^-(0)\rangle = \langle e^{-\frac{1}{2}\phi_1(x)-\frac{1}{2}\phi_2(x)}e^{+\frac{1}{2}\phi_1+\frac{1}{2}\phi_2}\rangle, \tag{9.151}$$

where averages now involve Eq. (9.149). Using the procedure of Sutherland, one must first of all diagonalize the Hamiltonian, Eq. (9.149), via the 'rotation'

$$\phi_1(x) \mapsto \phi_1(x)\cosh\varphi - \phi_2(x)\sinh\varphi, \tag{9.152}$$

where $\tanh 2\varphi = 2J_z/\pi$. It follows that $\phi_1(x)+\phi_2(x) \mapsto e^{-\varphi}[\phi_1(x)+\phi_2(x)]$, and the Hamiltonian is just the sum of the '1' and '2' Boson energies. As a result, one can obtain

$$\begin{aligned}\exp&\left(\frac{1}{2}\theta\langle\phi_1^2 - \phi_1(x)\phi_1\rangle + \frac{1}{2}\theta\langle\phi_2^2 - \phi_2(x)\phi_2\rangle\right)\\ &= \exp\left(\frac{1}{2}\theta\frac{2\pi}{L}\sum_{k>0}\frac{e^{ikx}-1}{k} + \mathrm{H.c.}\right) \sim \left(\frac{s}{x}\right)^{\theta},\end{aligned} \tag{9.153}$$

where

$$2\theta = e^{-2\varphi} = \left(\frac{\pi - 2J_z}{\pi + 2J_z}\right)^{\frac{1}{2}}, \tag{9.154}$$

use has been made of Eq. (9.147) with $\alpha \sim s$, and $x \gg s$ has been assumed. The other correlation functions may be similarly obtained, yielding the results

$$s^2\langle S^y(x)S^y\rangle \sim \left(\frac{s}{x}\right)^{\theta} + \ldots, \tag{9.155a}$$

$$s^2\langle S^z(x)S^z\rangle \sim \left(\frac{s}{x}\right)^2 + e^{2ik_F x}\left(\frac{s}{x}\right)^{\theta^{-1}} + \ldots, \tag{9.155b}$$

where only dimensional factors are kept. For the free-particle case, $J_z = 0$ and $m_0 = 0$. Then the problem reduces to the isotropic XY model, studied by Lieb *et al.* (1961), and by McCoy (1968).

The multispin correlation functions in the asymptotic regime can also be obtained (including the time dependence also). The static functions involve only the exponent denoted below by θ, since the properties that averages of exponentiated free Boson operators in the free density matrix can be expressed as

$$\langle e^{-p\phi_1(x_1)}e^{p\phi_1(x_2)}e^{-p\phi_1(x_3)}\cdots e^{p\phi_1(x_{2n})}\rangle_1$$
$$= \exp\sum_{i<j}^{2n}(-1)^{i-j}p^2\log\left(\frac{x_j - x_i}{s}\right). \tag{9.156}$$

In Eq. (9.156), $x_j - x_i \gg x$, and the ordering $x_1 < x_2 < \ldots < x_{2n}$ has been assumed. LP make use of this property to derive the higher-order correlations in these models.

Starting from the noninteracting-particle case ($J_z = 0$) (McCoy, 1968), a nonvanishing value of m_0 is seen to introduce a BCS-like gap (or mass) term into the Fermion spectrum. In the general (interacting) case, the momentum sum in Eq. (9.153) will be modified for $k < m$, being m the renormalized gap or mass. Consequently, the correlation function will change its behaviour in the length scale as $m^{-\theta}$. The assumption of homogeneity then implies $\langle S^+(x)S^-\rangle \propto m^{-\theta}$, as $x \to \infty$, if long-range order exists. In the special case of the XY model with $J_z = 0$, one explicitly has $\theta = \frac{1}{2}$ and $m = m_0$, since there are no interactions to renormalize the free-particle gap.

When interactions are switched on, one expects that $m \sim m_0^{\nu}$. One way to justify such a behaviour is to start from the scaling assumption

$\langle \rho_1(x)\rho_1\rangle \sim x^{-2}f(mx)$, and invoke then perturbation theory to express the density-density correlation function as a power series in m_0 as

$$\langle \rho_1(x)\rho_1\rangle = \sum_n m_0^n \int_0^\infty dt_1 \cdots \int_0^{t_{n-1}} dt_n \int dx_1 \cdots \int dx_n \\ \times x \left\langle \rho_1(x)\rho_1 \left(e^{-\phi_1(t_1,x_1)-\phi_2(t_1,x_1)} + \text{H.c.} \right) \right. \\ \left. \cdots \left(e^{-\phi_1(t_n,x_n)-\phi_2(t_n,x_n)} + \text{H.c.} \right) \right\rangle, \quad (9.157)$$

which can be evaluated by repeated application of Eq. (9.156). A dimensional analysis of the integrals involved yields eventually

$$2\nu = (1-\theta)^{-1}. \quad (9.158)$$

The above value for ν allows to determine the long-range order critical exponent β as

$$2\beta = \theta(1-\theta)^{-1}, \quad (9.159)$$

where β is defined from the asymptotic ($x \to \infty$) behaviour of the correlation function $\langle S^+(x)S^-\rangle \sim \langle (S^y)^s\rangle \sim |m_0|^{2\beta}$, which should coincide with $\langle S^+(x)S^-\rangle \sim m^\theta$, in the same limit. The relation of these critical exponents with the critical exponents in the Ising or Baxter models will be discussed in Sec. 9.5.4 immediately below.

LP caution that it is not immediately obvious that the critical behaviour of the continuum generalization of the XYZ model should coincide with its lattice counterpart. In general, also processes involving states far from the Fermi points might be relevant, and their contributions should be evaluated appropriately. However, the noninteracting limit ($J_z = 0$), corresponding to the Ising model, is described exactly, as the absence of interaction does not give rise to any further microscopic length scale.

9.5.4 *Critical exponents*

Knowledge of the asymptotic behaviour of the correlation functions discussed above enables various critical exponents of the Baxter model to be found. Identifying θ in the notation of LP above with the critical exponent η associated with an order parameter, using Fisher's scaling relation $\gamma = \nu(2-\eta)$ and Eqs. (9.158) and (9.159) above, one finds the exponent γ as

$$\gamma = \frac{1}{2}\frac{2-\theta}{1-\theta}. \quad (9.160)$$

The exponent δ is readily obtained from the Hubbard-Schofield (1972) relation (see also Zhang and March, 2013) as

$$\delta = \frac{d + (2 - \eta)}{d - (2 - \eta)} \tag{9.161}$$

in d dimensions [cf. Eq. (9.106) above]. This yields immediately that $\delta(d = 2)$ for the Baxter model is given by $\delta = (4 - \eta)/\eta$.

Chapter 10

Relativistic fields

Here, we review several wave equations for relativistic particles, with their account of spin, including Dirac and Pauli equations. Some of the known exact solutions of the Dirac equation are revisited, special cases being those in a central electrostatic field, and in an external electromagnetic field, either uniform and constant, or in the form of a plane wave. Various limits are examined, both in the nonrelativistic and semiclassical cases. Relativistic effects as accounted by the Dirac equation are relevant for atomic spectra, in quantum electrodynamics, in the density functional theory of heavy-nuclei atoms, and more recently in the description of the low-energy electronic and transport properties of two-dimensional graphene and related compounds. The spinless Salpeter equation is briefly mentioned at the end of the Chapter.

More detailed information on this general topic can be found *e.g.* in the fairly recent monographs by Bagrov and Gitman (1990, 2014).

10.1 Dirac wave equation

In order to generalize Schrödinger's equation to include relativistic effects, one must start from the correct relativistic expression for the energy of a free particle, *viz.*

$$(\text{Energy})^2 = c^2p^2 + m^2c^4, \tag{10.1}$$

where m is the particle's rest mass, p its momentum, and c the velocity of light in vacuum. At small momenta, this yields

$$\text{Energy} = mc^2 + \frac{p^2}{2m} + \dots, \tag{10.2}$$

the first term being the energy associated with the rest mass according to Einstein's mass-energy equivalence, while the second term is just the non-relativistic kinetic energy.

In order to make use of the correspondence principle of quantum mechanics, which identifies energy E with the Hamiltonian operator H, one then needs not the square of the energy, as in Eq. (10.1), but the energy itself. Dirac (1928) (see March, 1975b, for a review) therefore proposed that one writes the quantity $p^2 + m^2c^2$ appearing on the right-hand side of Eq. (10.1) as a perfect square, *i.e.*

$$p^2 + m^2c^2 = p_x^2 + p_y^2 + p_z^2 + m^2c^2 = (\alpha_x p_x + \alpha_y p_y + \alpha_z p_z + \alpha_m mc)^2. \quad (10.3)$$

It is evident that the α's defined in Eq. (10.3) cannot be ordinary scalars. Performing the square on the right-hand side of Eq. (10.3), and equating coefficients with the left-hand side, one therefore obtains that

$$\alpha_x^2 = \alpha_y^2 = \alpha_z^2 = \alpha_m^2 = 1 \quad (10.4a)$$

$$\begin{aligned} \alpha_x\alpha_y &= -\alpha_y\alpha_x, & \alpha_x\alpha_z &= -\alpha_z\alpha_x, & \alpha_x\alpha_m &= -\alpha_m\alpha_x, \\ \alpha_y\alpha_z &= -\alpha_z\alpha_y, & \alpha_y\alpha_m &= -\alpha_m\alpha_y, & \alpha_z\alpha_m &= -\alpha_m\alpha_z. \end{aligned} \quad (10.4b)$$

The α's cannot be c-numbers, as they do not commute, but they must rather be operators. Provided they satisfy Eqs. (10.4), one may write the Hamiltonian H for a particle moving in a potential field V as

$$H = -c\alpha_x p_x - c\alpha_y p_y - c\alpha_z p_z - \alpha_m mc^2 + V. \quad (10.5)$$

Here, we have chosen the negative sign in taking the square root of Eq. (10.1). This is a pure convention, and does not alter the physical content of Eq. (10.5). Finally, one makes use of the operator substitution $\mathbf{p} \mapsto (\hbar/i)\nabla$ to obtain the Dirac Hamiltonian for an electron moving in a field of potential energy $V(\mathbf{r})$ as

$$H = V + ic\hbar\alpha_x\frac{\partial}{\partial x} + ic\hbar\alpha_y\frac{\partial}{\partial y} + ic\hbar\alpha_z\frac{\partial}{\partial z} - \alpha_m mc^2. \quad (10.6)$$

While it is not strictly required by the theory that a specific form for the α's be chosen, it is usually more practical to do so. As in the Pauli spin theory, it is convenient to choose the form of such operators as matrices. It can then be proved that the smallest order with which a suitable set of matrices can satisfy Eqs. (10.4) is four. With Dirac, one then usually adopts the following (not unique) choice:[1]

$$\alpha_x = \begin{pmatrix} 0&0&0&1 \\ 0&0&1&0 \\ 0&1&0&0 \\ 1&0&0&0 \end{pmatrix}, \qquad \alpha_y = \begin{pmatrix} 0&0&0&-i \\ 0&0&i&0 \\ 0&-i&0&0 \\ i&0&0&0 \end{pmatrix},$$
$$\alpha_z = \begin{pmatrix} 0&0&1&0 \\ 0&0&0&-1 \\ 1&0&0&0 \\ 0&-1&0&0 \end{pmatrix}, \quad \alpha_m = \begin{pmatrix} 1&0&0&0 \\ 0&1&0&0 \\ 0&0&-1&0 \\ 0&0&0&-1 \end{pmatrix}. \tag{10.7}$$

Identifying then H with the time-evolution operator, $i(\partial/\partial t) \mapsto H$, as for the Schrödinger equation, the wave function must be a 4-component column vector, *viz.* $\Psi \equiv (\psi_1, \psi_2, \psi_3, \psi_4)^\top$. The stationary Dirac's equation is then equivalent to the system of coupled equations

$$\frac{i}{\hbar}\left(\frac{E-V}{c} + mc\right)\psi_1 + \left(\frac{\partial}{\partial x} - i\frac{\partial}{\partial y}\right)\psi_4 + \frac{\partial \psi_3}{\partial z} = 0, \tag{10.8a}$$

$$\frac{i}{\hbar}\left(\frac{E-V}{c} + mc\right)\psi_2 + \left(\frac{\partial}{\partial x} + i\frac{\partial}{\partial y}\right)\psi_4 - \frac{\partial \psi_3}{\partial z} = 0, \tag{10.8b}$$

$$\frac{i}{\hbar}\left(\frac{E-V}{c} - mc\right)\psi_3 + \left(\frac{\partial}{\partial x} - i\frac{\partial}{\partial y}\right)\psi_4 + \frac{\partial \psi_3}{\partial z} = 0, \tag{10.8c}$$

$$\frac{i}{\hbar}\left(\frac{E-V}{c} - mc\right)\psi_4 + \left(\frac{\partial}{\partial x} + i\frac{\partial}{\partial y}\right)\psi_4 - \frac{\partial \psi_3}{\partial z} = 0. \tag{10.8d}$$

10.2 Central field solutions of the Dirac's equation

When the potential is spherically symmetric, *i.e.* $V(\mathbf{r}) = V(r)$, with $r = |\mathbf{r}|$, as is the case of the electrostatic potential generated by a point charge, such as an isolated atom, one usually takes advantage of spherical coordinated (r, θ, ϕ). One finds

$$\frac{\partial}{\partial x} \pm i\frac{\partial}{\partial y} = e^{\pm i\phi}\left(\sin\theta\frac{\partial}{\partial r} + \frac{\cos\theta}{r}\frac{\partial}{\partial\theta} \pm \frac{i}{r\sin\theta}\frac{\partial}{\partial\phi}\right), \tag{10.9a}$$

$$\frac{\partial}{\partial z} = \cos\theta - \frac{\sin\theta}{r}\frac{\partial}{\partial\theta}. \tag{10.9b}$$

[1] For later reference [cf. Sec. 10.4], the Dirac α matrices are related to the Dirac matrices (γ^0, γ^i), $i = 1, 2, 3$, by $\alpha_m = \gamma^0$, $\alpha_i = \gamma^0\gamma^i$. One further finds $\gamma^0 = \begin{pmatrix} 1 & 0 \\ 0 & -1 \end{pmatrix}$ and $\gamma^i = \begin{pmatrix} 0 & \sigma_i \\ -\sigma_i & 0 \end{pmatrix}$, where 1 and σ_i here denote the 2×2 identity matrix and the Pauli matrices, respectively.

Operating the above substitutions in Eqs. (10.8), one finds that the resulting equations can then be separated, as in the non-relativistic case. However, since the electron spin and magnetic moment is now included in the Dirac equation itself, unlike the Schrödinger case, where the Pauli spin is not included, the total angular momentum **J** plays the role for the Dirac equation of the angular momentum **L** for the Schrödinger equation. One therefore looks for simultaneous solutions also of

$$J^2\Psi = j(j+1)\hbar^2\Psi, \qquad j = \frac{1}{2}, \frac{3}{2}, \ldots \tag{10.10a}$$

$$J_z\Psi = m\hbar\Psi, \qquad m = \pm\frac{1}{2}, \pm\frac{3}{2}, \ldots, \text{ with } |m| \leq j. \tag{10.10b}$$

The solutions Ψ for a given j and m then take the form

$$\psi_1 = f(r)\left(\frac{j+1-m}{2j+2}\right)^{\frac{1}{2}} Y_{j+\frac{1}{2},m-\frac{1}{2}}(\theta,\phi) + g(r)\left(\frac{j+m}{2j}\right)^{\frac{1}{2}} Y_{j-\frac{1}{2},m-\frac{1}{2}}(\theta,\phi), \tag{10.11a}$$

$$\psi_2 = -f(r)\left(\frac{j+1+m}{2j+2}\right)^{\frac{1}{2}} Y_{j+\frac{1}{2},m+\frac{1}{2}}(\theta,\phi) + g(r)\left(\frac{j-m}{2j}\right)^{\frac{1}{2}} Y_{j-\frac{1}{2},m+\frac{1}{2}}(\theta,\phi), \tag{10.11b}$$

$$\psi_3 = F(r)\left(\frac{j+1-m}{2j+2}\right)^{\frac{1}{2}} Y_{j+\frac{1}{2},m-\frac{1}{2}}(\theta,\phi) + G(r)\left(\frac{j+m}{2j}\right)^{\frac{1}{2}} Y_{j-\frac{1}{2},m-\frac{1}{2}}(\theta,\phi), \tag{10.11c}$$

$$\psi_4 = -F(r)\left(\frac{j+1+m}{2j+2}\right)^{\frac{1}{2}} Y_{j+\frac{1}{2},m+\frac{1}{2}}(\theta,\phi) + G(r)\left(\frac{j-m}{2j}\right)^{\frac{1}{2}} Y_{j-\frac{1}{2},m+\frac{1}{2}}(\theta,\phi), \tag{10.11d}$$

where $Y_{\ell m}(\theta,\phi)$ are spherical harmonics, which are related to the associated Legendre functions by (Gradshteyn and Ryzhik, 1994)

$$Y_{\ell m}(\theta,\phi) = (-1)^m \left(\frac{(2\ell+1)(\ell-m)!}{4\pi(\ell+m)!}\right)^{\frac{1}{2}} P_\ell^m(\cos\theta)e^{im\phi}. \tag{10.12}$$

The angular and spin dependence of Eqs. (10.11) can in turn be also expressed in terms of the so-called spinor spherical harmonics, which are basically proportional to the spherical harmonics, through appropriate Clebsch-Gordan coefficients (Kleinert, 2016). Substituting these results into Eqs. (10.8) written in spherical polar coordinates, one can show that the four radial functions f, g, F, and G satisfy the radial coupled equations

$$\frac{df}{dr} + \left(j + \frac{3}{2}\right)\frac{f}{r} + \frac{i}{\hbar c}[E - V(r) - mc^2]G = 0, \tag{10.13a}$$

$$\frac{dG}{dr} - \left(j - \frac{1}{2}\right)\frac{G}{r} + \frac{i}{\hbar c}[E - V(r) + mc^2]f = 0, \tag{10.13b}$$

$$\frac{dF}{dr} + \left(j + \frac{3}{2}\right)\frac{F}{r} + \frac{i}{\hbar c}[E - V(r) + mc^2]g = 0, \tag{10.13c}$$

$$\frac{dg}{dr} - \left(j - \frac{1}{2}\right)\frac{g}{r} + \frac{i}{\hbar c}[E - V(r) - mc^2]F = 0. \tag{10.13d}$$

Given a spherically symmetric potential $V(r)$, Eqs. (10.13) are then coupled first-order ordinary differential equations. Whenever a solution is not available analytically (as in Sec. 10.2.1 immediately below), a viable numerical method may be iterative, in the Hartree sense.

10.2.1 *Dirac equation for a hydrogenic atom*

In the case of a Coulomb field for a hydrogenic atom, $V(r) = -Ze^2/r$, Eqs. (10.13) can be solved analytically, their solution yielding the celebrated Sommerfeld formula for the fine structure energy of an atom [Eq. (10.26) below]. One looks for solutions of the first pair of Eqs. (10.13), the second pair being solved in essentially the same way. It is then convenient to perform the change of variables

$$\zeta = \frac{Ze^2}{\hbar c}, \quad \epsilon = \frac{E}{mc^2}, \quad \lambda = \frac{mc}{\hbar}\sqrt{1-\epsilon^2}, \quad \rho = 2\lambda r, \tag{10.14a}$$

$$f(r) = i\sqrt{1-\epsilon}e^{-\rho/2}\rho^{\gamma-1}[f_1(\rho) - f_2(\rho)], \tag{10.14b}$$

$$G(r) = i\sqrt{1+\epsilon}e^{-\rho/2}\rho^{\gamma-1}[f_1(\rho) + f_2(\rho)]. \tag{10.14c}$$

Then, the first pair of Eqs. (10.13) read

$$\frac{df_1}{d\rho} = \left[1 - \frac{1}{\rho}\left(\gamma + \frac{\zeta\epsilon}{\sqrt{1-\epsilon^2}}\right)\right] f_1 + \frac{1}{\rho}\left[j + \frac{1}{2} - \frac{\zeta}{\sqrt{1-\epsilon^2}}\right] f_2, \tag{10.15a}$$

$$\frac{df_2}{d\rho} = \frac{1}{\rho}\left[j + \frac{1}{2} + \frac{\zeta}{\sqrt{1-\epsilon^2}}\right] f_1 - \frac{1}{\rho}\left[\gamma - \frac{\zeta\epsilon}{\sqrt{1-\epsilon^2}}\right] f_2. \tag{10.15b}$$

Assuming for $f_1(\rho)$ and $f_2(\rho)$ power series

$$f_1(\rho) = \sum_{s=0}^{\infty} c_s\rho^s, \qquad f_2(\rho) = \sum_{s=0}^{\infty} d_s\rho^s, \tag{10.16}$$

substituting in Eqs. (10.15), and equating powers of ρ^{s-1}, one obtains the recurrence relations for the coefficients

$$\left(\gamma + s + \frac{\zeta\epsilon}{\sqrt{1-\epsilon^2}}\right) c_s - \left(j + \frac{1}{2} - \frac{\zeta}{\sqrt{1-\epsilon^2}}\right) d_s = c_{s-1}, \quad (10.17a)$$

$$\left(j + \frac{1}{2} + \frac{\zeta}{\sqrt{1-\epsilon^2}}\right) c_s - \left(\gamma + s - \frac{\zeta\epsilon}{\sqrt{1-\epsilon^2}}\right) d_s = 0. \quad (10.17b)$$

Since we are considering only positive powers of ρ, for $s = 0$ one can take $c_{-1} = 0$, to obtain

$$\left(\gamma + \frac{\zeta\epsilon}{\sqrt{1-\epsilon^2}}\right) c_0 - \left(j + \frac{1}{2} - \frac{\zeta}{\sqrt{1-\epsilon^2}}\right) d_0 = 0, \quad (10.18a)$$

$$\left(j + \frac{1}{2} + \frac{\zeta}{\sqrt{1-\epsilon^2}}\right) c_0 - \left(\gamma - \frac{\zeta\epsilon}{\sqrt{1-\epsilon^2}}\right) d_0 = 0. \quad (10.18b)$$

This is a homogeneous system of linear equations, which admits non-vanishing solutions if and only if

$$\gamma = \sqrt{\left(j + \frac{1}{2}\right)^2 - \zeta^2}, \quad (10.19)$$

the negative root leading to an unnormalizable, and therefore unacceptable, wave function. Since $j + \frac{1}{2} \geq 1$ and $\zeta = Z(e^2/\hbar c) \simeq Z/137 < 1$, it follows that γ is always a real quantity.

Eliminating d_s from Eqs. (10.17), one gets the recurrence relation

$$c_s = \frac{\gamma + s - \zeta\epsilon/\sqrt{1-\epsilon^2}}{(\gamma+s)^2 - \left(j+\frac{1}{2}\right)^2 + \zeta^2} c_{s-1}. \quad (10.20)$$

As in the non-relativistic case, the infinite series Eq. (10.16) for $f_1(\rho)$ will actually be truncated, and $f_1(\rho)$ will be a polynomial

$$f_1(\rho) = \sum_{s=0}^{n'-1} c_s \rho^s, \quad (10.21)$$

in which the last term is non-zero, if

$$\gamma + n' - \frac{\zeta\epsilon}{\sqrt{1-\epsilon^2}} = 0, \quad (10.22)$$

which can be solved for ϵ to yield

$$\epsilon = \pm \frac{\gamma + n'}{\sqrt{\zeta^2 + (\gamma + n')^2}}. \quad (10.23)$$

Taking the positive root, and defining the principal quantum number n as

$$n = n' + j + \frac{1}{2}, \quad (10.24)$$

one finds the result

$$E = mc^2 \left(1 + \frac{\zeta^2}{\left(n - j - \frac{1}{2} + \gamma \right)^2} \right)^{-\frac{1}{2}} . \tag{10.25}$$

Substituting γ from Eq. (10.19), one finds the celebrated Sommerfeld fine structure formula

$$\frac{E'}{mc^2} = \left[1 + \frac{Z^2 \alpha^2}{\left(n - j - \frac{1}{2} + \sqrt{\left(j + \frac{1}{2} \right)^2 - Z^2 \alpha^2} \right)^2} \right]^{-\frac{1}{2}} - 1, \tag{10.26}$$

where $\alpha = e^2/\hbar c \simeq 1/137$ is the fine structure constant ($\alpha = e^2/4\pi\epsilon_0 \hbar c$, in the International System of units). For small values of $Ze^2/\hbar c$, this reduces to the nonrelativistic formula

$$E' = -\frac{Z^2}{2n^2} \frac{e^2}{a_0}, \tag{10.27}$$

with a_0 the Bohr radius.

Solutions of Dirac equation in the spherically symmetric potential generated by heavy nuclei have been employed to study the inhomogeneous electron liquid in elemental atoms, where relativistic corrections are believed to be important. In particular, based on Dirac's relativistic wave equation, an approximate propagator solution using a WKB-like treatment of this equation by Linderberg for central fields has been obtained by Leys *et al.* (2004). This approach leads back to the exact single-particle kinetic energy of the uniform electron gas and hence to the virial relations when the central field is switched off. The electron density and kinetic energy density in the semi-infinite inhomogeneous electron-gas limit were also studied by Baltin and March (1987) using Dirac's relativistic wave equation.

10.3 Dirac equation in a magnetic field

We start by rewriting Dirac's equation for a free fermion field as (Itzykson and Zuber, 1980; Kleinert, 2016)

$$(i\hbar \not{\partial} - mc)\Psi = 0, \tag{10.28}$$

where the Feynman 'slash' notation has been used, with $\not{\partial} \equiv \gamma^\mu \partial_\mu$, and more generally $\not{a} \equiv \gamma^\mu a_\mu$, for any 4-vector a_μ. The Dirac matrices γ^μ

have been defined in note 1 on page 195, and satisfy the anticommutation relation

$$\{\gamma^\mu, \gamma^\nu\} \equiv \gamma^\mu \gamma^\nu + \gamma^\nu \gamma^\mu = 2g^{\mu\nu}, \tag{10.29}$$

where $g^{\mu\nu}$ is the metric tensor, which allows them to define a matrix representation of a Clifford algebra. One also defines $\gamma^5 = i\gamma^0\gamma^1\gamma^2\gamma^3$.

The standard way in which a spin-$\frac{1}{2}$ charged particle, with charge e, is coupled to an external electromagnetic field described a 4-vector potential A_μ is via the minimal coupling

$$\partial_\mu \mapsto \partial_\mu + i\frac{e}{\hbar c}A_\mu \equiv D_\mu. \tag{10.30}$$

Dirac's equation (10.28) in the presence of such an electromagnetic field then becomes

$$(i\hbar \not{D} - mc)\Psi \equiv \left(i\hbar \not{\partial} - \frac{e}{c}\not{A} - mc\right)\Psi = 0. \tag{10.31}$$

The minimal coupling prescription ensures that both the Maxwell equations for the electromagnetic field,

$$\partial_\mu F^{\mu\nu} = \frac{1}{c} j^\nu, \tag{10.32}$$

with $F^{\mu\nu} = \partial^\mu A^\nu - \partial^\nu A^\mu$ the electromagnetic tensor and $j^\mu \equiv (c\rho, \mathbf{A})$ the current density 4-vector, and Dirac's equation (10.28) remain invariant under the gauge transformation

$$A^\mu(x) \mapsto A^\mu(x) + \partial^\mu \Lambda(x), \tag{10.33a}$$

$$\Psi(x) \mapsto e^{-i\frac{e}{c}\Lambda(x)}\Psi(x), \tag{10.33b}$$

where $\Lambda(x)$ is an arbitrary (differentiable) function of the 4-vector x. D_μ is called a covariant derivative, not because it is covariant with respect to a Lorentz transformation (although it is a covariant 4-vector also in that sense), but because $D_\mu\Psi$ transforms in the same way as Ψ does under a local gauge transformation, *i.e.* $D_\mu\Psi(x) \mapsto e^{-i\frac{e}{c}\Lambda(x)} D_\mu\Psi(x)$ (Kleinert, 2016).

10.3.1 *Pauli equation*

The first experimental evidence of (half-integer) spin was provided by the Stern-Gerlach experiment (Gerlach and Stern, 1922b,a), where a beam of (originally silver) atoms is bent by a nonuniform external magnetic field into an (even) number of parts, depending on their 'intrinsic angular momentum'. This was then interpreted by Uhlenbeck and Goudsmit (1925,

1926) in terms of an intrinsic quantum variable, analogous to angular momentum, having no classical counterpart, *viz.* the spin. A phenomenological Schrödinger-like equation, but now for a 'spinor' wavefunction, *i.e.* a 2-components vector, had to be devised, including a suitable coupling between spin and the external magnetic field. Dirac's equation (10.28) was first to provide a microscopic interpretation of that equation, which naturally occurs in a matrix form, thereby laying the theoretical foundations of the spin, as a quantum-relativistic observable, and of a relativistic formulation of the quantum theory (see Itzykson and Zuber, 1980; Kleinert, 2016, for a review).

In order to obtain the relativistic version of Schrödinger equation for a spinor wavefunction, one acts with $i\hbar\not{D} + mc$ on Eq. (10.31) to obtain the Pauli equation (Pauli, 1927)

$$\left[-\left(\hbar\partial_\mu + i\frac{e}{c}A_\mu\right)^2 - \frac{1}{2}\frac{e\hbar}{c}\sigma^{\mu\nu}F_{\mu\nu} - m^2c^2\right]\Psi = 0. \tag{10.34}$$

Here, use has been made of the identity $\gamma^\mu\gamma^\nu = \frac{1}{2}\{\gamma^\mu,\gamma^\nu\} + \frac{1}{2}[\gamma^\mu,\gamma^\nu] = g^{\mu\nu} - i\sigma^{\mu\nu}$, via Eq. (10.29), where $\sigma^{\mu\nu} \equiv \frac{i}{2}[\gamma^\mu,\gamma^\nu]$. In the Pauli equation, Eq. (10.34), charge appears to be coupled non-minimally to the electromagnetic field through the double tensor product in the spin term, $\frac{1}{2}\sigma^{\mu\nu}F_{\mu\nu}$.

In the chiral representation, where the 4-spinor Ψ is represented as an upper and lower 2-spinor components as $\Psi = \begin{pmatrix}\varphi\\ \chi\end{pmatrix}$, Eq. (10.34) translates into

$$\left[-\left(\hbar\partial_\mu + i\frac{e}{c}A_\mu\right)^2 + \boldsymbol{\sigma}\cdot(\mathbf{H} \pm i\mathbf{E}) - m^2c^2\right]\begin{pmatrix}\varphi\\ \chi\end{pmatrix} = 0. \tag{10.35}$$

In the nonrelativistic limit, the large energy mc^2 is the driving term in Eq. (10.35). Removing the fast oscillations from the upper spinor through the position $\varphi(x) = e^{-imc^2t/\hbar}\tilde{\varphi}(\mathbf{x},t)/\sqrt{2m}$, one finds the nonrelativistic Pauli equation

$$\left[i\hbar\partial_t + \frac{\hbar^2}{2m}\left(\nabla - i\frac{e}{\hbar c}\mathbf{A}\right)^2 + \frac{e}{2mc}\boldsymbol{\sigma}\cdot\mathbf{H} - eA^0(\mathbf{x},t)\right]\tilde{\varphi}(\mathbf{x},t) = 0, \tag{10.36}$$

where the magnetic interaction Hamiltonian

$$H_{\text{mag}} = -\frac{e\hbar}{2mc}\boldsymbol{\sigma}\cdot\mathbf{H} \equiv -\boldsymbol{\mu}\cdot\mathbf{H} \tag{10.37}$$

explicitly makes its appearance, with

$$\boldsymbol{\mu} = \frac{e}{mc}\frac{\hbar}{2}\boldsymbol{\sigma} \equiv g\frac{e}{2mc}\mathbf{S} \tag{10.38}$$

the particle's intrinsic magnetic moment, here related to the particle's spin $\mathbf{S} = \frac{\hbar}{2}\boldsymbol{\sigma}$. Within Dirac's theory, one finds the value $g = 2$ for the gyromagnetic ratio of an electron, the experimental value being very close to it. Any small deviation from Dirac's value is referred to as an anomalous magnetic moment, and is a consequence of the quantum nature of the electromagnetic field (radiative corrections, cf. Itzykson and Zuber, 1980; Kleinert, 2016).

Further corrections to the nonrelativistic limit of Dirac's equation can be systematically investigated by means of the Foldy-Wouthuysen transformation (Foldy and Wouthuysen, 1950). This consists in determining a possibly time-dependent unitary transformation of the Dirac spinor field, which decouples the small and large components thereof. This results in a hierarchy of equations, which have to be solved to the desired order in $1/mc^2$, in a perturbative sense. Such a transformation can be constructed exactly in the free case (Itzykson and Zuber, 1980; Kleinert, 2016).

10.3.2 *Dirac propagator*

Exact solutions of Dirac's equation other than in the presence of a Coulomb potential (Sec. 10.2.1), or in the nonrelativistic limit (Sec. 10.3.1), are most conveniently expressed in terms of the associated Green's function, or propagator. This is basically the kernel of the integral operator equivalent to the inverse of the Hamiltonian (a differential operator, in fact).

10.3.2.1 *Dirac propagator: free case*

In the free case, this is defined (in natural units, $\hbar = c = 1$) as

$$(i\partial\!\!\!/_2 - m)S_{\mathrm{F}}(x_2 - x_1) = \delta^{(4)}(x_2 - x_1), \tag{10.39}$$

where ∂_2 is the 4-gradient associated with x_2. Such a propagator describes *(i)* the appearance of an electron at x_1, *(ii)* its free propagation from x_1 to x_2, *(iii)* its disappearance at x_2. As long as the electron has a positive energy, this process is physically meaningful for $t_2 > t_1$. Otherwise, one should regard the vanishing of an electron with negative energy as the appearance of a positron, and vice versa. Likewise, the intermediate step should then be regarded as the free propagation of a positron from x_2 to x_1, which conversely is physically meaningful only if $t_2 < t_1$. A full calculation shows that (Stueckelberg, 1941; Feynman, 1949; Itzykson and Zuber, 1980)

$$S_{\mathrm{F}}(x) = \int \frac{d^4p}{(2\pi)^4} e^{-ip\cdot x} \frac{p\!\!\!/ + m}{p^2 - m^2 + i\epsilon}, \tag{10.40}$$

where the positive infinitesimal $\epsilon > 0$ prescribes the contour along which the integration should be performed on the real p^0 axis. Fourier transforming, such a result can also be expressed as

$$S_{\rm F}(p) = \frac{1}{\not{p} - m + i\epsilon}. \tag{10.41}$$

Then, given the positive and negative frequency components of a solution as $\Psi^{(+)}(t_1, \mathbf{x})$ and $\Psi^{(-)}(t_2, \mathbf{x})$, respectively, at given times t_1 and t_2, with $t_1 < t_2$, and all $\mathbf{x}$, Feynman's propagator, Eq. (10.40), allows one to determine the solution $\Psi(t, \mathbf{x})$ at intermediate times t as

$$\begin{aligned}\Psi(t, \mathbf{x}) = i \int d^3\mathbf{x}' [S_{\rm F}(t - t_1, \mathbf{x} - \mathbf{x}')\gamma^0 \Psi^{(+)}(t_1, \mathbf{x}') \\ - S_{\rm F}(t - t_2, \mathbf{x} - \mathbf{x}')\gamma^0 \Psi^{(-)}(t_2, \mathbf{x}')]. \end{aligned} \tag{10.42}$$

10.3.2.2 *Dirac propagator: arbitrary external electromagnetic field*

In the presence of an external electromagnetic field described by a 4-vector potential A^μ, the equation for the Dirac propagator $S_A(x_2, x_1)$ reads

$$[i\not{\partial}_2 - e\not{A}(x_2) - m] S_A(x_2, x_1) = \delta^{(4)}(x_2 - x_1), \tag{10.43}$$

where $S_A(x_2, x_1)$ is in general a function of its two arguments, separately, and not of their difference, as in the free case (no translational invariance being, in this case, guaranteed). For an arbitrary external potential, it is not always possible to find a closed expression for $S_A(x_2, x_1)$. However, for cases in which the term $e\not{A}$ is 'small', in the perturbative sense, one may express $S_A(x_2, x_1)$ as an asymptotic series in $e\not{A}$. To that aim, multiply both sides of the above equation by the free Feynman propagator $S_{\rm F}(x_3 - x_2)$ and integrate with respect to x_2, to obtain

$$S_{\rm F}(x_3 - x_1) = \int d^4x_2 S_{\rm F}(x_3 - x_2)[i\not{\partial}_2 - e\not{A}(x_2) - m] S_A(x_2, x_1). \tag{10.44}$$

Since the Hermitian conjugate of Eq. (10.39) implies that $S_{\rm F}(x_3 - x_2)(-i\overleftarrow{\not{\partial}}_2 - m) = \delta^{(4)}(x_3 - x_2)$, one eventually finds an integral equation for $S_A(x_3, x_1)$ to read

$$S_A(x_3, x_1) = S_{\rm F}(x_3 - x_1) + e \int d^4x_2 \not{A}(x_2) S_A(x_2, x_1), \tag{10.45}$$

which can be solved iteratively.

10.3.2.3 *Fock-Schwinger proper time method*

An elegant method, due to Fock (1937) and Schwinger (1951, 1954a,b) (see Itzykson and Zuber, 1980; Kleinert, 2016, for a review), allows to express the propagators associated to an Hamiltonian whose functional dependence on the impulse variable is that of a polynomial, in terms of the proper time evolution associated to it. This will be useful to derive the Dirac propagator explicitly in two special cases, involving a constant and uniform magnetic field and an oscillating field, as that associated with a plane wave, respectively.

The problem to be addressed in general is that of finding the Green function $G(x,x')$ associated with a Hamiltonian $H(x,i\partial_x)$, defined by

$$H(x,i\partial_x)G(x,x') = \delta^{(4)}(x-x'). \tag{10.46}$$

Letting $i\partial_\mu = p_\mu$, with p_μ a canonically conjugate 4-momentum to x_μ, so that $[x^\mu,p^\nu] = -ig^{\mu\nu}$, then one may ask oneself the equivalent question of determining the propagator $\langle x|x'\rangle = \delta^{(4)}(x-x')$ associated to such a system. Introducing the unitary evolution operator $U(x,x';\tau)$ along the proper time τ as

$$i\partial_\tau U(x,x';\tau) = H(x,p)U(x,x';\tau), \tag{10.47a}$$

obeying boundary conditions

$$\lim_{\tau\to 0} U(x,x';\tau) = \delta^{(4)}(x-x'), \tag{10.47b}$$

$$\lim_{\tau\to -\infty} U(x,x';\tau) = 0, \tag{10.47c}$$

one formally finds

$$U(x,x';\tau) = \langle x|e^{-iH\tau}|x'\rangle \equiv \langle x|U(\tau)|x'\rangle, \tag{10.48}$$

$$G(x,x') = -i\int_{-\infty}^{0} d\tau U(x,x';\tau). \tag{10.49}$$

Differentiating Eq. (10.48) with respect to τ and conveniently inserting the identity written as $1 = U(\tau)U^\dagger(\tau)$, one finds

$$i\partial_\tau\langle x|U(\tau)|x'\rangle = \langle x|H(x,p)U|x'\rangle = \langle x|U(\tau)U^\dagger(\tau)H(x,p)U(\tau)|x'\rangle, \tag{10.50}$$

which is to say

$$i\partial_\tau\langle x(\tau)|x'(0)\rangle = \langle x(\tau)|H\left(x(\tau),p(\tau)\right)|x'(0)\rangle. \tag{10.51}$$

The problem is then shifted to that of finding $x(\tau)$ and $p(\tau)$, as solutions of Heisenberg-like equations, for a given Hamiltonian $H(x,p)$.

Going back to Eq. (10.43) defining the Dirac propagator in the presence of an external electromagnetic 4-vector potential, which we here rewrite as

$$[i\partial\!\!\!/ - eA\!\!\!/(x) - m]S_A(x,x') = \delta^{(4)}(x-x'), \tag{10.52}$$

one can relate $S_A(x,x')$ to the propagator $G_A(x,x')$ of the Pauli equation, Eq. (10.34), as

$$S_A(x,x') = [i\partial\!\!\!/ - eA\!\!\!/(x) + m]G_A(x,x'), \tag{10.53}$$

where

$$HG_A \equiv \left[(i\partial - eA)^2 - m^2 - \frac{e}{2}\sigma_{\mu\nu}F^{\mu\nu}\right] G_A(x,x') = \delta^{(4)}(x-x'). \tag{10.54}$$

Let $\pi_\mu = p_\mu - eA_\mu$ denote the momentum corresponding to the covariant derivative. Then the Ehrenfest equations for $x(\tau) = U^\dagger(\tau)xU(\tau)$ and $\pi(\tau) = U^\dagger(\tau)\pi U(\tau)$ read

$$\dot{x}_\mu(\tau) = i[H, x_\mu] = -2\pi_\mu, \tag{10.55a}$$

$$\dot{\pi}_\mu(\tau) = i[H, \pi_\mu] = -2eF_{\mu\rho}\pi^\rho - ie\partial^\rho F_{\mu\rho} - \frac{e}{2}\partial_\mu F_{\rho\nu}\sigma^{\rho\nu}, \tag{10.55b}$$

where use has been made of the fact that $[\pi_\mu, \pi_\nu] = -ieF_{\mu\nu}$.

10.3.2.4 *Dirac propagator: constant and uniform field*

In the quantum nonrelativistic limit, the study of the Schrödinger equation for an electron interacting with a constant and uniform magnetic field gives rise to Landau levels. When relativistic effects are included, one has to make recourse to the Dirac equation.

In the case of a constant and uniform field, Eqs. (10.55) above reduce to $\dot{x}_\mu(\tau) = -2\pi_\mu$, $\dot{\pi}_\mu(\tau) = -2eF_{\mu\rho}\pi^\rho$, which can be readily integrated, using the notation for functions of matrices (cf. *e.g.* Angilella, 2011), as

$$\pi(\tau) = e^{-2eF\tau}\pi(0), \tag{10.56a}$$

$$x(\tau) - x(0) = \left(\frac{e^{-2eF\tau} - 1}{eF}\right)\pi(0). \tag{10.56b}$$

Inverting the latter equation in favour of $\pi(0)$ and substituting in the former equation, one finds

$$\pi(\tau) = -\frac{1}{2}eFe^{-2eF\tau}[\sinh(eF\tau)]^{-1}\left(x(\tau) - x(0)\right), \tag{10.57}$$

and, on exploiting the antisymmetry of F,

$$\pi^2(\tau) = \left(x(\tau) - x(0)\right) K \left(x(\tau) - x(0)\right), \tag{10.58}$$

where

$$K = \frac{1}{4}e^2F^2[\sinh(eF\tau)]^{-2}. \tag{10.59}$$

Rearranging the terms in Eq. (10.58) involves the commutator for different components of the covariant 4-impulse at different proper times

$$[\pi_\mu(\tau), \pi_\nu(0)] = i\left(\frac{e^{-2eF\tau}-1}{eF}\right)_{\mu\nu}. \tag{10.60}$$

One finally gets the Pauli Hamiltonian in terms of $x(\tau)$ and $x(0)$ as

$$\begin{aligned} H &= x(\tau)Kx(\tau) - 2x(\tau)Kx(0) + x(0)Kx(0) \\ &\quad - \frac{i}{2}\,\mathrm{tr}\left[eF\coth(eF\tau)\right] - \frac{e}{2}\sigma_{\mu\nu}F^{\mu\nu} - m^2, \end{aligned} \tag{10.61}$$

while Eq. (10.47a) correspondingly becomes

$$\begin{aligned} i\partial_\tau U(x,x';\tau) &= \left[-\frac{e}{2}\sigma_{\mu\nu}F^{\mu\nu} - m^2 + (x-x')K(x-x')\right. \\ &\quad \left. - \frac{i}{2}\,\mathrm{tr}\left[eF\coth(eF\tau)\right]\right] U(x,x';\tau). \end{aligned} \tag{10.62}$$

This is now a (matrix) ordinary differential equation, which can be integrated to yield

$$\begin{aligned} U(x,x';\tau) &= C(x,x')\tau^{-2}\exp\left(-\frac{1}{2}\,\mathrm{tr}\log\left[(eF\tau)^{-1}\sinh(eF\tau)\right]\right) \\ &\times \exp\left[\frac{i}{4}(x-x')eF\coth(eF\tau)(x-x') + \frac{i}{2}\sigma_{\mu\nu}F^{\mu\nu}\tau + im^2\tau\right]. \end{aligned} \tag{10.63}$$

The function $C(x,x')$ is to be determined so that $U(x,x';\tau)$ satisfies the correct relations between position and momentum, *e.g.*

$$[i\partial_\mu^x - eA_\mu(x)]\langle x(\tau)|x'(0)\rangle = \langle x(\tau)|\pi_\mu(\tau)|x(0)\rangle, \tag{10.64a}$$

$$[-i\partial_\mu^{x'} - eA_\mu(x')]\langle x(\tau)|x'(0)\rangle = \langle x(\tau)|\pi_\mu(0)|x(0)\rangle. \tag{10.64b}$$

In this case, these are equivalent to

$$\left[i\partial_\mu^x - eA_\mu(x) - \frac{e}{2}F_{\mu\nu}(x-x')^\nu\right] C(x,x') = 0, \tag{10.65a}$$

$$\left[i\partial_\mu^{x'} - eA_\mu(x') - \frac{e}{2}F_{\mu\nu}(x-x')^\nu\right] C(x,x') = 0, \tag{10.65b}$$

whose solution can be expressed as

$$C(x,x') = C(x')\exp\left(-ie\int_{x'}^{x} d\xi\cdot\left[A(\xi) + \frac{1}{2}F(\xi-x')\right]\right), \tag{10.66}$$

where the integration is to be performed along an arbitrary path in spacetime between x' and x. That this does not depend on the integration path follows from the fact that the integrand has a vanishing curl. In particular, one may choose the straight line connecting x' and x, thus showing that the second term does not give any contribution, owing to the antisymmetry of $F_{\mu\nu}$, so that

$$C(x, x') = C \exp\left(-ie \int_{x'}^{x} d\xi \cdot A(\xi)\right), \tag{10.67}$$

where the constant C is finally determined from Eq. (10.47b) as

$$C = -\frac{i}{(4\pi)^2}. \tag{10.68}$$

Summarizing, the propagator for the Dirac equation in a constant and uniform field reads

$$S_A(x, x') = -i[i\partial\!\!\!/_x - eA\!\!\!/(x) + m] \int_{-\infty}^{0} d\tau U(x, x'; \tau), \tag{10.69}$$

where

$$\begin{aligned} U(x, x'; \tau) = -\frac{i}{(4\pi)^2}\frac{1}{\tau^2}\exp\bigg(&-ie\int_{x'}^{x} d\xi^\mu \cdot A_\mu(\xi) \\ &-\frac{1}{2}\operatorname{tr}\log\left[(eF\tau)^{-1}\sinh(eF\tau)\right] + \frac{i}{4}(x-x')eF\coth(eF\tau)(x-x') \\ &+\frac{i}{2}\sigma_{\mu\nu}F^{\mu\nu}\tau + i(m^2 - i\epsilon)\tau\bigg), \end{aligned} \tag{10.70}$$

with $m^2 \mapsto m^2 - i\epsilon$, with $\epsilon > 0$ a positive infinitesimal, to satisfy Eq. (10.47c). One may observe that the role of the phase $\exp(-ie \int d\xi \cdot A)$ provided by Eq. (10.67) is that of making $U(x, x'; \tau)$ gauge covariant, *i.e.* $U(x, x'; \tau) \mapsto e^{-ie\Lambda(x)} U(x, x'; \tau) e^{+ie\Lambda(x')}$ when $A_\mu(x) \mapsto A_\mu(x) + \partial_\mu \Lambda(x)$.

10.3.2.5 *Dirac propagator: oscillating, plane-wave field*

Let us now consider the electromagnetic field associated with a plane wave, travelling in spacetime along the direction n_μ, with $n^2 = 0$, and linearly polarized along ε_μ, with $\varepsilon^2 = -1$ and $\varepsilon_\mu n^\mu = 0$. Such a plane wave will be characterized by a 4-vector potential $A_\mu = \varepsilon_\mu f(\xi)$, where $\xi = n_\mu x^\mu$ is the spatial coordinate along n, and f is a sufficiently regular, but otherwise arbitrary function, defining the shape of the travelling wave. One also has $\partial_\mu A_\nu = n_\mu \varepsilon_\nu f'(\xi)$, and $F_{\mu\nu} = \phi_{\mu\nu} f'(\xi)$, with $\phi_{\mu\nu} = n_\mu \varepsilon_\nu - n_\nu \varepsilon_\mu$, and $\partial^\rho F_{\mu\rho} = 0$.

The Ehrenfest equations, Eqs. (10.55), in this case explicitly read

$$\dot{x}_\mu(\tau) = -2\pi_\mu, \tag{10.71a}$$

$$\dot{\pi}_\mu(\tau) = -2e\phi_{\mu\rho}\pi^\rho f'(\xi) - \frac{e}{2}n_\mu \phi_{\rho\nu}\sigma^{\rho\nu} f''(\xi). \tag{10.71b}$$

Observing that $\partial_\tau(\pi \cdot n) = 0$, $\dot{\xi} = -2\pi \cdot n$, and $[\xi, \dot{\xi}] = 0$, one first obtains the coordinate ξ in the direction of propagation as a function of proper time as

$$\xi(\tau) = n \cdot x(\tau) = \xi(0) - 2\pi \cdot n\tau, \tag{10.72}$$

and then

$$\phi^{\nu\mu}\pi_\mu = en^\nu f(\xi) + C^\nu, \tag{10.73}$$

with C^ν a constant operator commuting with $\pi \cdot n$. Inserting this back into Eqs. (10.71), after successive integrations one gets

$$-\frac{1}{2}\dot{x}_\mu(\tau) = \pi_\mu(\tau) = \frac{1}{2\pi \cdot n}\left(2eC_\mu f(\xi) + e^2 n_\mu f^2(\xi) + \frac{e}{2}n_\mu \phi_{\rho\nu}\sigma^{\rho\nu} f'(\xi)\right) + D_\mu, \tag{10.74a}$$

with D_μ another constant operator commuting with $\pi \cdot n$,

$$\begin{aligned}\pi_\mu(\tau) = {} & \frac{x_\mu(\tau) - x_\mu(0)}{2\tau} \\ & + \frac{\tau}{(\xi(\tau) - \xi(0))^2}\int_{\xi(0)}^{\xi(\tau)} d\xi \left(2eC_\mu f(\xi) + e^2 n_\mu f^2(\xi) + \frac{e}{2}n_\mu \phi_{\rho\nu}\sigma^{\rho\nu} f'(\xi)\right) \\ & - \frac{\tau}{\xi(\tau) - \xi(0)}\left(2eC_\mu f[\xi(\tau)] + e^2 n_\mu f^2[\xi(\tau)] + \frac{e}{2}n_\mu \phi_{\rho\nu}\sigma^{\rho\nu} f'[\xi(\tau)]\right).\end{aligned} \tag{10.74b}$$

This allows to determine the constant C_μ as

$$C_\mu = -\frac{1}{2\tau}\phi_{\mu\rho}\left(x^\rho(\tau) - x^\rho(0)\right) - \frac{en_\mu}{\xi(\tau) - \xi(0)}\int_{\xi(0)}^{\xi(\tau)} d\xi f(\xi). \tag{10.75}$$

After computation of the various relevant commutators as $[\xi(\tau), x_\mu(0)] = 0$, $[\xi(0), x_\mu(\tau)] = 2in_\mu\tau$, $[x_\mu(\tau), x^\mu(0)] = -8i\tau$, one can eventually express the Pauli Hamiltonian in terms of $x(\tau)$ and $x(0)$ [cf. Eq. (10.61) for the uniform and constant field] as

$$\begin{aligned}H = {} & \frac{1}{4\tau^2}\left(x^2(\tau) - 2x(\tau) \cdot x(0) + x^2(0)\right) - \frac{2i}{\tau} \\ & - \left(e^2\langle \delta f^2 \rangle + m^2\right) - \frac{e}{2}\phi_{\rho\nu}\sigma^{\rho\nu}\frac{f[\xi(\tau)] - f[\xi(0)]}{\xi(\tau) - \xi(0)},\end{aligned} \tag{10.76}$$

where

$$\langle \delta f^2 \rangle = \int_{\xi(0)}^{\xi(\tau)} d\xi \frac{f^2(\xi)}{\xi(\tau) - \xi(0)} - \left(\int_{\xi(0)}^{\xi(\tau)} d\xi \frac{f(\xi)}{\xi(\tau) - \xi(0)} \right)^2 . \tag{10.77}$$

Following analogous steps to those for the Dirac propagator in a uniform and constant electromagnetic field, one arrives at the expression for the evolution operator with respect to proper time

$$U(x, x'; \tau) = C(x, x')\tau^{-2} \exp\left(\frac{i}{4\tau}(x - x')^2 + i\tau \left(e^2 \langle \delta f^2 \rangle + m^2\right) + \frac{ie}{2}\tau \phi_{\rho\nu}\sigma^{\rho\nu} \frac{f(\xi) - f(\xi')}{\xi - \xi'} \right), \tag{10.78}$$

with

$$C(x, x') = C(x') \exp\left(-ie \int_{x'}^{x} dy_\mu \left[A^\mu(y) - \phi^{\mu\rho} \frac{y_\rho - x'_\rho}{n \cdot y - \xi'} \times \int_{\xi'}^{n \cdot y} du \frac{f(u)}{n \cdot y - \xi'} - f(n \cdot y) \right] \right), \tag{10.79}$$

where the line integral with respect to y is path independent. Choosing again to integrate along the straight line between x' and x results in a phase term ensuring gauge covariance, with which the evolution operator reads

$$U(x, x'; \tau) = -\frac{i}{(4\pi)^2}\tau^{-2} \exp\left(\frac{i}{4\tau}(x - x')^2 + i\left[e^2 \langle \delta f^2 \rangle + m^2 + \frac{e}{2}\phi_{\rho\nu}\sigma^{\rho\nu} \frac{f(\xi) - f(\xi')}{\xi - \xi'} - i\epsilon\right]\tau - ie \int_{x'}^{x} dy_\mu A^\mu(y) \right). \tag{10.80}$$

For a periodic function $f(\xi)$, the term proportional to $\phi_{\rho\nu}\sigma^{\rho\nu}$ becomes negligible after averaging over a few periods, and the net effect amounts to a mass renormalization

$$m_{\text{eff}}^2 = m^2 + e^2 \langle \delta f^2 \rangle. \tag{10.81}$$

This corresponds to a relative mass shift (Itzykson and Zuber, 1980; Kleinert, 2016)

$$\frac{\Delta m^2}{m^2} = 2\alpha \lambda\!\!\!^{-}_e{}^2 \lambda \rho, \tag{10.82}$$

where $\lambda\!\!\!^{-}_e = \hbar/mc$ is the Compton wavelength of the electron, λ is the wavelength of the electromagnetic wave, and ρ the photon number density (α is the fine structure constant). Present lasers would not be able to produce an observable mass shift in Dirac electrons interacting with intense fields (see Piraux *et al.*, 1993, see also Brown and Kibble (1964); Varró (2013), and the introduction of Braun *et al.* (1999)). However, recent experiments report of a sizeable renormalization of the effective band mass of Dirac-like electrons in semimetal bismuth (Minami *et al.*, 2015).

10.3.2.6 *Eigenfunction method*

Among other methods to obtain the propagator of Dirac's equation in an external electromagnetic field exactly, the eigenfunction method should be mentioned (Ritus, 1970, 1972, 1974, 1978, 1979). This is based on the diagonalization of the Dirac Hamiltonian in the basis of the eigenfunctions of the operator $\not{\pi}^2 = \pi^2 - \frac{e}{2}\sigma^{\mu\nu}F_{\mu\nu}$, where $\pi_\mu = p_\mu - eA_\mu$ is the canonical momentum operator.

In the presence of an electromagnetic potential 4-vector, the definition, Eq. (10.43), for the Dirac propagator can then be written as

$$(\not{\pi} - m)S_A(x, x') = \delta^{(4)}(x - x'). \tag{10.83}$$

Since $\not{\pi}$ does not commute with the momentum operator, it is not convenient to expand $S_A(x, x')$ in plane waves, so that its Fourier representation cannot be cast in diagonal form, as in the free case, Eq. (10.41). However, it will be diagonal in the basis of the simultaneous eigenvectors of $\not{\pi}$, and of all the operators commuting with it. These turn out to be $\not{\pi}^2$, $\sigma^{\mu\nu}F_{\mu\nu}$, and the dual field strength tensor $\tilde{F}_{\mu\nu} = \frac{1}{2}\epsilon_{\mu\nu\rho\sigma}F^{\rho\sigma}$, where $\epsilon_{\mu\nu\rho\sigma}$ is the 4-rank completely antisymmetric Levi-Civita tensor. In particular, let $u_p(x)$ denote the simultaneous eigenvectors (spinors) of $\not{\pi}$ and $\not{\pi}^2$, with eigenvalues as defined by the equations

$$\not{\pi}^2 u_p(x) = p^2 u_p(x), \tag{10.84a}$$

$$\not{\pi} u_p(x) = u_p(x)\not{\bar{p}}. \tag{10.84b}$$

These will constitute a complete set, in terms of which one may expand both $\delta^{(4)}(x - x')$ and $S_A(x, x')$ respectively as

$$\delta^{(4)}(x - x') = \int dp\, u_p(x)\bar{u}_p(x'), \tag{10.85a}$$

$$S_A(x, x') = \int dp\, u_p(x) S_A(p) \bar{u}_p(x'), \tag{10.85b}$$

where $\bar{u}_p \equiv u_p^\dagger \gamma^0$ is the conjugate spinor. Then, applying $\not{\pi} - m$ to both sides of Eq. (10.85b), one finds

$$S_A(p) = \frac{1}{\not{\bar{p}} - m + i\epsilon}. \tag{10.86}$$

In other words, in the basis of eigenfunctions defined by Eqs. (10.84), the Dirac propagator assumes a diagonal form, much similar to that of the free case, Eq. (10.41), but where now $\bar{p}$ denotes the eigenvalue with respect to $\not{\pi}$. The Dirac propagator in real space can of course be obtained by inverting Eq. (10.85b).

The eigenfunction method has been originally used to derive the Dirac propagator in the case of crossed fields, *i.e.* for perpendicular electric and magnetic fields of equal magnitude, which is a special case of the electromagnetic plane wave discussed above (Volkov, 1935, 1937) (see Bagrov and Gitman, 1990; Kuznetsov and Mikheev, 2013; Bagrov and Gitman, 2014, for a review). The density matrix of an electronic plasma in a magnetic field with a fixed number of Landau levels is also reviewed by Kuznetsov and Mikheev (2013).

10.3.2.7 *Dirac propagator in a magnetic field: reduced dimensionality*

The Feynman propagator for the Dirac equation in a uniform and constant external magnetic field, and for other field configurations, has been recently studied also in $2+1$ dimensions by Murguía *et al.* (2010). Such exact results in reduced dimensionality may prove relevant in the massless limit ($m=0$) for two-dimensional systems, such as graphene (Novoselov *et al.*, 2004, 2005a,b), where the low-energy electron excitations are described by a massless, linear dispersion relation, which could then be described by the formalism of quantum electrodynamics (Katsnelson *et al.*, 2006; Katsnelson, 2006, 2007; Katsnelson and Novoselov, 2007; Katsnelson, 2012), but at much smaller energy scales (see Castro Neto *et al.*, 2009, for a review).

10.4 Semiclassical limit of Dirac's relativistic wave equation

Our aim in this Section is to discuss the semiclassical limit (Planck's constant $h \to 0$: velocity of light in vacuum c remains finite) of Dirac's one-electron equation discussed at some length above (cf. Sec. 10.1).

Below, we shall follow closely the study of Rafanelli and Schiller (1964) (see also Keller and Oziewicz, 1997; Keller, 2002). Rafanelli and Schiller (1964) demonstrate that the semiclassical limit of the Dirac equation leads to a particle under the Lorentz force. The trajectory and average spin of such a particle satisfy a set of relativistic equations due to Bargmann *et al.* (1959, BMT, below):

$$m\dot{v}_\mu = -\frac{e}{c} F_{\mu\nu} v_\nu, \tag{10.87a}$$

$$\dot{S}_\mu = \frac{e}{mc}\left[-(1+g_1) F_{\mu\nu} S_\nu + \frac{1}{c^2} g_1 F_{\rho\sigma} S_\rho v_\sigma v_\mu \right]. \tag{10.87b}$$

Here, $v_\mu = \dot{x}_\mu \equiv dx_\mu / d\tau$ is the particle quadrivelocity, S_μ its spin pseudo-quadrivector, $g_1 = \frac{1}{2}g - 1$ the anomalous gyromagnetic factor, $F_{\mu\nu}$ the electromagnetic quadritensor, and a dot denotes differentiation with respect to proper time τ. Eqs. (10.87), according to Della Selva *et al.* (1996) (see also Keller and Oziewicz, 1997, p. 71), are not limited regarding the velocity of the particle and appear valid for arbitrary strengths of the electromagnetic field. Rafanelli and Schiller (1964) then proceeds to derive Eqs. (10.87) as the semiclassical limit of Dirac's relativistic equation. Dirac's equation in the presence of an electromagnetic field (assumed homogeneous), and including Pauli's anomalous magnetic moment term, may be written as

$$\mathcal{D}\Psi \equiv \left[\pi_\mu \gamma_\mu - imc - \frac{g_1 e\hbar}{4mc^2} \gamma_\mu \gamma_\nu F_{\mu\nu} \right] \Psi = 0. \tag{10.88}$$

Here, $\pi_\mu = -i\hbar\partial_\mu + \frac{e}{c}A_\mu$ is the canonical momentum (including minimal coupling to the electromagnetic quadripotential A_μ), γ_μ are the usual Dirac matrices ($\mu = 1, 2, 3, 4$), and Ψ is the particle spinor. Rafanelli and Schiller (1964) then seek a solution of Eq. (10.88) such that

$$\left[\pi_\rho \gamma_\rho + imc + \frac{g_1 e\hbar}{4mc^2} \gamma_\rho \gamma_\sigma F_{\rho\sigma} \right] \Phi = 2mc\Psi, \tag{10.89}$$

with Φ an auxiliary spinor. Applying the Dirac operator $\mathcal{D}$ to both sides of the above equation, and neglecting terms of order $\hbar^2$, Rafanelli and Schiller (1964) derive the following second order equation for Φ:

$$\left[\pi_\mu \pi_\mu + m^2c^2 + (1 + g_1) \frac{e\hbar}{2ic} \gamma_\mu \gamma_\nu F_{\mu\nu} + \frac{g_1 e\hbar}{mc^2} F_{\mu\nu} \gamma_\mu \pi_\nu \right] \Phi = 0. \tag{10.90}$$

Within the spirit of Wentzel-Kramers-Brillouin (WKB) approach, solutions to Eq. (10.90) can be expanded as a power series in $\hbar$ as

$$\Phi_{\text{WKB}} = e^{iS/\hbar} \sum_{n=0}^{\infty} (-i\hbar)^n \phi_n, \tag{10.91}$$

where S is an action (a scalar function of spacetime), and ϕ_n are auxiliary spinors. To leading order in $\hbar$, Rafanelli and Schiller (1964) then show that S and ϕ_0 are solutions to following set of coupled equations:

$$mv_\mu = \partial_\mu S + \frac{e}{c} A_\mu, \tag{10.92a}$$

$$v_\mu v_\mu + c^2 = 0, \tag{10.92b}$$

$$\begin{aligned} 2imv_\mu \partial_\mu \phi_0 + im(\partial_\mu v_\mu)\phi_0 = {} & (1 + g_1) \frac{e}{2ic} \gamma_\rho \gamma_\sigma F_{\rho\sigma} \phi_0 \\ & + \frac{g_1 e}{c^2} F_{\alpha\beta} \gamma_\alpha \gamma_\beta \phi_0. \end{aligned} \tag{10.92c}$$

The WKB solution of Eq. (10.90) is to be found by solving Eqs. (10.92) above. Fock (1937) also shows that Eq. (10.92c) can be expressed as an ordinary differential equation with the proper time τ acting as an independent variable. One then writes the lowest order coefficient ϕ_0 entering the WKB expansion, Eq. (10.91), as

$$\phi_0 = R\phi, \tag{10.93}$$

ϕ here being a unit spinor ($\bar{\phi}\phi = 1$, with $\bar{\phi} = \phi^\dagger \gamma_4$), and R a scalar quantity whose square satisfies the equation of continuity

$$\partial_\mu (R^2 v_\mu) = 0. \tag{10.94}$$

Then, if one has a solution of the relativistic Hamilton-Jacobi equation (10.92a–10.92b), then a solution of Eq. (10.94) is given by

$$R^2 = ic \det \left(\frac{\partial^2 S}{\partial x^i \partial \alpha^k} \right) \dot{x}_4^{-1}. \tag{10.95}$$

Here, α^k in the (Van Vleck) determinant $\det(\partial^2 S/\partial x^i \partial x^k)$ $(i,k = 1,2,3)$ denote the three separation constants entering the complete solution of the relativistic Hamilton-Jacobi equation (see, *e.g.* Schiller, 1962).

Returning to Eq. (10.92c), when R^2 satisfies Eq. (10.94), then $v_\mu \partial_\mu \phi \equiv \dot{\phi}$, the latter quantity being obtained from

$$\dot{\phi} = -(1+g_1)\frac{e}{4mc}\gamma_\rho \gamma_\sigma F_{\rho\sigma}\phi - \left(\frac{ig_1 e}{2mc^2}\right) F_{\rho\sigma} v_\rho \gamma_\sigma \phi. \tag{10.96}$$

Likewise, one has

$$\dot{\bar{\phi}} = (1+g_1)\frac{e}{4mc}\bar{\phi}\gamma_\rho \gamma_\sigma F_{\rho\sigma} - \left(\frac{ig_1 e}{2mc^2}\right) \bar{\phi} F_{\rho\sigma} v_\rho \gamma_\sigma. \tag{10.97}$$

In the special case of uniform constant electromagnetic field, Eqs. (10.96) and (10.97) turn into ordinary differential equations. This is due to the fact that the velocity $v_\mu(\tau)$ is a solution of Eq. (10.87a), and therefore a known function of τ.

One is now left with Eqs. (10.87a), (10.96), and (10.97) as the semi-classical equations of motion associated with the squared Dirac equation, Eq. (10.90). Eq. (10.87a) determines the trajectory of the particle, while the unit spinor ϕ entering Eq. (10.96) describes its spin. Before finding the classical spin equation, we follow Rafanelli and Schiller (1964) by analyzing the WKB solution of the squared Dirac equation, *viz.*

$$\Phi_{\text{WKB}} = R\phi \exp(iS/\hbar). \tag{10.98}$$

The fact that $\bar{\phi}\phi = 1$ evidently preserves the possibility of interpreting the wave function Φ as a probability amplitude even in the asymptotic limit of the wave theory. If $(\bar{\Phi}\Phi)_{\mathrm{WKB}}$ is to determine the relative numbers of particles in a given four-volume element, then in the classical limit $(\bar{\Phi}\Phi)_{\mathrm{WKB}}$ must satisfy an equation of continuity:

$$\partial_\mu[v_\mu(\bar{\Phi}\Phi)_{\mathrm{WKB}}] = 0. \tag{10.99}$$

Rafanelli and Schiller (1964) note that since R^2 satisfies this same equation, then ϕ must be a unit spinor.

However, it follows from Eqs. (10.96) and (10.97) that

$$\frac{d}{d\tau}(\bar{\phi}\phi) = -\frac{ig_1 e}{mc^2} F_{\alpha\beta}\bar{\phi}\gamma_\alpha\phi v_\beta, \tag{10.100}$$

so that ϕ is a unit spinor only if $\bar{\phi}\gamma_\alpha\phi \propto v_\alpha$, otherwise the right-hand side of the above equation for $d(\bar{\phi}\phi)/d\tau$ will not vanish. Since v_α is a time-like quadrivector, Eq. (10.92b), one requires that

$$ic\bar{\phi}\gamma_\alpha\phi = v_\alpha, \quad \text{so that } \bar{\phi}\phi = 1. \tag{10.101}$$

To show that the above constraint is consistent, we note that $ic\bar{\phi}\gamma_\alpha\phi$ can be shown to satisfy the same equation for v_α, *viz.* Eq. (10.87a). In addition to the requirement that $\bar{\phi}\phi = 1$, Rafanelli and Schiller (1964) note that one may also make $\bar{\phi}\gamma_5\phi = 0$, since it is a constant of motion:

$$\frac{d}{d\tau}(\bar{\phi}\gamma_5\phi) = 0, \tag{10.102}$$

so that if this value is chosen initially, it is maintained at all time.

Under the assumption of Eq. (10.101), the classical spin pseudo-quadrivector S_μ can be related to the bilinear form $\bar{\phi}\gamma_5\gamma_\mu\phi$. If, by analogy with Eq. (10.101), one writes

$$S_\mu = i\hbar\bar{\phi}\gamma_5\gamma_\mu\phi, \tag{10.103}$$

then one finds the dynamical equation for S_μ, *viz.*

$$\dot{S}_\mu = -(1+g_1)\frac{e}{mc}F_{\mu\nu}S_\nu + \frac{g_1 e}{mc^3}F_{\alpha\beta}S_\alpha v_\beta v_\mu, \tag{10.104}$$

which coincides with the BMT equation (10.87b) of Bargmann *et al.* (1959). Rafanelli and Schiller (1964) note that its derivation needs to make use additionally of the following identity:

$$\frac{1}{2}\bar{\phi}(\gamma_\mu\gamma_\nu - \gamma_\nu\gamma_\mu)\gamma_5\phi = \frac{1}{c\hbar}(S_\mu v_\nu - S_\nu v_\mu), \tag{10.105}$$

v_μ and S_μ being given by Eqs. (10.101) and (10.103), respectively, while $\bar{\phi}\phi = 1$ and $\bar{\phi}\gamma_5\phi = 0$.

10.5 Spinless Salpeter equation

To conclude this Chapter, we record here the form of the so-called spinless Salpeter equation, as the conceptually simplest wave equation including some relativistic effects, but still allowing for bound states (see, *e.g.* Brau, 2005, and references therein). In natural units ($\hbar = c = 1$), this equation has the form

$$\left[\sqrt{p^2+m^2}+V(\mathbf{r})\right]\Psi(\mathbf{r}) = M\Psi(\mathbf{r}), \tag{10.106}$$

where m denotes the mass of the particle, and M is the mass of the eigenstate ($M = m + E$, E being the binding energy). Salpeter equation follows from the Bethe-Salpeter equation (Salpeter and Bethe, 1951), when all timelike variables are eliminated (Salpeter, 1952), and spin is neglected, as well as negative energy solutions. Eq. (10.106) is suitable to study kinetic relativistic effects on bosons, or whenever the spin of the particles involved can be neglected. Interactions can be introduced in the free equation via the substitution $M \mapsto M - V(\mathbf{r})$, rather than via the minimal coupling substitution $\mathbf{p} \mapsto \mathbf{p} - \mathbf{A}(\mathbf{r})$, which is ruled out by the derivation of Eq. (10.106) from the Bethe-Salpeter equation (Brau, 2005).

Unfortunately, because of the pseudodifferential nature of the kinetic energy operator, few exact results exist which characterize this equation. Such results are mostly for a Coulomb potential, and include, *e.g.* a partial wave analysis of its solution (Castorina *et al.*, 1984), and upper and lower bounds on its energy levels (Hall *et al.*, 2003, and refs. therein), or on the number of possible bound states (Daubechies, 1983; Brau, 2003).

The propagator associated with the free version of Eq. (10.106) ($V \equiv 0$), defined by

$$\left(\sqrt{p^2+m^2}-m\right)G_m(\mathbf{r}-\mathbf{r}') = \delta^{(3)}(\mathbf{r}-\mathbf{r}'), \tag{10.107}$$

can be evaluated explicitly as (Brau, 2005)

$$G_m(\mathbf{r}) = \frac{m}{4\pi r}\left(1+\frac{2}{\pi}F(mr)\right), \tag{10.108a}$$

$$F(\xi) = \frac{\pi}{2} + K_1(\xi) - \int_\xi^\infty d\zeta\, K_0(\zeta), \tag{10.108b}$$

where $K_\nu(\zeta)$ is a modified Bessel function (Gradshteyn and Ryzhik, 1994). The full Salpeter equation, Eq. (10.106), can then be equivalently recast into the integral form

$$\Psi(\mathbf{r}) = -\int d^3\mathbf{r}'\, G_m(\mathbf{r}-\mathbf{r}')V(\mathbf{r}')\Psi(\mathbf{r}'). \tag{10.109}$$

Restricting to spherically symmetric potentials, $V(\mathbf{r}) \equiv V(r)$, so that $\Psi(\mathbf{r}) = r^{-1}u_\ell(r)Y_{\ell\mathrm{m}}(\hat{\mathbf{r}})$, reduces such an integral equation into a one-dimensional integral equation

$$u_\ell(r) = -\int_0^\infty dr'\, G_m^{(\ell)}(r,r')V(r')u_\ell(r'), \tag{10.110}$$

where

$$G_m^{(\ell)}(r,r') = \frac{1}{2}mrr' \int_0^\pi d\theta'\, \sin\theta' \frac{1}{|\mathbf{r}-\mathbf{r}'|}\left(1+\frac{2}{\pi}F(mr)\right)P_\ell(\cos\theta'), \tag{10.111}$$

with $P_\ell(\cos\theta')$ the Legendre polynomials.

Further exact expressions can be obtained in the ultrarelativistic regime ($m = 0$), in which $\lim_{m\to 0} m\left(1 + 2\pi^{-1}F(mr)\right) = 2(\pi r)^{-1}$. In this limit, the angular projected propagator $G_{m=0}^{(\ell)}(r,r')$ can be expressed as

$$G_{m=0}^{(\ell)}(r,r') = \frac{1}{\pi}Q_\ell\left(\frac{r^2+r'^2}{2rr'}\right), \tag{10.112a}$$

with $Q_\ell(\zeta)$ Legendre functions of the second kind, particular expressions at low angular momentum ℓ being

$$G_{m=0}^{(0)}(r,r') = \frac{1}{\pi}\log\left|\frac{r+r'}{r+r'}\right|, \tag{10.112b}$$

$$G_{m=0}^{(1)}(r,r') = \frac{1}{\pi}\left(\frac{r^2+r'^2}{2rr'}\log\left|\frac{r+r'}{r+r'}\right| - 1\right). \tag{10.112c}$$

Finally, for a discussion of the spinless Salpeter equation in the semirelativistic limit, we refer the reader to the review by Lucha and Scöberl (1999).

Chapter 11

Towards quantum gravity

Soon after the foundations of both quantum mechanics and general relativity were laid, during the first two decades of the 20th century, it became immediately clear to both scientific communities that the two theories claimed for some sort of unification. It was then felt as conceptually natural to endeavour at a quantization of general relativity, owing perhaps to the already overwhelming experimental confirmations of quantum mechanics over general relativity, for which only observational evidence could be available at that time (as it is mostly the case nowadays).[1] This quickly excited an intense research program aiming at a consistent formulation of a quantum theory of general relativity, with prominent names such as those of Rosenfeld (1930a,b) and Fierz (1939); Pauli and Fierz (1939); Fierz and Pauli (1939) immediately taking part in it (see Rovelli, 2000, for a historical perspective on quantum gravity).

It was quite clear at the very beginning that quantum mechanics and general relativity are fundamentally, let alone also technically, different in many respects. While general relativity is a deterministic theory, in the same sense as classical mechanics, quantum mechanics is inherently probabilistic (see Hardy, 2007, for a discussion of these aspects).[2] Besides, while quantum mechanics is essentially *linear* (the quantum states of a system belong to a linear space, *viz.* the Hilbert space, and the quantum superposition principle holds, leading to the celebrated Schrödinger's 'cat' paradox), the equations of general relativity are inherently *nonlinear*: while their linearized version admits 'gravitational waves' (yet to be observed), Ein-

[1]This tendency, which surmises some kind of superiority of quantum mechanics *vs* general relativity, is today debated by some workers, including Penrose (2014a,b), who plainly calls for a *gravitization* of quantum mechanics, instead. See Sec. 11.2.

[2]The apparent irreducibility of the two frameworks is captured by Albert Einstein's famous quote 'God does not play dice'.

stein's complete equations are also characterized by the existence of exact solutions in the form of solitons (Belinskiĭ and Zakharov, 1978), as is typical of nonlinear equations in general.

Another intrinsic difference between the two theories is the role of time, which is part of the spacetime continuum in general relativity, whereas it stands as the (preferential) direction along which 'evolution' takes place in quantum systems governed by an Hamiltonian operator.

One of the first formal and coherent formulations of a 'quantum gravidynamics', the so-called *canonical quantum gravity,* was that contained in the 'trilogy' of papers by DeWitt (1967a,b,c), which of course synthesized earlier contributions by Feynman (1963), as well as by Bergmann (1949); Bergmann and Brunings (1949), Dirac (1958, 1959), Misner (1957), and others (see DeWitt, 2009, for a posthumous account of the development of those ideas). In the canonical theory, spacetime is described as the foliation of space-like 3-dimensional submanifolds. The metric element is written as (DeWitt, 1967a)

$$ds^2 = g_{\mu\nu} dx^\mu dx^\nu = (-\alpha^2 + \beta_k \beta^k) dt^2 + 2\beta_i dx^i dt + \gamma_{ij} dx^i dx^j, \qquad (11.1)$$

where Roman indices $i, j, k = 1, 2, 3$ run over space coordinates, and γ_{ij} is therefore the metric tensor over each space-like submanifold of the foliation. Momenta π^{kl} are then extracted from the conventional Einstein Lagrangian density, as canonically conjugate to the 3-metric elements γ_{ij}, which act as the 'fields' of the theory. A Hamiltonian density is thus constructed in terms of the fields and its conjugate momenta, which is finally quantized. This results in what DeWitt (1967a) called the 'Einstein-Schrödinger equation', but that immediately afterwards was universally acknowledged as the Wheeler-DeWitt equation (Rovelli, 2000). This can be written as (DeWitt, 1967a)

$$\left(\frac{1}{2\sqrt{\gamma}} G_{ijkl} \pi^{ij} \pi^{kl} - \sqrt{\gamma}\, {}^{(3)}R \right) \Psi[\gamma] = 0. \qquad (11.2)$$

(See also Peres, 1962, for its equivalent formulation in the Hamilton-Jacobi framework). In Eq. (11.2), $\gamma = \det(\gamma_{ij})$, $G_{ijkl} = \gamma_{ik}\gamma_{jl} + \gamma_{il}\gamma_{jk} - \gamma_{ij}\gamma_{kl}$, ${}^{(3)}R$ is the intrinsic curvature of the submanifold (where contractions have been performed via γ_{ij}), depending only on γ_{ij} of that hypersurface, $\pi^{ij} = -i\hbar\delta/\delta\gamma_{ij}$ involves *functional* variation with respect to the 3-metric tensor, and $\Psi[\gamma]$ is a wave-*functional* of the 3-metric $\gamma_{ij} \equiv \gamma_{ij}(t, x^k)$. At variance with Schrödinger equation, which is a partial differential equation, the Wheeler-DeWitt equation, Eq. (11.2), is in fact a *functional differential equation.* Apart from the technical difficulties that this implies, an important aspect of Eq. (11.2) is that it cannot be interpreted as an evolution equation for the quantum state, but rather as a *primary constraint, viz.* the Hamiltonian constraint. This property is known as timelessness, and leads to the problem of time (Peres, 1968; Anderson, 2012), as it seems to imply that the Universe is 'stationary' or even static, and that time would only be a variable observable only intrinsically. A possible solution, which is still under discussion (Moreva *et al.*, 2014), to this open problem has been recently put forward by Page and Wootters (1983) and relies on the concept of quantum entanglement.

Other severe problems haunt quantum gravity theories, such as their non-renormalizability ('t Hooft, 1973; Deser and van Nieuwenhuizen, 1974a,b). More generally, any theory of quantum gravity has serious difficulties in proposing experiments through which it could be 'falsified', as these experiments should involve phenomena at or below the Planck length,

$$\ell_{\mathrm{P}} = \sqrt{\frac{\hbar G}{c^3}} \approx 10^{-35}\ \mathrm{m}, \tag{11.3}$$

or equivalently at or above the Planck energy scale, $\approx 10^{19}$ GeV (see Ellis and Silk, 2014; Castelvecchi, 2015, for a more general discussion).

The horizon encompassed by the title of the present Chapter is wider than a single book may contain. The general topic of quantum gravity is indeed wide and rapidly evolving, with several alternatives to the canonical theory being actively studied, including string theory (Polchinski, 1998a,b; Green *et al.*, 2012a,b) with its correspondence to quantum field theory within the AdS/CFT holographic duality (Maldacena, 1999; Gubser *et al.*, 1998; Witten, 1998, see also Sec. 9.4.1), spin foams and 'sum over histories' models (Reisenberger and Rovelli, 1997), and loop quantum gravity (Rovelli, 2004) (cf. Section 11.3 at the end of the Chapter). We refer the reader *e.g.* to Horowitz (2000); Booß-Bavnbek *et al.* (2010), and especially to Gracia-Bondía (2010), for a fairly recent account of the recent advances and open problems in this rapidly evolving field.

In the following, after a brief account of the graviton as laid out by Feynman *et al.* (1999), we shall limit ourselves at reviewing the so-called Schrödinger-Newton equation, and its possible implications towards a link between quantum mechanics and gravitation.

11.1 The graviton and its relevance for quantum gravity

There is, already, the book *Feynman lectures on gravitation* (Feynman *et al.*, 1999), which however summarizes some of R. P. Feynman's thinking in the 1962-63 academic year. But still, as always, the physical clarity of Feynman's thinking is worth, at least in part, attempting to recapture below.

Thus, on p. 11 of the above book, Feynman already emphasizes that a prediction of the quantum theory of gravitation had to be that the force would be mediated by the virtual exchange of some particle, termed the graviton. But half a century ago, Feynman could add 'There is apparently no hope of ever observing a graviton.' Nevertheless, this quantum aspect of gravitational waves has some basic physical properties. Thus, below we classify mass, spin and polarization of this quantum particle: the graviton.

11.1.1 *Mass and spin of the quantum of gravitation*

Feynman *et al.* (1999, p. 29: lecture 3) stress the first implication of the fact that gravity has a long range: namely that the field is carried by the exchange of the above particle, the graviton. Then to have the force proportional to $1/r^2$, the mass of this quantum particle has to be $m = 0$.

But to write down a field theory, it is essential also to know the spin of the graviton. The difference between electromagnetism and gravitation, as Feynman stresses, implies beyond reasonable doubt that the spin of the graviton must be some integer in the sequence 0, 1, 2 etc. Any of these spins would give an interaction potential proportional to $1/r$, since any radial dependence is exclusively determined by the mass.

Feynman rejects spin-zero theories of gravitation on the basis of the gravitational behaviour of the binding energies. Spin 1 has the consequence that likes repel and unlikes attract: in fact he stresses this as a property of all odd-spin theories. Conversely, it is also found that even spins lead to attractive forces, so that one needs to consider only spins 0 and 2 (he adds 4, if 2 fails), but at the time of writing our book, spin 2 is widely accepted. Indeed, it was recognized that an early fillip for string theory was that it contained within its fundamental framework a massless particle having spin 2.

11.1.2 *Comments on polarization*

Feynman turns in his third lecture to construct a spin-2 theory in analogy to the other field theories that exist, while he argues that at this point one could switch to Einstein's viewpoint on gravitation. However, Feynman instead attempts to arrive at a correct theory in analogy with electrodynamics.

He then stresses that in the theories of scalar, vector, and tensor fields (which he notes as another way of denoting spins 0, 1, and 2), the fields are described by scalar, vector, or tensor potential functions:

spin 0	χ	scalar potential,
spin 1	A_μ	vector potential,
spin 2	$h_{\mu\nu}$	symmetric tensor potential.

At this point, Feynman notes that while another theory would result from assuming that the tensor is antisymmetric, this would not lead to a description of gravity, but rather to something resembling electromagnetism.

This leads Feynman into a discussion of amplitudes for exchange of a graviton, but we must refer the interested reader to (Feynman *et al.*, 1999) for Feynman's account of this (see also Feynman, 1963).

The expression of the relativistic propagator for a particle with arbitrary spin, thereby also including the graviton, was derived by Fierz (1939); Pauli and Fierz (1939); Fierz and Pauli (1939).[3] The cross section for a graviton-graviton scattering in the tree approximation was evaluated by De Witt and Wesley (1971).

As emphasized by B. Hatfield in the Feynman lectures on gravitation (Feynman *et al.*, 1999, p. xxxi; see especially p. xi), a reformulation of general relativity in terms of new variables (Ashtekar, 1986, 1987) has led to a new loop representation of quantum general relativity (Jacobson and Smolin, 1988), where the equations are much easier to solve, and this has allowed some progress to be made.

[3]These authors pointed out that the most immediate approach, proposed by Dirac (cf. Chapter 10), leads to inconsistent equations as soon as the spin is greater than unity. To make this clear, Fierz and Pauli (1939) considered immediately Dirac's equation for a particle of spin $\frac{3}{2}$, and found such inconsistencies.

11.2 The Schrödinger-Newton equation

The so-called Schrödinger-Newton equation is a particular instance of the nonlinear Schrödinger equation, describing the quantum evolution of a self-gravitating mass distribution (Diósi, 1984; Penrose, 1996, 1998)

$$i\hbar\frac{\partial}{\partial t}\psi(\mathbf{r},t) = \left(-\frac{\hbar^2}{2m}\nabla^2 - Gm^2\int d\mathbf{r}'\frac{|\psi(\mathbf{r}',t)|^2}{|\mathbf{r}-\mathbf{r}'|}\right)\psi(\mathbf{r},t). \tag{11.4}$$

This equation belongs to a class of Schrödinger-Poisson problems (see Lange *et al.*, 1995, for a review), as it may be cast into the form

$$i\hbar\frac{\partial}{\partial t}\psi(\mathbf{r},t) = \left(-\frac{\hbar^2}{2m}\nabla^2 + mV(\mathbf{r},t)\right)\psi(\mathbf{r},t), \tag{11.5a}$$

$$\nabla^2 V(\mathbf{r},t) = 4\pi Gm|\psi(\mathbf{r},t)|^2. \tag{11.5b}$$

The stationary Schrödinger-Newton equation is then obtained from Eq. (11.4) assuming $\psi(\mathbf{r},t) \equiv \psi(\mathbf{r})^{-iEt/\hbar}$, which implies $V(\mathbf{r},t) \equiv V(\mathbf{r})$ in Eqs. (11.5), and it reads

$$\left(-E - \frac{\hbar^2}{2m}\nabla^2 - Gm^2\int d\mathbf{r}'\frac{|\psi(\mathbf{r}')|^2}{|\mathbf{r}-\mathbf{r}'|}\right)\psi(\mathbf{r}) = 0. \tag{11.6}$$

The Schrödinger-Newton equation was originally proposed by Ruffini and Bonazzola (1969) to describe a boson star, where the constituent particles are subject to their own mutual gravitational attraction, and more recently revived by Diósi (1984) and by Penrose (1996, 1998) to describe the collapse of the quantum wave function, *i.e.* the process by which a quantum system, initially prepared in a (linear) superposition of two or more quantum states, is reduced to a single, well-defined quantum state, as a consequence of a measurement process ('observation': *e.g.* a wave packet reducing to a coordinate eigenstate, as a result of a position measurement). While time evolution is described by the Schrödinger equation, which is a linear equation and therefore preserves any linear superposition of states, quantum collapse should then be due to some *nonlinear,* and thus presumably non-quantum, cause, which is usually assumed as the thermodynamically irreversible interaction of the quantum system and the environment (treated as classical). In the Schrödinger-Newton equation, this nonlinear cause would be brought about by the gravitational interaction.

Specifically, according to Penrose (1996, 1998), a superposition of two or more quantum states, ψ_1 and ψ_2, say, characterized by a significant mass distribution between the states, ought to be unstable and reduce to one of the states within a finite time, which can be estimated as $\tau_G \approx \hbar/E_G$, where $E_G \propto \frac{1}{2} \int d\mathbf{r} |\nabla(\psi_1 - \psi_2)|^2$, if a gravitational attraction between such mass distribution is accounted for, according to Eq. (11.4).

Moreover, a self-consistent nonlinear term in the Schrödinger equation would solve the apparent paradox that the centre of mass of a macroscopic, classical object, in the absence of any external force, either moves uniformly along a straight line or, in the inertial frame co-moving with the object, stays fixed, whereas a quantum wave packet (which is the closest approximation to a 'particle') widens in time according to Schrödinger equation, and is characterized by a steady increase of the indeterminacy of its centre-of-mass position. To quote Diósi's words, "the gravitational interaction possibly could prevent the unbounded quantum-mechanical spreading of the centre-of-mass position of macroobjects, at least in certain quantum states. If this interaction is included, it destroys the linearity of quantum mechanics" (Diósi, 1984). Indeed, Eq. (11.4) allows to estimate the characteristic width of the centre-of-mass wavefunction as $a_0 \approx \hbar^2/Gm^3$, for a point-like object of mass m, and as $a_0^{(R)} \approx a_0^{1/4} R^{3/4}$, for a homogeneous spherical mass distribution with radius R (Diósi, 1984). The condition $a_0^{(R)} = R$ would then yield a criterion for establishing a 'critical' size, below which objects behave quantum-mechanically. This can be estimated as $R_c \approx 10^{-5}$ cm for objects with normal densities (Diósi, 1984).

The idea that gravity could or should somehow influence quantum mechanical processes actually dates back to the 1960s (Károlyházy, 1966), when a semiclassical version of Einstein's equations was proposed (Møller, 1962; Rosenfeld, 1963)

$$R_{\mu\nu} + \frac{1}{2} g_{\mu\nu} R = \frac{8\pi G}{c^4} \langle \psi | \hat{T}_{\mu\nu} | \psi \rangle, \tag{11.7}$$

where $R_{\mu\nu}$ denotes the Ricci tensor associated with the spacetime metrics $g_{\mu\nu}$, and R is the curvature. In Eq. (11.7), the energy-momentum tensor for a given mass distribution, which in Einstein's equations is usually regarded as the 'source' term for spacetime curvature, has been replaced by a quantum operator, averaged in the quantum state ψ. Thus, while matter and its properties are treated quantum mechanically, spacetime curvature is treated classically. In other words, gravity is *not* quantized.

Such an assumption is of course quite controversial, and has been criticized by several authors from various respects. One constraint on the quantum state would of course be that the average of the energy-momentum operator does not violate energy-momentum conservation, *i.e.* $\partial^\mu \langle \psi | \hat{T}_{\mu\nu} | \psi \rangle = 0$. However, there are no specific restrictions to this condition (see Bahrami *et al.*, 2014, and references therein). It can then be proved that Schrödinger-Newton equation, Eq. (11.4), can be derived from the semiclassical Einstein's equations, Eq. (11.7), under quite general assumptions (Bahrami *et al.*, 2014).

That gravity should not be necessarily quantized, but rather quantum mechanics *gravitized,* is indeed a (provocative) proposal by Penrose (2014a,b). Several experiments have been proposed to test the hypothesis that spacetime could or should be treated as classical (*i.e.* non-quantum), when looking at the quantum evolution of many-particle systems (Penrose, 2000; Yang *et al.*, 2013; Bahrami *et al.*, 2014). These may be feasible nowadays, with the recourse to modern state-of-the-art optomechanics (Jones *et al.*, 2015).

11.2.1 *Solutions of the Schrödinger-Newton equation*

Despite its possible fundamental significance discussed above, and mainly because of its nonlinear character, few exact results only are available for the Schrödinger-Newton equation, Eq. (11.4). The Schrödinger-Newton equation can be classified as a kind of nonlinear Schrödinger equation, where self-gravitation coupling is described via a Hartree-like, self-consistent term. Introducing natural scales for energy, time, and length as $\epsilon_0 = 2G^2 m^5/\hbar^5$, $\tau_0 = \hbar/\epsilon_0$, and $\ell_0^2 = \hbar^2/(2m\epsilon_0)$, respectively, and rescaling variables and functions in Eqs. (11.4) and (11.5) as $\mathbf{r} \mapsto \ell_0 \mathbf{r}$, $t \mapsto \tau_0 t$, $\psi \mapsto \ell_0^{-3/2}\psi$, $V \mapsto mV/\epsilon_0$, $E \mapsto \epsilon_0 E$, the Schrödinger-Newton equation acquires the nondimensional form

$$i\frac{\partial\psi(\mathbf{r},t)}{\partial t} = -\nabla^2\psi(\mathbf{r},t) - \int d\mathbf{r}' \frac{|\psi(\mathbf{r}',t)|^2}{|\mathbf{r}-\mathbf{r}'|}\psi(\mathbf{r},t), \tag{11.8}$$

or

$$i\partial_t\psi = (-\nabla^2 + V)\psi, \tag{11.9a}$$

$$\nabla^2 V = 4\pi|\psi|^2, \tag{11.9b}$$

or, for the associated stationary problem,

$$\left(-E - \nabla^2 - \int d\mathbf{r}' \frac{|\psi(\mathbf{r}')|^2}{|\mathbf{r}-\mathbf{r}'|}\right)\psi(\mathbf{r}) = 0, \tag{11.10}$$

where all variables and functions are now dimensionless. It is remarkable that the Schrödinger-Newton equation in nondimensional form, Eq. (11.8), is free from any tunable parameter, at variance with the 'pure' nonlinear Schrödinger equation, where such a parameter appears as a coupling constant in front of the quartic, self-interaction term. From now on, we will refer to the nondimensional Schrödinger-Newton equation, Eqs. (11.8), (11.9), and (11.10).

Like the nonlinear Schrödinger equation, the Schrödinger-Newton equation conserves probability, which fact can be expressed globally or locally as

$$\int d\mathbf{r}\, |\psi(\mathbf{r},t)|^2 = 1, \tag{11.11a}$$

$$\partial_t |\psi|^2 + \partial_i J^i = 0, \tag{11.11b}$$

with

$$J_i = -i\left(\psi^* \partial_i \psi - \psi \partial_i \psi^*\right). \tag{11.11c}$$

Among other conservation properties, the Schrödinger-Newton equation preserves a suitably defined 'conserved energy'

$$\mathcal{E} = T + \frac{1}{2}U \equiv \int d\mathbf{r}\, |\nabla\psi|^2 + \frac{1}{2}\int d\mathbf{r}\, V|\psi|^2, \tag{11.12}$$

while the total energy $\equiv T + U$ is not conserved. Exploiting a variational derivation of the Schrödinger-Newton equation, and making use of standard inequalities, Tod (2001) showed that $\mathcal{E} \geq -1/(54\pi^4)$ for the ground state. Other variational constraints for T and U in terms of the conserved energy $\mathcal{E}$ could be derived analytically (Arriola and Soler, 1999; Tod, 2001).

Lie point symmetries of the equation were obtained and studied by Robertshaw and Tod (2006).

If $(\psi, V, \mathbf{r}, t)$ is a solution of the Schrödinger-Newton equation as a Schrödinger-Poisson problem, Eqs. (11.9), then another solution can be obtained via the re-scaling $(\psi, V, \mathbf{r}, t) \mapsto (\lambda^2\psi, \lambda^2 V, \lambda^{-1}\mathbf{r}, \lambda^{-2}t)$, while the energy (and the conserved action) of the corresponding stationary problem is also re-scaled as $E \mapsto \lambda^2 E$.

The proof of the existence and uniqueness of the ground state of the Schrödinger-Newton is equivalent to that for the Choquard equation, describing the Coulomb problem of a one-component plasma of many charged particles, which was obtained by Lieb (1977) (see also Illner *et al.*, 1994, 1998). Moroz *et al.* (1998) specifically proved that the spherically symmetric stationary equation, Eq. (11.10), possesses infinite real solutions, corresponding to a discrete, negative spectrum, E_n say. It is conjectured that the number of zeroes of the nth eigenstate along the r axis is exactly $n-1$ (which could only be verified numerically), and that E_n scales with n as $E_n = -\alpha/(n+\beta)^\gamma$, with $\alpha = 0.096$, $\beta = 0.76$, and $\gamma = 2.00$ (best-fit values against numerically determined eigenvalues Bernstein *et al.*, 1998).

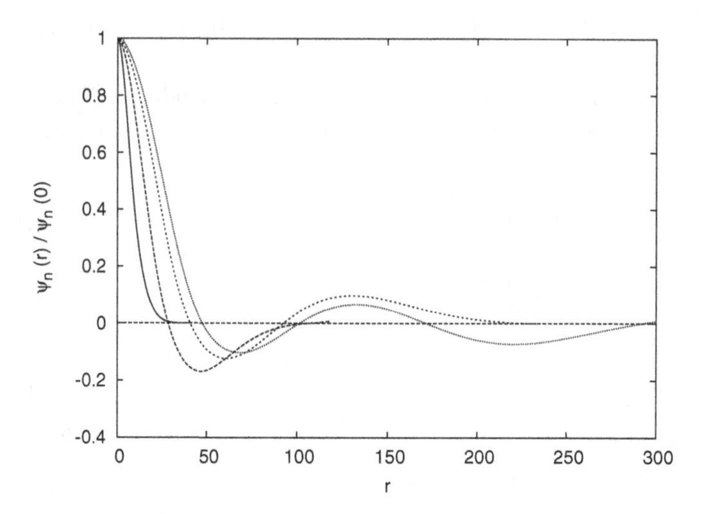

Fig. 11.1 Radial dependence of the stationary eigenstates $\psi_n(r)$ of the spherically symmetric, nondimensional Schrödinger-Newton equation, for $n = 1, 2, 3, 4$. Adapted from Harrison (2001) (see also Harrison *et al.*, 2003).

Numerical solutions of the Schrödinger-Newton equation have been obtained with various approximations, at various dimensionalities, and with different geometries (see *e.g.* Moroz *et al.*, 1998; Harrison, 2001; Harrison *et al.*, 2003). In Fig. 11.1, we reproduce the results of Harrison (2001); Harrison *et al.* (2003) for the radial part of the lowest four eigenstates of the spherically symmetric stationary Schrödinger-Newton equation.

11.3 Concluding remarks

In a paper by Clark (2016) the question is raised as to whether exploding black holes can reveal the 'true fabric of spacetime'. The experimental evidence cited there goes back to November 2012, when an intense burst of radio waves was observed above the Caribbean island of Puerto Rico. Apparently, radio astronomers had been waiting for a while for such a fast radio burst. Lasting but a few thousandths of a second, these super-bright pulses are thought to come from deep space, and are extremely rare.

Haggard and Rovelli (2015) (referred to by Clark, 2016) invoked the model known as loop quantum gravity (Rovelli, 2004). This proposed that spacetime is made of interlocking loops that form a fabric akin to chain mail. Looked at from afar, this fabric appears to be smooth and continuous, but viewed close up it is made from tiny indivisible pieces. These loops would represent the fundamental quanta of spacetime: nothing could be smaller.

When the above authors considered what loop quantum gravity would mean for black holes, they concluded that a black hole would eventually attain a density at which such loops of spacetime would shrink no further. They concluded there would be no singularity. Instead, they proposed that the loops would generate an outward pressure, resulting in a 'quantum bounce'—an explosion that they asserted would destroy the black hole. Haggard and Rovelli's bouncing black holes would, it appears, create a white hole. This is a massive object that emits particles but never absorbs them. Haggard and Rovelli's idea is that all the matter collapsing to form a black hole singularity can never actually reach that stage. As it nears the size of an individual spacetime loop, the probability that the entire black hole undergoes a quantum tunnelling event becomes greater and greater, until it suddenly becomes a white hole!

However, it has to be said, in concluding this Chapter, that there are those who think the process is too much of a conceptual leap. We expect this will be one area where either such a leap will be confirmed, or a plausible alternative explanation should be offered.

In fact, on September 14, 2015, the gravitational waves emitted by the merger of two massive black holes have been observed, thereby certainly initiating a new era of observational astrophysics. The observation resulted from the joint efforts of the Laser Interferometer Gravitational-wave Observatory (LIGO) in the USA and of the Virgo interferometer in Italy Abbott, B. P., *et al.* (LIGO Scientific Collaboration and Virgo Collaboration) (2016). From the analysis of the signal, these researchers inferred that the gravitational waves were produced by the merger of two black holes each with a mass exceeding 25 times the mass of our Sun. The catastrophic event would have released an energy equivalent to that of three solar masses. In addition to providing evidence for the existence of gravitational waves, *i.e.* oscillatory solutions to the equations of general relativity, something that had been predicted by Einstein exactly one century earlier (Einstein, 1916), the LIGO–Virgo experiment is probably the first nearly direct observation of a black hole itself, or even better of a black hole binary system. The observation of such events is so elusive, that it involved the detection of length variations of the order of 10^{-21}. In fact, the 'legs' of the Michelson interferometer employed in the LIGO experiment were as long as 4 km, with a sensitivity of less than the diameter of an atomic nucleus. It should be mentioned that is was theorist Felix Pirani (1965) who first pointed out that a connection between Newton's Second Law and the equations of geodesic deviation could be exploited to detect a gravitational wave.[4] Pirani (1965) also recognized the necessity of a quantum theory of gravitation (see discussion in Berti, 2016). After the LIGO–Virgo result, more observatories are planned to be built, both on the Earth, and in space stations, which will form a network of detectors for events of this kind, which will presumably enable not only to shed light on the physics of black holes, but of the Universe itself.

During an interview, Caltech physicist and general relativist Kip Thorne stated that 'a new window onto the Universe has been opened'.[5]

---

[4]Sadly, F. Pirani passed away on December 31, 2015, *i.e.* a few weeks after the official announcement of the observation of the gravitational wave was given, at a momentous conference transmitted world-wide also in webcast, on February 11, 2016.

[5]Recently, yet another, albeit indirect, possible evidence of the existence of gravitational vacuum fluctuations has been proposed by Mari *et al.* (2016). This relies on the possibility of entangling a particle with quantum gravitational radiation. Other condensed matter systems, such as optomechanical oscillators, have been recently proposed to test several quantum gravity scenarios, which imply the occurrence of nonlocal dynamics below the Planck scale (Belenchia *et al.*, 2016).

Appendix A

Feynman propagator for an arbitrary DFT-like one-body potential

March and Murray (1960b) analyzed central-field problems characterized by a local spherical potential $V(r)$, with $r = |\mathbf{r}|$, using the so-called canonical or Bloch density matrix. In terms of the (normalized) eigenfunctions $\psi_i(\mathbf{r})$ and corresponding eigenvalues ϵ_i generated by this potential,

$$[T(\mathbf{r}) + V(\mathbf{r})]\psi_i(\mathbf{r}) = \epsilon_i \psi_i(\mathbf{r}), \tag{A.1a}$$

$$T(\mathbf{r}) = -\frac{1}{2}\nabla_{\mathbf{r}}^2 = -\frac{1}{2}\sum_{\alpha=1}^{d} \frac{\partial^2}{\partial r_\alpha^2} \tag{A.1b}$$

(in d dimensions, in atomic units), the Bloch density matrix $C(\mathbf{r}, \mathbf{r}_0; \beta)$ at inverse temperature $\beta = (k_{\mathrm{B}}T)^{-1}$ is defined by

$$C(\mathbf{r}, \mathbf{r}_0, \beta) = \sum_{\text{all } i} \psi_i(\mathbf{r})\psi_i^*(\mathbf{r}_0)e^{-\beta\epsilon_i}. \tag{A.2}$$

On the other hand, the corresponding propagator (Feynman and Hibbs, 1965) is defined by replacing β in the right-hand side of Eq. (A.2) by imaginary time it. In the March and Murray (1960b) central-field case, this is a one-dimensional problem, and these authors derived a differential equation for the so-called Slater sum

$$S(\mathbf{r}, \beta) \equiv C(\mathbf{r}, \mathbf{r}_0 = \mathbf{r}, \beta), \tag{A.3}$$

which is essentially defined as the diagonal element of the Bloch density, Eq. (A.2). From this definition, and from Eq. (A.2), it is clear that the Slater sum is a generalization of the partition function $Z(\beta)$, defined by

$$Z(\beta) = \sum_{\text{all } i} e^{-\beta\epsilon_i}, \tag{A.4}$$

to embrace inhomogeneity, *i.e.* a spatial dependence. For a spherically symmetric potential $V(r)$, one finds that also the Slater sum is spherically symmetric, $S(\mathbf{r},\beta) \equiv S(r,\beta)$. In the one-dimensional case, $V(r) \equiv V(x)$, March and Murray (1960b) derived the following differential equation for the Slater sum (in $d = 1$ dimension, in atomic units):

$$\frac{1}{8}\frac{\partial^3 S(x,\beta)}{\partial x^3} = \frac{\partial^2 S(x,\beta)}{\partial x \partial \beta} + V(x)\frac{\partial S(x,\beta)}{\partial x} + \frac{1}{2}V'(x)S(x,\beta). \qquad \text{(A.5)}$$

Here, we will present a generalization of Eq. (A.5) to higher dimensions for the Slater sum $S(\mathbf{r},\beta)$ associated to a given local potential $V(\mathbf{r})$.

A.1 Generalization of the March-Murray equation for the Slater sum to $d > 1$

Following Holas and March (2006), we start by recalling the so-called differential virial equation Holas and March (1995, see also Section 1.1). The electron density $n(\mathbf{r})$ of an electron gas at energy E in an arbitrary one-body external potential $V(\mathbf{r})$ in d dimensions obeys the differential virial equation

$$-n(\mathbf{r})\nabla V(\mathbf{r}) = \mathbf{z}(\mathbf{r}) - \frac{1}{4}\nabla\nabla^2 n(\mathbf{r}). \qquad \text{(A.6)}$$

Here, the kinetic vector

$$\mathbf{z}(\mathbf{r}) = \boldsymbol{\mathcal{D}}(\mathbf{r}_1,\mathbf{r}_2)\,\rho(\mathbf{r}+\mathbf{r}_1,\mathbf{r}+\mathbf{r}_2;E)|_{\mathbf{r}_1=\mathbf{r}_2=\mathbf{0}} \qquad \text{(A.7)}$$

is defined in terms of derivatives of the Dirac density matrix

$$\rho(\mathbf{r},\mathbf{r}';E) = \sum_{\text{all } j} \Theta(E-\epsilon_j)\psi_j(\mathbf{r})\psi_j^*(\mathbf{r}'), \qquad \text{(A.8)}$$

with $\rho(\mathbf{r},\mathbf{r};E) \equiv n(\mathbf{r})$ at energy E, and $\Theta(z)$ is the usual Heaviside function. The differential operator $\boldsymbol{\mathcal{D}}$ has components (Holas and March, 1995, $\alpha = 1,\ldots d$)

$$\mathcal{D}_\alpha(\mathbf{r}_1,\mathbf{r}_2) = \frac{1}{2}\sum_{\beta=1}^{d}\left(\frac{\partial}{\partial r_{1\beta}} + \frac{\partial}{\partial r_{2\beta}}\right)\left(\frac{\partial^2}{\partial r_{1\alpha}\partial r_{2\beta}} + \frac{\partial^2}{\partial r_{1\beta}\partial r_{2\alpha}}\right), \qquad \text{(A.9)}$$

which can be further simplified, when acting on a real, symmetric density matrix, *i.e.* when ρ, Eq. (A.8), is constructed in terms of real orbitals only.

The Dirac density matrix, Eq. (A.8), is related to the Bloch density matrix, Eq. (A.2) by the Laplace transform (March and Murray, 1960b)

$$\beta^{-1}C(\mathbf{r}_1,\mathbf{r}_2,\beta) = \int_0^\infty dE\,\rho(\mathbf{r}_1,\mathbf{r}_2;E)e^{-\beta E}, \qquad \text{(A.10)}$$

which in turn relates the Slater sum $S(\mathbf{r},\beta)$, Eq. (A.3), to the density $n(\mathbf{r})$ at energy E. Making use of Eqs. (A.1), it is straightforward to show that the Bloch density matrix satisfies the Bloch equation

$$[T(\mathbf{r}_1)+V(\mathbf{r}_1)]C(\mathbf{r}_1,\mathbf{r}_2,\beta)=-\frac{\partial C(\mathbf{r}_1,\mathbf{r}_2,\beta)}{\partial\beta}. \tag{A.11}$$

Setting $\mathbf{r}_1=\mathbf{r}_2=\mathbf{r}$, one thus derives the β derivative of the Slater sum as

$$\frac{\partial S(\mathbf{r},\beta)}{\partial\beta}=-t(\mathbf{r},\beta)-V(\mathbf{r})S(\mathbf{r},\beta), \tag{A.12}$$

where the canonical kinetic energy density is defined as

$$t(\mathbf{r},\beta)=\frac{1}{2}[T(\mathbf{r}_1)+T(\mathbf{r}_2)]\ C(\mathbf{r}+\mathbf{r}_1,\mathbf{r}+\mathbf{r}_2,\beta)|_{\mathbf{r}_1=\mathbf{r}_2=\mathbf{0}}\,. \tag{A.13}$$

Finally, taking the Laplace transform of both sides of the differential virial equation, Eq. (A.6), one finds a differential equation for the Slater sum:

$$-S(\mathbf{r},\beta)\nabla V(\mathbf{r})=\mathbf{r}z(\mathbf{r},\beta)-\frac{1}{4}\nabla\nabla^2 S(\mathbf{r},\beta) \tag{A.14}$$

where now

$$\mathbf{z}(\mathbf{r},\beta)=\boldsymbol{\mathcal{D}}(\mathbf{r}_1,\mathbf{r}_2)\ C(\mathbf{r}+\mathbf{r}_1,\mathbf{r}+\mathbf{r}_2,\beta)|_{\mathbf{r}_1=\mathbf{r}_2=\mathbf{0}} \tag{A.15}$$

is in fact the Laplace transform of the kinetic vector $\mathbf{z}(\mathbf{r})$, Eq. (A.7).

It is then immediate (Holas and March, 2006) to show that Eq. (A.14) reduces to Eq. (A.5), when $d=1$. The merit of both Eqs. (A.5) and (A.14) is that it can be brought to a form where the β-derivative is eliminated, at the expense of involving spatial derivatives of the diagonal elements of the canonical density matrix C. Specific examples include the isotropic harmonic oscillator (Amovilli and March, 1995; Holas and March, 2006), for which $V(r)=\frac{1}{2}\omega^2 r^2$.

Appendix B

Phonons in Wigner crystals near melting

B.1 Jellium model

An extremely versatile, yet powerful, model to describe Coulomb correlations in globally neutral quantum fluids is the jellium model. In its simplest version, it describes the quantum dynamics of N electrons in a volume V, coulombically interacting with one another and with a positive background of equal charge, which is regarded as classical and uniform, *i.e.* having density $n_b(\mathbf{r}) \equiv n$, where $n = N/V$ is the electron density (see *e.g.* March *et al.*, 1995; March and Tosi, 2002; Fetter and Walecka, 2004; Giuliani and Vignale, 2005, for an introduction). Moreover, the background's dynamics is assumed 'frozen', within Born-Oppenheimer approximation (think of an electron liquid flowing through a lattice of positive ions fixed in their equilibrium positions). The model's Hamiltonian then reads

$$H = H_e + H_{eb} + H_b, \tag{B.1a}$$

$$H_e = \sum_{i=1}^{N} \frac{p_i^2}{2m} + \frac{1}{2} \sum_{i \neq j} \frac{e^2}{|\mathbf{r}_i - \mathbf{r}_j|}, \tag{B.1b}$$

$$H_{eb} = -e^2 \sum_{i=1}^{N} \int d\mathbf{r} \frac{n_b(\mathbf{r})}{|\mathbf{r} - \mathbf{r}_i|}, \tag{B.1c}$$

$$H_b = \frac{e^2}{2} \int d\mathbf{r} \int d\mathbf{r}' \frac{n_b(\mathbf{r}) n_b(\mathbf{r}')}{|\mathbf{r} - \mathbf{r}'|}, \tag{B.1d}$$

where only the electron variables $\mathbf{r}_i$, $\mathbf{p}_i$ $(i = 1, \ldots N)$ are treated at the quantum level. Although charge neutrality is manifestly fulfilled, the jellium model is ill-conditioned in the thermodynamic limit, as all three contributions to Eq. (B.1a) tend to infinity, as $V \to \infty$. This annoying drawback is effectively regularized by cutting-off the Coulomb interaction at large distances via a (physically consistent) effective Yukawa-like potential, say $V(r) = -e^2/r \mapsto -e^2 e^{-\kappa r}/r$, where the phenomenological parameter κ is required to satisfy $V\kappa^d \to \infty$, as $V \to \infty$ and $\kappa \to 0$, where $d = 3$ is the dimensionality of the system under consideration. This is equivalent to keep the range of the interaction well confined within the system's boundaries. One then observes (Fetter and Walecka, 2004) that the diverging contributions to each term in Eq. (B.1a) come from the long-wavelength ($q \to 0$) limit, and that these terms cancel exactly under the assumption of charge neutrality, in the thermodynamic limit. Passing to Fourier space and performing second quantization, Eq. (B.1a) then becomes (in $d = 3$) (Fetter and Walecka, 2004)

$$H \equiv T + V = \sum_{\mathbf{k}\sigma} \epsilon_{\mathbf{k}\sigma} a^\dagger_{\mathbf{k}\sigma} a_{\mathbf{k}\sigma} + \frac{e^2}{2V} \sum_{\mathbf{kpq}}{}' \sum_{\sigma_1\sigma_2} \frac{4\pi}{q^2} a^\dagger_{\mathbf{k}+\mathbf{q},\sigma_1} a^\dagger_{\mathbf{p}-\mathbf{q},\sigma_2} a_{\mathbf{p}\sigma_2} a_{\mathbf{k}\sigma_1}, \tag{B.2}$$

where $a^\dagger_{\mathbf{k}\sigma}$ ($a_{\mathbf{k}\sigma}$) creates (annihilates) an electron with wavevector $\mathbf{k}$ and spin projection σ, $\epsilon_{\mathbf{k}\sigma}$ is the dispersion relation of the electron liquid under consideration ($\epsilon_{\mathbf{k}\sigma} = \hbar^2 k^2/2m$ for free electrons, or electrons described within a plane-wave basis), and the prime in the summation excludes long-wavelength contributions $q \to 0$ from the Coulomb potential.

It is then convenient to parametrize the electronic density by means of the (dimensionless) Wigner-Seitz radius r_s, defined as the radius (in a.u.) of a sphere whose volume is equal to the mean volume per particle (Ashcroft and Mermin, 1976), *viz.* $\frac{4}{3}\pi(r_s a_0)^3 N = V$, in $d = 3$, where $a_0 = \hbar^2/me^2$ is Bohr radius. Then, standard scaling analysis shows (Fetter and Walecka, 2004) that the two terms in the Hamiltonian, Eq. (B.2), scale as $T \sim r_s^{-2}$ and $V \sim r_s^{-1}$, respectively. This suggested that the potential term could be treated perturbatively in the high-density limit, $r_s \to 0$. Indeed, within standard first-order perturbation theory around the ground state, *i.e.* the Fermi sea, the ground-state energy per particle in $d = 3$ is found to be (Fetter and Walecka, 2004)

$$\left.\frac{E_0}{N}\right|_{r_s \to 0} = \frac{1}{r_s^2}(a + br_s + \ldots) \text{ [Ry]}, \tag{B.3}$$

where $1 \text{ Ry} = e^2/2a_0 \approx 13.6$ eV, and $a = 2.21$, $b = -0.916$ for $d = 3$. Such an expression has a local minimum at a finite value of r_s, whose density and energy compare well with the observed values for the lighter alkalis, say Na, under normal conditions (Fetter and Walecka, 2004; Ashcroft and Mermin, 1976).

The remaining terms in Eq. (B.3) are denoted as the correlation energy of the jellium model. Higher order perturbation theory however shows that the remaining contributions to Eq. (B.3) are generally not simple powers of r_s, the next leading term $\approx -0.094 + 0.0622 \log r_s$ due to exchange being logarithmic (Macke, 1950; Gell-Mann and Brueckner, 1957; Sawada, 1957), thus showing that the model cannot be treated perturbatively (see Fetter and Walecka, 2004, for a comprehensive review). (See also Giuliani and Vignale, 2005, for a comprehensive review also in systems with reduced dimensionality, $d = 2$ and $d = 1$.)

B.2 Low-density limit: Wigner crystallization

As the dimensionless length r_s increases (*i.e.* as the electron density decreases), the kinetic energy contribution to Eq. (B.2) becomes less and less important relative to the potential energy associated with Coulomb correlations. The latter eventually dominates, as the interparticle separation is always shorter than the interaction range, which is infinite, in the case of Coulomb repulsion.[6] It is therefore the kinetic, rather than the potential, term that might be treated as a perturbation, in the low-density limit.

Rather than attacking the problem perturbatively, however, Wigner (1934) proposed that a system of interacting electrons at low-density, as described by the jellium model, should crystallize at fixed positions, thereby forming a lattice, which has since been termed a Wigner crystal. This phase would correspond to a spontaneous breaking of the continuous translational symmetry endowed by the jellium Hamiltonian, Eq. (B.1a). As this takes place for some critical value of r_s, intermediate between high and low density, and at zero temperature, it can be viewed as an instance of a quantum phase transition.

The structure of a Wigner crystal lattice cannot be determined a priori, as is the case for ionic lattices. One may compare the electrostatic potential energies corresponding to several given possible alternative structures, and determine which is lowest in energy, at a given density. This of course does not exclude the occurrence of other symmetries. It turns out that in a uniform background a Wigner crystal forms with body-centred cubic (bcc) structure in $d = 3$, and with hexagonal structure in $d = 2$. The stability of a Wigner crystal, *i.e.* the existence of a rigorous lower bound for its

[6]This makes a charged system inherently different from a system of neutral particles, for which the interaction range is finite, and often relatively short.

ground-state energy E_0 at all values of the Wigner-Seitz radius, has been proved by Lieb and Narnhofer (1975), who showed that (energies in Ry)

$$\frac{E_0}{N} > \frac{3}{5}\left(\frac{9\pi}{4}\right)^{2/3}\frac{1}{r_s^2} - \frac{9}{5}\frac{1}{r_s}. \tag{B.4}$$

Denoting by $\mathbf{R}_i$ ($i = 1, \ldots N$) the fixed positions occupied by the N electrons within a Wigner crystal at equilibrium, the configurational potential energy is given by the classical expression

$$U = \frac{1}{2}\sum_{i\neq j}\frac{e^2}{|\mathbf{R}_i - \mathbf{R}_j|} - \sum_i \int d\mathbf{r}\frac{ne^2}{|\mathbf{r} - \mathbf{R}_i|} + \frac{1}{2}\int d\mathbf{r}\int d\mathbf{r}'\frac{n^2e^2}{|\mathbf{r} - \mathbf{r}'|}. \tag{B.5}$$

Again, each term in Eq. (B.5) above diverges, their compensation being guaranteed by charge neutrality.

A rough estimate of the configurational potential energy U can be obtained by assuming the crystal as composed by neutral unit cells of spherical shapes, each of radius r_s (in a.u.), each containing exactly one electron at its centre. The electrostatic energy of a positive, uniformly charged sphere with a point negative charge at its centre can be readily estimated. An assembly of such globally neutral spheres do not interact in space (by Gauss' theorem), therefore the overall electrostatic energy is simply equal to the energy of a single cell times their number, yielding a result (in a.u.)

$$\frac{U}{N} \approx -\frac{1.8}{r_s}. \tag{B.6}$$

This definitely (*i.e.* for sufficiently large r_s) overwhelms the kinetic contribution, which is shown to scale as $r_s^{-3/2}$ for large r_s in $d = 3$.

An exact, although still numerically inconvenient, alternative expression for U was suggested by Overhauser, yielding (Giuliani and Vignale, 2005)

$$\frac{U}{N} = \frac{n}{2}\sum_{\mathbf{G}\neq\mathbf{0}} V(|\mathbf{G}|) - \frac{1}{2V}\sum_{\mathbf{q}} V(q). \tag{B.7}$$

Here, $\mathbf{G}$ denotes a reciprocal lattice vector, and $V(q) = 4\pi e^2/q^2$ is the Fourier transform of the Coulomb potential in $d = 3$. Even though again each term in Eq. (B.7) is separately divergent, one expects that the crystal structure with the lowest configurational potential energy corresponds to that having the first reciprocal lattice vector with the largest magnitude, so as to minimize the first term in Eq. (B.7). This is indeed the case for the bcc structure at a given electron density, among other cubic structures in $d = 3$ (see also Rapisarda and Senatore, 1996; Drummond *et al.*, 2004).

Following an original method due to Ewald (1921), Bonsall and Maradudin (1977) were able to express exactly the configurational potential energy of a Wigner crystal as the sum of two summations, both rapidly converging, one over the reciprocal lattice, and the other over the direct lattice:

$$\frac{U}{N} = -\frac{ne^2}{2}\left(\frac{\pi}{\eta}\right)^{(d-1)/2}\left(\frac{2}{d-1} - \sum_{\mathbf{G}\neq\mathbf{0}} \phi_{\frac{d-3}{2}}\left(\frac{G^2}{4\eta}\right)\right) - \frac{e^2}{2}\left(\frac{\eta}{\pi}\right)^{1/2}\left(2 - \sum_{\mathbf{R}\neq\mathbf{0}} \phi_{-\frac{1}{2}}(\eta R^2)\right). \quad \text{(B.8)}$$

Here, $\phi_\nu(z) \equiv \int_1^\infty dt\, e^{-zt} t^\nu$ denotes a Misra function (Gradshteyn and Ryzhik, 1994), and η is an adjustable parameter, usually taken to be $\eta = \pi n^{2/3}$ in $d = 3$.

A major advance in the theoretical study of the Wigner crystal, and of the electron liquid in general, was provided by the introduction of numerical Quantum Monte Carlo (QMC) methods to evaluate the ground-state energies of such a correlated quantum liquid (Ceperley and Alder, 1980). These studies in particular indicated that with increasing r_s the ground state of the jellium in $d = 3$ dimensions undergoes a continuous transition from the paramagnetic (spin-disordered) fluid state to a ferromagnetic (spin-aligned) fluid state, and then a first-order transition to a ferromagnetic crystal at $r_s \approx 65$. Similar studies of the jellium in $d = 2$ indicated a first-order transition from a paramagnetic to a ferromagnetic fluid, and crystallization into a triangular lattice at $r_s \approx 35$ Rapisarda and Senatore (1996). (For an extensive review, see also Giuliani and Vignale, 2005, and references therein.)

B.2.1 *Experimental observation of Wigner crystallization*

The first unambiguous observation of Wigner crystallization was in an electron liquid suspended on top of a ^{4}He droplet (Grimes and Adams, 1979). Metal-insulator transitions due to interparticle correlations have been subsequently reported in quasi-bidimensional systems of carriers in semiconductor structures subject to very strong magnetic fields (Buhmann *et al.*, 1991) or in the presence of disorder (Kravchenko *et al.*, 1994, 1996), and also in samples with high purity and high carrier mobility (Yoon *et al.*, 1999).

More recently, evidence of the formation of electron lattice structures has been observed in $d = 1$ quantum wires (Hew *et al.*, 2009; Meyer and Matveev, 2009).

A notable exception is that of graphene. In such atomically thin, and therefore 'perfectly' $d = 2$ system, the linear low-energy quasiparticle dispersion relation conjures against the formation of a spatially ordered structure with broken translational symmetry, since the kinetic and potential energies both scale identically with the density of carriers. However, Wigner crystal phases have been predicted theoretically in the presence of a magnetic field (Zhang and Joglekar, 2007).

B.3 Elementary excitations in Wigner crystals

Wigner crystallization occurs at low temperatures, *i.e.* for $k_{\mathrm{B}}T \ll e^2/(r_s a_0)$. Above some critical temperature, on quite general grounds, a Wigner crystal should melt into an electron liquid. However, even within this admittedly restricted temperature range, one should expect quantum fluctuations of the electrons around their equilibrium positions, as for atoms and ions in conventional crystalline solids. This leads to quantum corrections to the equilibrium Hamiltonian, which to quadratic order in the displacements $\mathbf{u}_i \equiv \mathbf{r}_i - \mathbf{R}_i$ from equilibrium can be written as

$$H \approx \sum_i \frac{p_i^2}{2m} + U(\{\mathbf{R}_i\}) + \frac{1}{2}\sum_{ij} u_{i\alpha} V_{ij,\alpha\beta} u_{j\beta}, \tag{B.9}$$

where $V_{ij,\alpha\beta}$ are the second derivatives of the potential energy function with respect to the electron position coordinates $r_{i\alpha}$ and $r_{j\beta}$, calculated at the equilibrium positions, $\mathbf{r}_i \equiv \mathbf{R}_i$. Standard quantization procedures then lead to phonon modes, characterized by a dispersion relation $\omega_{\mathbf{q}\lambda}$, with λ labelling the phonon branch under consideration for a given lattice structure. Since $V_{ij,\alpha\beta} \sim e^2/(r_s^3 a_0^3)$, in view of the Coulombic nature of the interaction between two electrons, one has $\hbar\omega \sim \sqrt{V_{ij,\alpha\beta}/m} \sim r_s^{-3/2}$ Ry, as anticipated. This leads to a quantum correction (Bonsall and Maradudin, 1977, in $d = 3$)

$$\left.\frac{\Delta U}{N}\right|_{\text{harmonic}} = \frac{1}{N}\sum_{\mathbf{q}\lambda} \frac{\hbar\omega_{\mathbf{q}\lambda}}{2} \approx \frac{2.66}{r_s^{3/2}}\,\text{Ry} \tag{B.10}$$

to the ground-state energy at equilibrium, due to phonon excitations within this harmonic approximation.

B.3.1 *Wigner crystals in 2DEG at high magnetic field*

In two-dimensional electron-gas (2DEG) systems, such as those formed in semiconductor inversion layers, *e.g.* a metal-oxide-semiconductor field-effect transistor (MOSFET), a strong magnetic field B applied in the direction perpendicular to the plane can help stabilizing a Wigner crystal of electrons. In a magnetic field, an appropriate basis of the Hamiltonian is provided by the Landau levels (LLs), whose energy is quantized as $\epsilon_n = \hbar\omega_c(n + \frac{1}{2})$, where $\omega_c = eB/mc$ is the cyclotron frequency. In the strong-magnetic-field limit, all electrons have the minimum quantized kinetic energy, *i.e.* they are all in the lowest Landau level (LLL). The large quantized kinetic energy allows the electrons to be localized to a length comparable to the classical Larmor radius $\ell = \sqrt{\hbar c/eB}$ of the cyclotron orbit without any further cost in kinetic energy. With increasing field B, the magnetic length ℓ becomes small compared to the typical distance between electrons, and crystallization occurs. In zero field the Wigner crystal can only exist below some critical density. For any density, however, crystallization will occur for a sufficiently strong magnetic field. In particular, the high field limit $\hbar\omega_c \gg e^2/a_0$ excludes LL mixing due to electron-electron interactions.

In the presence of a magnetic field, a Wigner crystal phase, characterized by a broken translational symmetry, is in competition with incompressible fluid states, as induced by the fractional quantum Hall effect (Tsui *et al.*, 1982; Laughlin, 1983a,b). This is stable when the LL filling factor $\nu = 2\pi\ell^2 n = nhc/eB$, with n the electron number per unit area, is the reciprocal of an odd integer. Apart from those special values, a Wigner crystal has been claimed to be stable for $\nu \leq \frac{1}{4}$ (Andrei *et al.*, 1988; Goldman *et al.*, 1988) (see also Lea *et al.*, 1991, and references therein).

The dispersion relations of the normal modes of a Wigner crystal in $d = 2$ dimensions in the presence of a magnetic field have been calculated both in the classical (Bonsall and Maradudin, 1977) and in the quantum limit (Fukuyama, 1975), within the harmonic approximation. For a high applied magnetic fields, these findings suggest the existence of two phonon branches, the low-frequency one being nearly transverse and with an unusual $\sim k^{3/2}$ dispersion, and the high-frequency one being nearly longitudinal, with a frequency close to ω_c and corresponding to the classical magnetoplasmon in the long-wavelength limit.

The harmonic approximation is however justified only for sufficiently small filling factors even at low temperature, as an increasing ν would imply larger ℓ, and therefore larger zero-point fluctuations around the equilibrium positions. One may then expect that anharmonic effects may modify the magnetophonon dispersion relations at high magnetic field. Earlier treatments of anharmonic effects were equivalent to an effective harmonic approximation, where the force constant is averaged over the motion of the particles.

One successful approach, which does not treat perturbatively the electron displacement from the equilibrium position, consists in extracting the magnetophonon frequencies from the density-density response function within the time-dependent Hartree-Fock approximation (TDHFA), where they appear as poles corresponding to collective excitations (Côté and MacDonald, 1990, 1991). This approach is based on a mean-field picture of the lattice ground state in which each particle moves in a self-consistent potential which attracts the particles to the lattice sites. However, the particles are allowed to be itinerant, rather than assumed to be bound to a particular lattice site. For itinerant particles, it is no longer possible to make use of the displacement of a specific particle from its equilibrium position. Therefore, one is led to consider the density-density response function, which is also directly relevant experimentally, rather than the displacement-displacement response function, which appears naturally in the RPA-phonon theory.

Côté and MacDonald (1991) then find that, for filling factor $\nu \leq \frac{1}{2}$, the TDHFA phonons are only slightly different from those calculated by the usual harmonic approximation. For filling factor $\frac{1}{3} < \nu \leq 0.45$, however, exchange effects become important and modify considerably the dispersion relation until the crystal starts to soften at $\nu \approx 0.45$. Moreover, additional collective modes occur near multiples of the cyclotron frequency ω_c. These modes are absent in the harmonic approximation and are analogous to excitations which occur in the fluid state. Extensions of these results to finite temperature are not believed to be relevant for the melting of the Wigner crystal, which in $d = 2$ should take place according to a Kosterlitz-Thouless mechanism (Parrinello and March, 1976; Ferraz *et al.*, 1978, 1979; March and Tosi, 1985; Nagara *et al.*, 1987; March, 1988). (See, in particular, Chapter 9 of March, 1996, for a general review on Wigner crystal melting.)

The phonon spectrum of the 2D Wigner crystal in the presence of a strong magnetic field has been calculated also by Ferconi and Vignale (1992) using a density functional approach. Such a method could be used to relate the phonon dispersion curves in a crystal near melting to the structure factor of its liquid near freezing (Ferconi and Tosi, 1991a,b), and to extract vibrational and elastic properties of the Wigner crystal near melting (Tozzini and Tosi, 1993).

Appendix C

Lattice chains, Bethe Ansatz and polynomials for energies *E* of ground and excited states

Lieb and Wu (1968a,b) have solved exactly a short-range, one-band model in one dimension (1D). The account below follows closely their method. They first stress the widespread interest in the effect of correlation in a partially fixed energy band. A number of authors (Gutzwiller, 1963; Hubbard, 1963, 1964a,b, 1965, 1967a,b; Kemeny, 1965a,b) have considered such a case in 1D.

In Lieb and Wu; Lieb and Wu's model (LW, below), one considers electrons in such a narrow energy band as a hopping between Wannier states of neighbouring lattice sites (their number being N_a), a repulsive interaction energy being taken between two electrons of opposite spins occupying the same lattice site. LW point out that they appeal to a general theorem of Lieb and Mattis (1962): this theorem holds also for the Hamiltonian, Eq. (C.1), studied below. This theorem yields the result that the ground-state is always antiferromagnetic.

Turning to the Hamiltonian employed by Lieb and Wu (1968a,b), it takes the form

$$H = t \sum_{\langle ij\rangle,\sigma} c^{\dagger}_{i\sigma} c_{i\sigma} + U \sum_{i} c^{\dagger}_{i\uparrow} c^{\dagger}_{i\uparrow} c^{\dagger}_{i\downarrow} c^{\dagger}_{i\downarrow}. \tag{C.1}$$

In Eq. (C.1), $c^{\dagger}_{i\sigma}$ and $c_{i\sigma}$ are respectively creation and annihilation operators for an electron having spin σ in the Wannier state on the ith lattice site. The summation over $\langle ij\rangle$ is restricted to nearest-neighbour sites.

Lieb and Wu (1968a,b) then note that the energy spectrum under the sign change $t \mapsto -t$. Therefore, in the interests of simplicity, they assume $t = -1$ in appropriate units. Denote M and M' the number of down and upward spin electrons, respectively, with $M + M' = N$, the total number of electrons. As these are good quantum numbers, the ground-state energy corresponding to Eq. (C.1) can then be written as $E = E(M, M'; U)$.

Considering holes instead of electrons, the following relations can be proved to hold (Lieb and Wu, 1968a):

$$\begin{aligned} E(M, M'; U) &= -(N_a - M - M')U \\ &\quad + E(N_a - M, N_a - M'; U) &\text{(C.2a)} \\ &= MU + E(M, N_a - M'; -U) &\text{(C.2b)} \\ &= M'U + E(N_a - M, M'; -U). &\text{(C.2c)} \end{aligned}$$

Without loss of generality, they write

$$S_z = \frac{1}{2}(N - 2M) \geq 0 \text{ and } N \leq N_a, \tag{C.3}$$

i.e. less than a half-filled band. LW show that the maximum energy $G(M, M'; U)$ is related to the above ground state by

$$G(M, M'; U) = M'U - E(N_a - M, M'; U). \tag{C.4}$$

For a 1D chain, denote with $f(x_1, \ldots x_M, x_{M+1}, \ldots x_N)$ the amplitude of the wave-function for which the down spins are at the sites $x_1, \ldots x_M$, while the up spins are at $x_{M+1}, \ldots x_N$. We explicitly require that such an amplitude be separately antisymmetric in the first M and in the last $N - M$ variables. The eigenvalue equation for the wave-function then leads to

$$\begin{aligned} -\sum_{i=1}^{N} \sum_{s=\pm 1} & f(x_1, \ldots x_i + s, \ldots x_N) \\ & + U \sum_{i<j} \delta(x_i - x_j) f(x_1, \ldots x_N) = E f(x_1, \ldots x_N). \end{aligned} \tag{C.5}$$

LW further make the Ansatz that f has the form

$$f(x_1, \ldots x_M, x_{M+1}, \ldots x_N) = \sum_P c_{Q,P} \exp\left(i \sum_{j=1}^{N} k_{Pj} x_{Qj} \right) \tag{C.6}$$

in each region defined by $1 \leq x_{Q1} \leq \ldots x_{QN} \leq N$. Here, $P = (P1, \ldots PN)$, $Q = (Q1, \ldots QN)$ are two independent permutations of the integers $1, \ldots N$, $\{k_1, \ldots k_N\}$ is a set of real numbers, with $k_i \neq k_j$ for $i \neq j$, and $c_{Q,P}$ is a set of $N! \times N!$ coefficients to be determined. These are constrained by the requirement of continuity and single-valuedness of f, and by the condition that Eq. (C.6) be a solution of Eq. (C.5). For the total energy LW find then

$$E = -2 \sum_{j=1}^{N} \cos k_j, \tag{C.7}$$

while the coefficients $c_{Q,P}$ must satisfy

$$c_{Q,P} = Y^{ab}_{nm} c_{Q,P'}, \tag{C.8a}$$

$$Y^{ab}_{nm} = \frac{-\frac{1}{2}iU}{\sin k_n - \sin k_m + \frac{1}{2}iU} + \frac{\sin k_n - \sin k_m}{\sin k_n - \sin k_m + \frac{1}{2}iU} P^{ab}, \tag{C.8b}$$

where,

$$Qi = a = Q'j, \tag{C.9a}$$
$$Qj = b = Q'i, \quad \text{for } j = i + 1, \tag{C.9b}$$
$$Qk = Q'k, \quad \text{for all } k \neq i, j, \tag{C.9c}$$
$$Pi = m = P'j, \tag{C.9d}$$
$$Pj = n = P'i, \quad \text{for } j = i + 1, \tag{C.9e}$$
$$Pk = P'k, \quad \text{for all } k \neq i, j, \tag{C.9f}$$

and P^{ab} is an operator which exchanges $Qi = a$ and $Qj = b$. By making use of previous results by Gaudin (1967) and Yang (1967), LW determine the set of k_j as

$$N_a k_j = 2\pi I_j + \sum_{\beta=1}^{M} \theta(2 \sin k_j - 2\Lambda_\beta), \quad j = 1, \ldots N, \tag{C.10}$$

where

$$-\sum_{j=1}^{N} \theta(2\Lambda_\alpha - 2 \sin k_j) = 2\pi J_\alpha - \sum_{\beta=1}^{M} \theta(\Lambda_\alpha - \Lambda_\beta), \quad \alpha = 1, \ldots M, \tag{C.11a}$$

$$\theta(p) \equiv -2\,\mathrm{atan}(2p/U), \tag{C.11b}$$

with $\pi \leq \theta \leq \pi$, and I_j an integer (resp., an half-odd integer) for M even (resp., odd), and J_α an integer (resp., an half-odd integer) for M' odd (resp., even). It can be verified that

$$\sum_{j=1}^{N} k_j = \frac{1}{N_a} \left(\sum_j I_j + \sum_\alpha J_\alpha \right). \tag{C.12}$$

In the ground state, J_α and I_j are consecutive integers (or half-integers) centred around zero and obeying the condition $\sum_j k_j = 0$.

In the thermodynamic limit ($N \to \infty$, $N_a \to \infty$, $M \to \infty$, but N/N_a and M/N_a kept finite), k and Λ become real variables distributed continuously, with $-Q \leq k \leq Q \leq \pi$, and $-B \leq \Lambda \leq B \leq \infty$. From Eqs. (C.10),

(C.11a), their distributions $\rho(k)$, $\sigma(\Lambda)$ are then seen to obey the coupled integral equations

$$2\pi\rho(k) = 1 + \cos k \int_{-B}^{B} \frac{8U\sigma(\Lambda)d\Lambda}{U^2 + 16(\sin k - \Lambda)^2}, \quad \text{(C.13a)}$$

$$\int_{-Q}^{Q} \frac{8U\rho(k)dk}{U^2 + 16(\sin k - \Lambda)^2} = 2\pi\sigma(\Lambda) + \int_{-B}^{B} \frac{4U\sigma(\Lambda')d\Lambda'}{U^2 + 4(\Lambda - \Lambda')^2}, \quad \text{(C.13b)}$$

these distributions being normalized as $\int_{-Q}^{Q} \rho(k)dk = N/N_a$, $\int_{-B}^{B} \sigma(\Lambda)d\Lambda = M/N_a$. The ground state energy Eq. (C.7) can then be expressed as

$$E = -2N_a \int_{-Q}^{Q} \rho(k) \cos k dk, \quad \text{(C.14)}$$

and is of course an extensive quantity in the number of lattice sites N_a. The absolute ground state is antiferromagnetic with $S_z = 0$, $M/N = \frac{1}{2}$ for $B = \infty$. In the half-filled band ($N/N_a = 1$), corresponding to $Q = \pi$, the problem can be solved in closed form by means of Fourier transforms, the quantities of interest being

$$\sigma(\Lambda) = \frac{1}{2\pi} \int_0^\infty \operatorname{sech}\left(\frac{1}{4}\omega U\right) \cos(\omega\Lambda) J_0(\omega)\, d\omega, \quad \text{(C.15a)}$$

$$\rho(k) = \frac{1}{2\pi} + \frac{1}{\pi} \cos k \int_0^\infty \frac{\cos(\omega \sin k) J_0(\omega)\, d\omega}{1 + \exp(\frac{1}{2}\omega U}, \quad \text{(C.15b)}$$

$$E \equiv E(\frac{1}{2}N_a, \frac{1}{2}N_a; U)$$

$$= -4N_a \int_0^\infty \frac{J_0(\omega)J_1(\omega)\, d\omega}{\omega\left(1 + \exp(\frac{1}{2}\omega U)\right)}, \quad \text{(C.15c)}$$

where $J_0(\omega)$, $J_1(\omega)$ are Bessel functions of the first kind in the given arguments.

The chemical potential is then defined as the variation in ground-state energy by adding or subtracting one spin, *i.e.*

$$\mu_+ = E(M+1, M; U) - E(M, M; U), \quad \text{(C.16a)}$$

$$\mu_- = E(M, M; U) - E(M-1, M; U). \quad \text{(C.16b)}$$

Lieb and Wu (1968a) then characterize the system as conducting, if $\mu_+ = \mu_-$, or insulating, if $\mu_+ > \mu_-$. From Eqs. (C.10), (C.11a), one can compute μ_- by replacing M by $M - 1$ and N by $N - 1$, while allowing all the k's, Λ's and their distribution functions to alter slightly. If $N < \frac{1}{2}N_a$, one may compute μ_+ in the same manner, and one thereby finds that $\mu_+ = \mu_-$ for all values of the interaction strength U. Lieb and Wu (1968a) note however that if $N = \frac{1}{2}N_a$ exactly, then, in order to compute μ_+, one must resort to Eq. (C.2a). This leads to

$$\mu_+ = U - \mu_- \tag{C.17}$$

for a half-filled band. The calculation of μ_- for such a band yields the result that

$$\begin{aligned}\mu_- - 2 &= -4 \int_0^\infty \frac{J_1(\omega)\, d\omega}{\omega\,(1 + \exp(\frac{1}{2}\omega U)} \\ &= -4 \sum_{n=1}^{\infty} (-1)^n \left[\left(1 + \frac{1}{4} n^2 U^2\right)^{\frac{1}{2}} - \frac{1}{2} nU \right].\end{aligned} \tag{C.18}$$

One then immediately finds that $\mu_+ > \mu_-$ for $U > 0$, with $\lim_{U\to 0} \mu_\pm = 0$. Therefore, the ground-state of such a half-filled band is insulating for any nonzero value of the interaction strength U, and has conducting character for $U = 0$. Since temperature T has been kept equal to zero throughout the derivation and solution of the LW model, thermal fluctuations have no role in establishing the transition from the insulating to the conducting phase in this 1D model, and one may therefore speak of a 'quantum critical point' (QCP) at $U = 0$ (Sachdev, 2011b).

Lieb and Wu (1968a) also investigate the excitation spectrum $E(p)$ for a given total momentum $p = \sum_j k_j$ and a specified value of the spin component S_z. They find three types of excitations, *viz.* (i) a 'hole' state in the Λ distribution, (ii) a 'hole' state in the k distribution, and (iii) a 'particle' state in the k distribution. A spin-wave excited state with $S_z = 0$ can pertain to any of these types. On the other hand, an excited state with $S_z = 1$ can only be of type (i), and turns out to be characterized by the lowest energy, and by antiferromagnetic order. In the limit $U \to 0$, the energy spectrum of type-(i) excited states goes into $E(p) = |\sin p|$, while the energy spectra of type-(ii) and type-(iii) excited states go into $E(p) = |2 \sin \frac{1}{2}p|$. In all cases, one recovers a linear dispersion at small p.

Appendix D

Kondo lattices

D.1 The Kondo effect

As is known from early work, the electrical resistivity of metals falls of as T^5 at low temperatures. However, a resistance minimum is observed in good conductors such as Cu, Ag, and Au containing a very small concentration of impurities such as Cr, Mn, and Fe. As such impurities carry magnetic moments, the reason for the upturn of their resistivity is to be sought in the interaction of the conduction electrons with these impurities. Typically, there is the *s*-*d* interaction: specifically for Mn in Cu, the 3*d* electrons in the impurity have an exchange *s*-*d* interaction with the lowest-lying conduction electrons of Cu (see *e.g.* Isihara, 1991).

To evaluate electrical resistivity, interaction such as considered by Zener (1951) (see also Kasuya, 1956), Kondo (1964) arrived at the following formula for the resistivity, say $\Delta\rho$, due to the magnetic impurities

$$\Delta\rho = \frac{2\pi\Omega}{\hbar e^2 v_F^2 N_0} cJ^2 S(S+1)\left(1 + \frac{4Jg(\epsilon_F)}{N_0}\log T\right). \tag{D.1}$$

The total resistivity includes a residual resistivity contribution ρ_0, the T^5 term referred to above, and has therefore the overall form

$$\rho = \rho_0 + aT^5 \Delta\rho, \tag{D.2}$$

where

$$\Delta\rho = \Delta\rho_1 - bc|J|\log T. \tag{D.3}$$

In Eq. (D.3), $\Delta\rho_1$ is the temperature-independent term of $\Delta\rho$, and bc denote the coefficient of the logarithmic term in $\Delta\rho$. It is to be noted that $b > 0$, because J is negative. The resistance minimum thus occurs at

$$T_{\mathrm{K}} = \left(\frac{bc|J|}{5a}\right)^{1/5}. \tag{D.4}$$

This so-called *Kondo temperature* is seen to be proportional to $c^{1/5}$, as found by experiment. The temperature variation of $\Delta\rho$ also agrees with existing measurements.

We conclude by recording that exactly solvable models exist, and the solutions were in fact due respectively to the independent studies of Andrei (1980); Andrei *et al.* (1983) and of Wiegmann *et al.* (1980); Wiegmann and Tsvelick (1981) (see Hewson, 1993, for a review).

D.2 Kondo lattices

The Kondo lattice model describes the interaction of conduction electrons in a band with localized spins via a spin-exchange term. This model has been especially studied to account for the rich phase diagram of several magnetic materials, including the so-called heavy fermion systems, examples being $CeAl_3$, UPt_3, and UBe_{13} (cf. Table D.1). In these materials, anomalously large values of the Sommerfeld coefficient γ in the T-linear contribution to the specific heat at low temperature T, and of the Pauli spin susceptibility χ_s have been observed, yet with a relative Sommerfeld-Wilson ratio $R = \pi^2 k_{\mathrm{B}}^2 \chi_s / (3\mu_{\mathrm{eff}}^2 \gamma)$ of order unity. Such a behaviour can still be described in terms of Landau theory of Fermi liquids, but with a remarkably high value of the effective quasiparticle mass, with $m^*/m_e \approx 400$ for $CeCu_2Si_2$ (Steglich *et al.*, 1979). Since these compounds almost invariably include lanthanides or actinides, $4f$ or $5f$ atomic orbitals are involved in their electronic band structure (Stewart, 1984; Fulde, 1999).

Using c ($c^\dagger$) and f ($f^\dagger$) as the annihilation (creation) operators for conducting and localized spins, respectively, one starts by considering the periodic Anderson model

$$H = -t \sum_{\langle ij \rangle \tau} (c^\dagger_{i\tau} c_{j\tau} + \mathrm{H.c}) + \epsilon_f \sum_{i\tau} n^f_{i\tau} + U \sum_i n^f_{i\uparrow} n^f_{i\downarrow} + V \sum_{i\tau} (c^\dagger_{i\tau} f_{i\tau} + \mathrm{H.c}), \quad \text{(D.5)}$$

where $n^f_{i\tau} = f^\dagger_{i\tau} f_{i\tau}$ is the number operator associated with f spin-orbitals on site i with energy ϵ_f, U measures the Hubbard-like Coulomb repulsion among f electrons with opposite spins on the same site i, t is the hopping energy for conduction electrons between nearest-neighbour sites $\langle ij \rangle$, and V is the matrix element of the hybridization potential between spin-orbitals c and f on the same site.

Table D.1 Selected physical properties for uranium and cerium based heavy fermion systems. Where available, multiple entries separated by slashes refer to properties along different crystallographic directions. T_c is the superconducting critical temperature, T_N is the magnetic ordering (Néel) temperature, ξ is the superconducting correlation length, m^* the quasiparticles effective mass, γ denotes Sommerfeld's specific-heat coefficient. After Angilella *et al.* (2005), and references therein (see also Angilella *et al.*, 2000, 2001).

Compound	T_c (K)	T_N (K)	ξ (Å)	m^*/m_e	γ (J mol^{-1} K^{-2})
UPt_3	0.52, 0.48	5.0	100/120	180	0.450
UBe_{13}	0.87	—	100	260	1.100
UNi_2Al_3	1.0	4.3–4.6	240	48	0.120
UPd_2Al_3	2.0	14.5	85	66	0.145
URu_2Si_2	1–1.5	17–17.5	100/150	140	0.065–0.18
$CeCu_2Si_2$	0.65	1.3	90	380	0.73–1.1
$CeRh_2Si_2$	0.35	35–36	370	220	0.08
$CePd_2Si_2$	0.4	10.2	150		0.13
$CeCu_2Ge_2$	0.64	4.1			
$CeNi_2Ge_2$	0.22				0.4
$CeRu_2Ge_2$	7.40	8.55			
$CePt_3Si$	0.75	2.2	81–97		0.39
$CeNiGe_2$	—	3, 4	—		0.22
$CeNiGe_3$	0.48	5.5	130		0.034
$CeCoIn_5$	2.3		58, 35/82	83, 5/49/87	0.29
$CeRhIn_5$	2.1	3.8	57		0.40
$CeIrIn_5$	0.40	0	241	140, 20/30	0.72–0.75
$CeIn_3$	0.25	10			≤ 0.13
$CePd_3$	—		—	36	0.037

When only one f spin-orbital is considered on a single site, the periodic Anderson model reduces to the Anderson model for a single impurity in a metal (Anderson, 1958, cf. Chapter 8). In that limit, in the spirit of Cooper pairing, but now for localized spins, a singlet Ansatz for the ground state has been assumed (Yosida, 1966; Varma and Yafet, 1976), consisting of the superposition of the Fermi sea of the conduction electrons, and of states obtained by transferring a single conduction electron to the localized state.

In the Kondo regime, each f spin-orbital is occupied by a single electron, with a given spin projection, while empty and doubly occupied sites take part in the problem only as virtual states. Therefore the physics described by the periodic Anderson Hamiltonian, Eq. (D.5), can be captured by a low-energy effective model, where the f electrons are represented by localized spins. Formally, this can be achieved by using a second-order perturbation with respect to V, yielding the Kondo lattice Hamiltonian (Schrieffer and

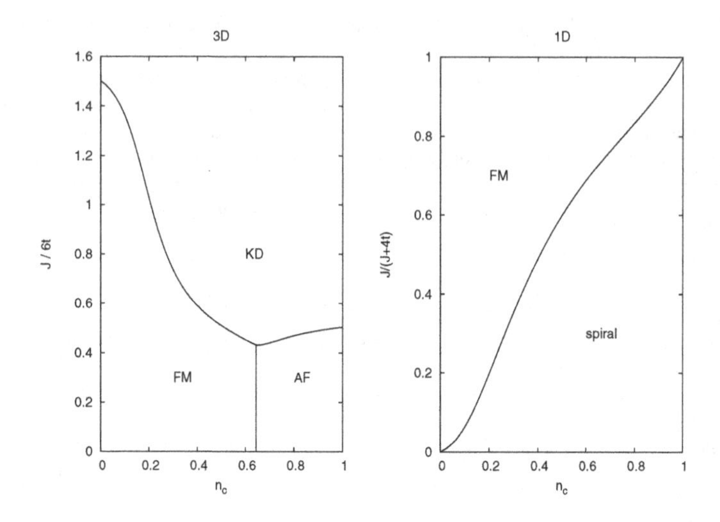

Fig. D.1 *Left panel:* Mean-field phase diagram for the Kondo lattice model in three dimensions, as a function of conduction electron concentration n_c and spin correlation to hopping ratio $J/6t$. Labels indicate the ferromagnetic (FM), antiferromagnetic (AF), and Kondo disordered (KD) phases. Redrawn after Lacroix and Cyrot (1979) (see also Tsunetsugu *et al.*, 1997). *Right panel:* Phase diagram for the Kondo lattice model in one dimension. Magnetic spiral order is favoured at low spin correlation. Redrawn after Fazekas and Müller-Hartmann (1991) (see also Tsunetsugu *et al.*, 1997).

Wolff, 1966)

$$H = -t \sum_{\langle ij \rangle \tau} (c^\dagger_{i\tau} c_{j\tau} + \text{H.c}) + J \sum_i \mathbf{s}_i \cdot \mathbf{S}_i. \tag{D.6}$$

Here, $\mathbf{S}_i = \frac{1}{2} \sum_{\tau\tau'} f^\dagger_{i\tau} \boldsymbol{\sigma}_{\tau\tau'} f_{i\tau'}$ and $\mathbf{s}_i = \frac{1}{2} \sum_{\tau\tau'} c^\dagger_{i\tau} \boldsymbol{\sigma}_{\tau\tau'} c_{i\tau'}$ are the spin operators on site i associated with the localized and conduction electrons, respectively, with $\boldsymbol{\sigma}$ being the (vector of the) Pauli matrices, and $J > 0$ is the Kondo exchange coupling constant. Under the symmetric condition $\epsilon_{\text{F}} - \epsilon_f = \epsilon_f + U - \epsilon_{\text{F}}$, one finds $J = 8V^2/U$. While electron-electron interaction is neglected within each individual band, the Kondo coupling induces hybridization between the conduction and localized electrons, which becomes operative at temperatures below the Kondo temperature, $T_{\text{K}} = \epsilon_{\text{F}} \exp(-1/JN(\epsilon_{\text{F}}))$.

The three-dimensional Kondo lattice model has been studied within the Hartree-Fock (mean-field) approximation by (Lacroix and Cyrot, 1979). They determined the magnetic phase diagram of this model as a function of the conduction electron density n_c and of the correlation-to-hopping ratio J/t, by comparing the relative stability of the ferromagnetic, antiferromagnetic, and magnetically disordered Kondo singlet states. While in the weak-coupling regime magnetic ordering is always favoured, with ferromagnetic order prevailing at low electron density and Néel order being obtained somewhat above half-filling, the nonmagnetic Kondo singlet phase was the ground state in the large-J regime (Fig. D.1). More refined treatments in reduced-dimensional systems partially modifies the phase diagram, with magnetically ordered phases being favoured at any electron density (Fig. D.1 Fazekas and Müller-Hartmann, 1991) (cf. Tsunetsugu *et al.*, 1997; Gulácsi, 2006, and references therein).

Through bosonization of the Kondo lattice model in one dimension (a Kondo 'chain' or 'necklace'), Zachar *et al.* (1996) were able to derive some exact results under appropriate limits. Their treatment and formalism are similar in spirit to those employed in the context of Luttinger liquids in 1D models (cf. Appendix E). In particular, they find a second energy scale, corresponding to a characteristic coherence temperature $T_{\rm coh}$, which increases with increasing impurity concentration, and is robust with respect of the character of the impurity lattice, whether commensurate, incommensurate, or weakly disordered. While a single-impurity Kondo scenario is appropriate for temperatures $T_{\rm coh} < T < T_{\rm K}$ in the dilute limit ($T_{\rm coh} \ll T_{\rm K}$), and coherence between impurities sets in at $T \sim T_{\rm coh}$, in the dense limit ($T_{\rm coh} \gg T_{\rm K}$) coherence and Kondo quenching of the impurity spins occur simultaneously, in much the same way as pairing and condensation occur simultaneously in BCS superconductors. Thus, a 'pseudogap' phase would be ruled out in the latter case, as is observed in the underdoped regime of the cuprate superconductors (cf. discussion on p. 175). Zachar *et al.* (1996) also find that the coherent low-temperature state is characterized by the appearance of long-range order of a non-local order parameter, analogous to the hidden (or topological) order found in integer spin-chains, or in systems exhibiting fractional quantum Hall effect (Girvin and MacDonald, 1987). These results may be relevant for the study of competing orders in the cuprate superconductors (Fradkin *et al.*, 2015), and provide some rigorous foundation towards recent proposals of topological insulating states and quantum phase transitions in the Kondo lattices (Tsvelik and Yevtushenko, 2015; Mezio *et al.*, 2015).

Appendix E

Luttinger liquid: spinons and holons

In three dimensions (3D), Landau theory of Fermi liquids applies to a wide class of many-body fermion assemblies, and enables one to describe these as a system of non-interacting quasiparticles, but with physical properties such as mass renormalized because of interparticle interaction, in the limit of low-energy excitations (*i.e.* energy close to the Fermi level) and low temperature. In lower dimensions, the paradigm of Landau-Fermi liquids may break down and, in one dimension (1D), the so-called Luttinger liquid may prevail (Haldane, 1981). Below, we closely follow Carmelo and Ovchinnikov (1991, CO, below), who deal in particular with the 1D Hubbard model.

CO introduce a novel representation of the Bethe Ansatz for such 1D Luttinger liquids. This describes the spectral properties and asymptotic behaviour of the correlation functions of the 1D Hubbard model in terms of the (renormalized) interaction of charge and spin pseudoparticles. Their study of the low-lying eigenstates can then be reduced to familiar energy band theory and Fermi liquid descriptions. CO obtain results which provide understanding of the decoupling of charge and spin degrees of freedom in 1D interacting assemblies. Their results turn out to be consistent with the ideas of Anderson (1990) that the Luttinger liquid is a fixed point of the same renormalization group which, in some 3D systems, leads to the Landau-Fermi liquid (see, *e.g.* Jones and March, 1986a) as a unique fixed point.

E.1 1D Hubbard model in an arbitrary magnetic field

CO consider specifically the 1D Hubbard Hamiltonian in an external magnetic field H

$$\mathcal{H} = -t \sum_{j,\sigma} (c_{j+1,\sigma}^{\dagger} c_{j\sigma} + \text{H.c.}) + U \sum_{j} c_{j\uparrow}^{\dagger} c_{j\uparrow} c_{j\downarrow}^{\dagger} c_{j\downarrow} - \mu_0 H \sum_{j,\sigma} \sigma c_{j\sigma}^{\dagger} c_{j\sigma}, \quad \text{(E.1)}$$

where as customary $c_{j\sigma}^{\dagger}$ ($c_{j\sigma}$) is the creation (annihilation) operator for an electron with spin σ at site j, t is the nearest-neighbour hopping, and U the Hubbard on-site repulsion energy. The system comprises N electrons on N_a sites, with periodic boundary conditions. Below, the notation employed takes the dimensionless on-site repulsion $u = U/4t$, density $n = N/N_a$, $k_{\text{F}} = \pi n/2$, spin density $s = (k_{\text{F}\uparrow} - k_{\text{F}\downarrow})/2\pi$, $k_{\text{F}\sigma}$ being $\pi N_\sigma / N_a$, and $N_\uparrow = M'$ and $N_\downarrow = M$ are the numbers of up and down spins, respectively.

Lieb and Wu (1968a,b, see also Appendix C) earlier employed the Bethe Ansatz approach to reduce the eigenvalue problem presented by the above Hamiltonian, Eq. (E.1), to the solution of coupled algebraic equations. The crystal momentum P and the energy E are given by

$$P = \sum_{j=1}^{N} q_j + \sum_{\alpha=1}^{M} p_\alpha, \quad \text{(E.2a)}$$

$$E = -2t \sum_{j=1}^{N} \cos K_j - 2\mu_0 N_a H s, \quad \text{(E.2b)}$$

where the notation employed is $q_j = (2\pi/N_a) I_j$, and $p_\alpha = (2\pi/N_a) J_\alpha$. An eigenstate of the many-body system under discussion is specified by a choice of the quantum numbers $\{I_j\}$, $\{J_\alpha\}$, or the pseudomomentum distributions $\{q_j\}$, $\{q_\alpha\}$. In the ground state, I_j and J_α are consecutive integers (or half odd integers) centred around zero (Lieb and Wu, 1968a). After choosing $\{q_j\}$ and $\{q_\alpha\}$, the Lieb-Wu algebraic equations determine the change and spin quantities $K_j = K_j(q_j)$ and $S_\alpha = S_\alpha(q_\alpha)$, respectively ($S_\alpha = \Lambda_\alpha / u$ (Lieb and Wu, 1968a).

Taking next the thermodynamic limit ($N_a \to \infty$, n fixed), the roots $K_j = K_j(q_j)$ and $S_\alpha = S_\alpha(q_\alpha)$ proliferate on the real axis and the Lieb-Wu equations lead to

$$K(q) = q + \frac{1}{\pi}\int_{-k_{\mathrm{F}\downarrow}}^{k_{\mathrm{F}\downarrow}} dp'\, N_\downarrow(p')\,\mathrm{atan}\left[S(p') - u^{-1}\sin K(q)\right], \quad \text{(E.3a)}$$

$$p = \frac{1}{\pi}\int_{-\pi}^{\pi} dq'\, M_c(q')\,\mathrm{atan}\left[S(p) - u^{-1}\sin K(q')\right]$$
$$-\frac{1}{\pi}\int_{-k_{\mathrm{F}\uparrow}}^{k_{\mathrm{F}\uparrow}} dp'\, N_\downarrow(p')\,\mathrm{atan}\left[\frac{1}{2}\left(S(p) - S(p')\right)\right], \quad \text{(E.3b)}$$

$$E = \frac{N_a}{2\pi}\int_{-\pi}^{\pi} dq'\, M_c(q')[-2t\cos K(q')] - 2\mu_0 N_a H s, \quad \text{(E.3c)}$$

$$s = \frac{1}{2\pi}\left(\frac{1}{2}\int_{-\pi}^{\pi} dq'\, M_c(q') - \int_{-k_{\mathrm{F}\uparrow}}^{k_{\mathrm{F}\uparrow}} dp'\, N_\downarrow(p')\right). \quad \text{(E.3d)}$$

The quantities $M_c(q)$ and $N_\downarrow(p)$ appearing in the above equations can be interpreted as pseudomomentum distributions of charge and spin pseudoparticle, respectively. One has also $K(\pm\pi) = \pm\pi$ and $S(\pm k_{\mathrm{F}\uparrow}) \pm \infty$. In the case of the ground state at fixed magnetization, $K(q)$ and $S(p)$ are odd functions such that $K(2k_{\mathrm{F}}) = Q$ and $K(k_{\mathrm{F}\downarrow}) = B/u$, where Q and B denote the cut-off of the Lieb-Wu equations. In this case, the distributions $M_c(q)$ and $N_\downarrow(p)$ take the form

$$M_c^0(q) = \Theta(2k_{\mathrm{F}} - q), \quad \text{(E.4a)}$$

$$N_\downarrow^0(p) = \Theta(k_{\mathrm{F}\downarrow} - |p|). \quad \text{(E.4b)}$$

Just as in Fermi liquid theory, these distributions do not depend on the interaction. In the present two-fluid-like Landau liquid, the charge and spin pseudo-Fermi surfaces are defined as the set of points $\{q = \pm 2k_{\mathrm{F}}, p = \pm k_{\mathrm{F}\downarrow}\}$ separating occupied from unoccupied regions. Also it is to be noted that the limits of the pseudo-Brillouin zones of the charge and spin pseudoparticles are $\{q = \pm\pi\}$, $\{p = \pm k_{\mathrm{F}\uparrow}\}$, respectively. The restrictions on I_j and J_α imply that each pseudomomentum value cannot be occupied by more than one pseudoparticle, *i.e.* demonstrating the Fermionic nature of the pseudoparticles. These, for instance, can be identified with the pseudo-Fermions considered by Carmelo and Baeriswyl (1988a,b), and then related to holons and spinons (Anderson, 1990).

In a little more detail, the spin pseudoparticles are of the same nature as those of the Heisenberg chain, being intimately related to the spin-$\frac{1}{2}$ spin waves introduced by Faddeev and Takhtajan (1981). They represent many-body collective modes, and the above class of pseudoparticles cannot exist outside the many-body assembly for any value of the bare interaction.

CO then consider small pseudomomentum fluctuations around the ground-state distributions, Eqs. (E.4), as

$$M_c(q) = M_c^0(q) + \delta_c(q), \tag{E.5a}$$

$$N_\downarrow(p) = N_\downarrow^0(p) + \delta_\downarrow(p). \tag{E.5b}$$

The departures $\delta_c(q)$, $\delta_\downarrow(p)$ from the ground-state values control the low-energy physics of the model considered above. This is true for both elementary excitations. CO emphasize that it is important to consider $K(q)$, $S(p)$, and E in Eqs. (E.3) as functionals of the pseudoparticle excitations. Provided that these involve a small number of pseudoparticles, an expansion of the energy as

$$E = E_0 + E_1 + E_2 + \cdots \tag{E.6}$$

can be performed to arbitrary order in the fluctuations, the leading order corrections being given by (Carmelo and Ovchinnikov, 1991)

$$E_1 = \frac{N_a}{2\pi}\int_{-\pi}^{\pi} dq\, \delta_c(q)\epsilon_c(q) + \frac{N_a}{2\pi}\int_{-k_{F\uparrow}}^{k_{F\uparrow}} dp\, \delta_\downarrow(p)\epsilon_s(p), \tag{E.7a}$$

$$\begin{aligned} E_2 = {} & \frac{N_a}{4\pi^2}\int_{-\pi}^{\pi} dq \int_{-\pi}^{\pi} dq'\, \delta_c(q)\delta_c(q')\frac{1}{2}f_{cc}(q,q') \\ & + \frac{N_a}{4\pi^2}\int_{-k_{F\uparrow}}^{k_{F\uparrow}} dp \int_{-k_{F\uparrow}}^{k_{F\uparrow}} dp'\, \delta_\downarrow(p)\delta_\downarrow(p')\frac{1}{2}f_{ss}(p,p') \\ & + \frac{N_a}{4\pi^2}\int_{-\pi}^{\pi} dq \int_{-k_{F\uparrow}}^{k_{F\uparrow}} dp\, \delta_c(q)\delta_\downarrow(p)f_{cs}(q,p). \end{aligned} \tag{E.7b}$$

Here, $\epsilon_c(q)$ and $\epsilon_s(q)$ determine the charge and spin quasiparticle dispersion relations, respectively, while $f_{cc}(q,q')$, $f_{ss}(p,p')$, $f_{cs}(q,p)$ describe their mutual interactions, as in the Landau theory of Fermi liquids. In the presence of a magnetic field, the change and spin bands can be expressed as

$$\epsilon_c(q) = \int_{Q}^{K(q)} dk'\, 2t\eta_c(k'), \tag{E.8a}$$

$$\epsilon_s(p) = \int_{B/u}^{S(p)} dv'\, 2t\eta_s(v'), \tag{E.8b}$$

where $K(q)$, $S(p)$ are implicitly defined by Eqs. (E.3a), (E.3b), while $2t\eta_c(k)$, $2t\eta_s(v)$ are solutions of coupled integral equations of similar form to the Lieb and Wu equations (1968a). The pseudoparticle group velocities are then defined as usual as $v_c = d\epsilon_c(q)/dq$, $v_s = d\epsilon_s(p)/dp$.

Appendix F

States with fractional Fermion charge in both condensed matter and relativistic field theories

Su, Schrieffer, and Heeger (1979; 1980; 1983; SSH, below) studied a coupled electron-phonon model for the quasi-one-dimensional (quasi-1D) conductor polyacetylene $(CH)_n$. They found a dynamical symmetry breaking of the assembly, leading to degenerate vacua and soliton formation. In the presence of a soliton, a c-number solution ψ_0 of the electron field, localized near the soliton, with energy at the centre of the gap, was found (see also Jackiw and Schrieffer, 1981). A consequence is that one-half a state of each spin orientation is removed from the sea in the vicinity of the soliton (see also Rice, 1979). Thus, neglecting the electron spin, the existence of the zero-energy state and Fermion number $\pm\frac{1}{2}$ are common to the two situations.

In the study of Jackiw and Schrieffer (1981), this above work is emphasized to have a counterpart in relativistic field theories. We start below by summarizing this discussion of Fermionic solitons in a relativistic field theory.

F.1 Relativistic field theory

Jackiw and Schrieffer (1981) studied the spectrum of a Dirac field coupled to a broken symmetry Bose field. These workers discovered, in a variety of models, the existence of zero-energy Fermion eigenstates, localized in the vicinity of a soliton of the Bose field. In their 1D model, Jackiw and Schrieffer (1981) focus on a Hamiltonian of the form

$$H = \int dx \left[\frac{1}{2}\Pi^2 + \frac{1}{2}\left(\frac{d}{dx}\Phi\right)^2 + V(\Phi) + \Psi^\dagger(\alpha p + \beta g \Phi)\Psi \right], \qquad \text{(F.1)}$$

where $p = -id/dx$, Π denotes the momentum field operator conjugate to Φ, and $V(\Phi)$ is a potential energy density for Φ. The Dirac spinors have

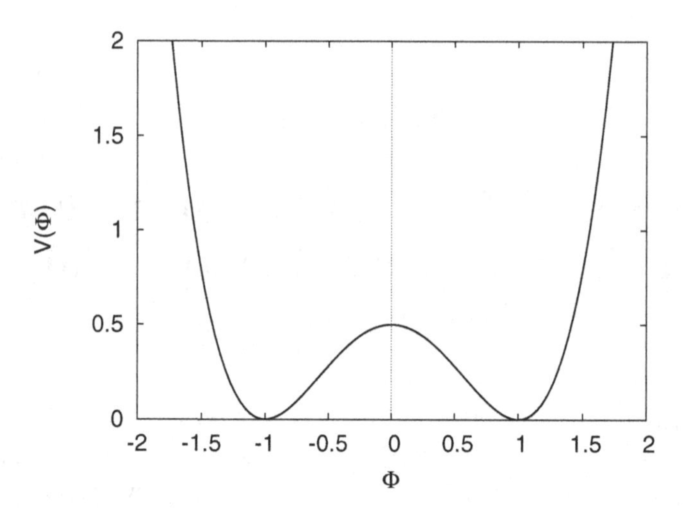

Fig. F.1 Double-well potential, Eq. (F.3).

two components: hence the Fermions carry no spin, which is a simplifying option which is only available in 1D. As a consequence, the two Dirac matrices α and β are two-dimensional:

$$\alpha = \sigma^2 = \begin{pmatrix} 0 & -i \\ i & 0 \end{pmatrix}, \tag{F.2a}$$

$$\beta = \sigma^1 = \begin{pmatrix} 0 & 1 \\ 1 & 0 \end{pmatrix}. \tag{F.2b}$$

The model is analyzed in an adiabatic framework. The ground state is then determined by minimizing $\int dx\, [\frac{1}{2}(d\Phi/dx)^2 + V(\Phi)]$, with constant Φ, while the soliton is obtained by minimizing with an x-dependent Φ, subject to the condition of finite total energy. The Dirac Hamiltonian is then quantized, with Φ taken as an external, prescribed c-number, background field. $V(\Phi)$ is taken as a symmetric, double-well potential, but having the simpler analytical form (Fig. F.1)

$$V(\Phi) = \frac{\lambda^2}{2\mu^2}(\mu^2 - \Phi^2)^2 = V(-\Phi). \tag{F.3}$$

F.2 Broken symmetry ground state

The minimum of Eq. (F.3) is at $\Phi = \pm\mu$. Thus the reflection symmetry $\Phi \mapsto -\Phi$ is spontaneously broken. The Dirac Hamiltonian describes a free particle with mass gap $\Delta = g\mu = m$. The Fermion modes satisfy the free Dirac equation

$$(\alpha p + \beta m)u^{(\pm)} = \pm|\epsilon|u^{(\pm)}, \qquad \text{(F.4)}$$

where $|\epsilon| = \sqrt{k^2 + m^2}$. Following Jackiw and Schrieffer (1981), charge conjugation is implemented by the σ^3 matrix. It takes the positive energy solutions of Eq. (F.4) into negative energy states, and vice versa. Quantization is achieved by expanding in modes

$$\Psi = \sum_k \left[e^{-i\epsilon_k t} b_k u_k^{(+)}(x) + e^{i\epsilon_k t} d_k^\dagger v_k^{(-)}(x) \right]. \qquad \text{(F.5)}$$

In Eq. (F.5), $u_k^{(+)}$ denotes a positive energy solution, and $v_k^{(-)}$ is the charge conjugate of the negative energy solution $u_k^{(-)}$, $v_k^{(-)} = \sigma^3 \left(u_k^{(-)} \right)^*$. The above operators $b_k^\dagger$ (b_k) create (annihilate) particles, while $d_k^\dagger$ (d_k) perform the same for antiparticles. In terms of these creation and annihilation operators, the charge operator

$$\begin{aligned} Q &= \int dx : \Psi^\dagger(x)\Psi(x): \\ &= \frac{1}{2} \int dx \sum_{i=1}^{2} \left(\Psi_i^\dagger(x)\Psi_i(x) - \Psi_i(x)\Psi_i^\dagger(x) \right) \end{aligned} \qquad \text{(F.6)}$$

can be expressed as

$$Q = \sum_k (b_k^\dagger b_k - d_k^\dagger d_k). \qquad \text{(F.7)}$$

Jackiw and Schrieffer (1981) stress that the spectrum is elementary, and that it can be built on either vacuum (A and B, say), corresponding to the positive ($\Phi = \mu$) and negative ($\Phi = -\mu$) minima of Eq. (F.3) (cf. Fig. F.1). Both vacua are charge neutral, in the sense that $Q|0\rangle = 0$.

F.3 Soliton excitations

The existence of two ground states leads to topological solitons which interpolate between them. To find the soliton shape of Φ, Jackiw and Schrieffer (1981) solve the equation

$$-\frac{d^2}{dx^2}\Phi(x) + V'(\Phi) = 0. \qquad \text{(F.8)}$$

For $V(\Phi)$ in Eq. (F.3), this reduces to the nonlinear Schrödinger equation, whose finite energy solutions are

$$\Phi(x) = \pm\mu \tanh(\lambda x), \tag{F.9}$$

where $x = 0$ at the chosen origin. The two signs refer to a soliton (S) and an antisoliton ($\bar{S}$), respectively. The formation energy of a soliton can be computed as

$$E_S = \int dx \left[\frac{1}{2} \left(\frac{d}{dx} \Phi \right)^2 + V(\Phi) \right] = \frac{4}{3}\mu^2\lambda. \tag{F.10}$$

In the presence of solitons, the Dirac modes now obey the modified Dirac equation

$$(\alpha p + \beta m \tanh \lambda x) U^{(\pm)} = \pm|\epsilon| U^{(\pm)}, \tag{F.11}$$

where again solutions corresponding to energies of different sign are connected by charge conjugation. Additionally, Eq. (F.11) admits a zero energy solution, given by

$$\psi_0(x) \propto \begin{pmatrix} 1 \\ 0 \end{pmatrix} \exp \left(-m \int_0^x dx' \tanh \lambda x' \right), \tag{F.12}$$

which is self-conjugate with respect to charge conjugation, $\psi_0^c = \sigma^3 \psi_0^* = \psi_0$. The occurrence of solitons thus induces the appearance of an isolated energy level at the centre of the gapped region of the spectrum. In analogy to Eq. (F.5), quantization proceeds by expanding solutions in terms of eigenmodes as

$$\Psi = a\psi_0(x) + \sum_k \left[e^{-i\epsilon_k t} B_k U_k^{(+)}(x) + e^{i\epsilon_k t} D_k^\dagger V_k^{(-)}(x) \right], \tag{F.13}$$

with obvious meaning of the notation. However, in addition to the particle and antiparticle in the soliton sector, and at variance with Eq. (F.5), the further operator a fills a state with the same energy, when operating on the soliton state. Hence, the two states are degenerate in energy. Denoting them by $|\pm, S\rangle$, or 'occupied' and 'unoccupied' states, the action of a and $a^\dagger$ is defined as

$$a|+, S\rangle = |-, S\rangle, \tag{F.14a}$$

$$a^\dagger|-, S\rangle = |+, S\rangle, \tag{F.14b}$$

$$a|-, S\rangle = 0, \tag{F.14c}$$

$$a^\dagger|+, S\rangle = 0. \tag{F.14d}$$

For the charge operator, one finds

$$Q = a^\dagger a - \frac{1}{2} + \sum_k (B_k^\dagger B_k - D_k^\dagger D_k), \tag{F.15}$$

and the occupied and unoccupied states are eigenstates thereof, with eigenvalues

$$Q|\pm, S\rangle = \pm|\pm, S\rangle, \tag{F.16}$$

i.e. each of the two solitons carry $\frac{1}{2}$ unit of charge.

F.4 SSH model for polyacetylene

Turning from the relativistic field theory above to the SSH model of Su, Schrieffer, and Heeger (Su *et al.*, 1979, 1980, 1983), their Hamiltonian reads

$$H = \sum_{n=1}^{N} \left(\frac{p_n^2}{2M} + V(u_n, u_{n+1}) \right) + \sum_{n=1}^{N} \sum_{s=\pm\frac{1}{2}} t_{n+1,n} (c_{n+1,s}^\dagger c_{n,s} + \text{H.c.}). \tag{F.17}$$

Here, u_n is a Bosonic operator describing the shift of the nth CH group in a linear chain of N elements with respect to its equilibrium position, p_n its conjugate momentum, and $c_{n,s}^\dagger$ ($c_{n,s}$) creates (destroys) an electron on site n with spin projection s. The lattice potential is taken as harmonic, *i.e.* in the form

$$V(u_n, u_{n+1}) = \frac{1}{2} k (u_{n+1} - u_n)^2. \tag{F.18}$$

The last term in Eq. (F.17) describes electron hopping between neighbouring sites, where the hopping integral is expanded to first order in the lattice displacement as

$$t_{n+1,n} = t_0 - \alpha(u_{n+1} - u_n), \tag{F.19}$$

thereby providing an effective electron-phonon interaction term. The various constants appearing in Eq. (F.17) have been estimated for polyacetylene as $t_0 = 2.5$ eV, $\alpha = 4.1$ eV/Å, $k = 21$ eV/Å^2 (Su *et al.*, 1979, 1980, 1983).

In the absence of electron-phonon coupling ($\alpha = 0$), the ground state is nondegenerate, and consists of a Fermi sea for electrons, with all states doubly occupied up to Fermi wave-vector k_F, and the phonon vacuum. When such an interaction is switched on ($\alpha \neq 0$), the ground state is characterized by broken reflection symmetry. Introducing the staggered displacement $\phi_n = (-1)^n u_n$, within the adiabatic (Born-Oppenheimer) approximation,

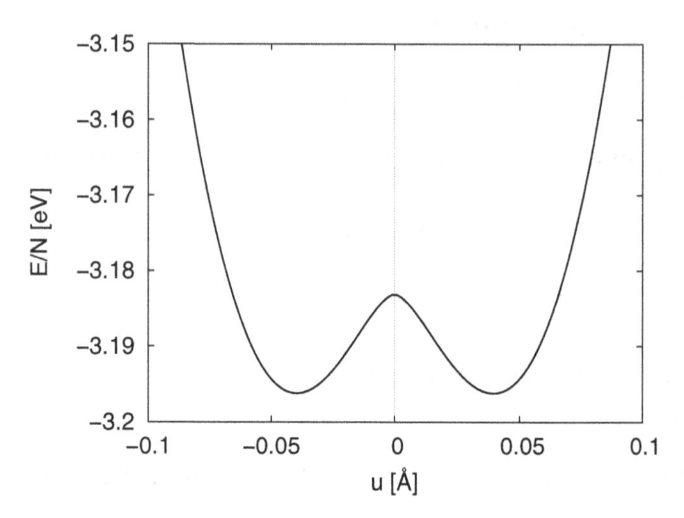

Fig. F.2 Ground state energy per site $E_0(u)/N$ for the SSH model, Eq. (F.20), as a function of lattice deformation u.

one sets $u_n = u =$ const. One then readily finds that the ground state energy is symmetric with respect to a sign change of u, $E_0(u) = E_0(-u)$, the equilibrium values $\pm u_0$ for u being determining by a variational procedure. One finds (Su *et al.*, 1979, 1980, 1983)

$$\frac{E_0(u)}{N} = -\frac{4t_0}{\pi} E\left(1 - \frac{4\alpha^2 u^2}{t_0^2}\right) + 2ku^2, \qquad \text{(F.20)}$$

where E is the elliptic integral (Gradshteyn and Ryzhik, 1994). Eq. (F.20) exhibits the typical double-well structure (Fig. F.2; cf. Fig. F.1), with two minima at $u = \pm u_0$ corresponding to the doubly degenerate ground state with broken reflection symmetry, and a local maximum at $u = 0$. This satisfies Peierls theorem (dating back to 1930), stating that any 1D perfectly periodic lattice is unstable with respect to local distortions (Peierls, 1955). With the numerical values for the various parameters given above, Su *et al.* (1979, 1980, 1983) find $u_0 = 0.042$ Å, in good agreement with quantum chemical calculations. The adiabatic approximation is also verified a posteriori, in that the energy gap $\Delta = 4\alpha|u_0|$ in the electron spectrum, Eq. (F.20), is much larger than the phonon energy scale $\sim \hbar\sqrt{k/M}$.

As a consequence of the ground state two-fold degeneracy, $\phi_n = \pm u_0$, topological solitons may arise, which interpolate smoothly between adjacent regions in the chain corresponding to the two distinct minima in Fig. F.2. Imposing periodic boundary conditions on the linear chain, there must be

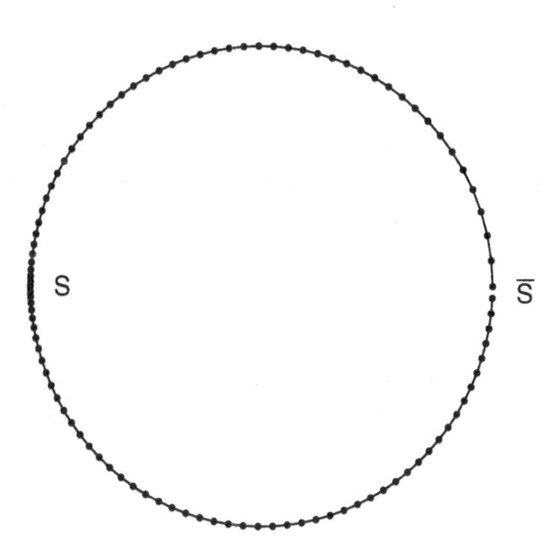

Fig. F.3 Soliton (S) and antisoliton ($\bar{S}$) pair in a 1D chain, with periodic boundary conditions.

an even number of such solitons, corresponding to an even number of sign changes in ϕ_n. The profile of ϕ_n can be found by minimizing the total energy, and requiring that the ground state approaches locally either minima in Fig. F.2, far from the solitons. In the case of a single soliton (S) and antisoliton ($\bar{S}$) pair, occurring at positions $n = n_1$ and $n = n_2$, respectively, Su *et al.* (1979, 1980, 1983) find

$$\phi_n = u_0 \tanh[(n - n_1)/\ell] - u_0 \tanh[(n - n_2)/\ell] - u_0, \tag{F.21}$$

where ℓ is the width of each soliton, with $\ell \approx 7$ in the case of polyacetylene, and $|n_2 - n_1| \gg \ell$, thus ensuring that each soliton is well separated from its antisoliton counterpart. Figure F.3 provides a schematic picture of a soliton/antisoliton pair.

In the presence of a soliton-antisoliton pair, the electronic spectrum is characterized by the presence of two discrete states $\psi_{0\pm}$, at energy levels symmetrically located with respect to the centre of the gap, $E = 0$. As $|n_2 - n_1| \to \infty$, the energy splitting between these two states vanishes as $\exp(-|n_2 - n_1|/\ell)$, so that it becomes energetically costless to form eigenstates ψ_{0S} and $\psi_{0\bar{S}}$, localized around S and $\bar{S}$, as linear combinations of $\psi_{0\pm}$. These are given by

$$\psi_0(n) = \frac{1}{\sqrt{\ell}} \operatorname{sech}[(n - n_i)/\ell] \cos[\frac{1}{2}\pi(n - n_i)], \tag{F.22}$$

where $i = 1, 2$ for S and $\bar{S}$, respectively.

F.5 Discussion

The two models treated above differ, of course, in detail. For instance, the SSH Hamiltonian is on a lattice, plus the fact that the Fermions have a spin degree of freedom. The relativistic field theory of (Jackiw and Rebbi, 1976) is in the continuum and the Fermions are spinless.

Nevertheless, in certain crucial respects, both models are similar. Both result in spontaneous breaking of the field reflection symmetry, a consequence of which being doubly degenerate ground states. They possess soliton excitations which interpolate between the degenerate vacua. Furthermore, the Fermion equation in both cases admits a localized zero-energy solution ψ_0, which in turn implies charge fractionalization: $\frac{1}{2}$ unit of charge being gained or lost, depending whether ψ_0 is filled or empty.

Appendix G

Haldane gap

G.1 Valence bond states for spin-$\frac{1}{2}$ chains

Majumdar and Ghosh (MG, 1969) first proposed a solvable model, in which the ground states could be expressed exactly in terms of valence bonds. Their model is for particles with spin $s = \frac{1}{2}$, and is characterized by the Hamiltonian (see Affleck, 1989)

$$
\begin{aligned}
H &= \sum_i (\mathbf{S}_i + \mathbf{S}_{i+1} + \mathbf{S}_{i+2})^2 \\
&= \sum_i (J_1 \mathbf{S}_i \cdot \mathbf{S}_{i+1} + J_2 \mathbf{S}_i \cdot \mathbf{S}_{i+2} + \text{const}),
\end{aligned}
\tag{G.1}
$$

with $J_2/J_1 = \frac{1}{2}$.

G.2 Higher spin chains: the Haldane gap

A rather different type of solvable model, but now for integer spin s, was found by Affleck *et al.* (1987, 1988). Again, as for the MG model, the ground state can be expressed simply in terms of valence bonds. Then, as a starting point, one requires a generalization of the idea of valence bonds to higher spin s. This can be achieved by regarding a spin-s operator as a symmetrized product of $2s$ spin-$\frac{1}{2}$ operators. Thus, for $s = 1$ (see Affleck, 1989), one can imagine two $s = \frac{1}{2}$ variables per site. One can then construct singlet states by suitably contracting these $s = \frac{1}{2}$ variables. Affleck stresses that any state obtained in terms of valence bonds must be symmetrized with respect to the two spins on each site. A convenient notation (see Affleck, 1989) is not to distinguish the two spins per site, but to draw two lines emerging from each site, both of which must terminate on other

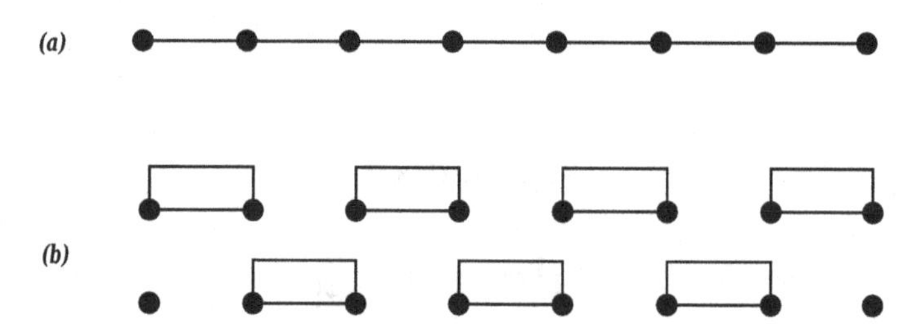

Fig. G.1 Valence bond states for a $s = 1$ chain: (a) the VBS; (b) the dimerized state. Redrawn after Affleck (1989).

sites. With this notation, as again Affleck stresses, the simple candidate ground state is almost obvious. One draws one line between each pair of neighbouring sites (Fig. G.1).

Since the structure of valence bonds mimics the lattice structure, Affleck terms this a valence-bond solid (VBS) state. This can also be described as the ground state of the Hamiltonian

$$H = \sum_i P^{(2)}(\mathbf{S}_i + \mathbf{S}_{i+1}) \propto \sum_i [\mathbf{S}_i \cdot \mathbf{S}_{i+1} + \frac{1}{3}(\mathbf{S}_i \cdot \mathbf{S}_{i+1})^2] + \text{const}, \quad \text{(G.2)}$$

where $P^{(j)}$ denotes the projection operator onto spin j. To see that the VBS state is indeed a ground state, Affleck observes that of the four $s = \frac{1}{2}$ variables associated with any pair of neighbouring sites, two of them are contracted to form a singlet. Symmetrizing or antisymmetrizing the other two gives spin 1 or 0, but never 2. Thus the ground state energy is zero.

An elegant algebraic representation of the VBS state can be traced back at least to Arovas *et al.* (1988). These authors used the Weyl representation of $SU(2)$. This allows to express the relevant components of a spin $\mathbf{S}$ operator in terms of polynomials of a unit vector $\mathbf{\Omega} = (u, v)$, with $|u|^2 + |v|^2 = 1$, as

$$S^+ = u\frac{\partial}{\partial v}, \quad \text{(G.3a)}$$

$$S^- = v\frac{\partial}{\partial u}, \quad \text{(G.3b)}$$

$$S^z = \frac{1}{2}\left(u\frac{\partial}{\partial u} - v\frac{\partial}{\partial v}\right). \quad \text{(G.3c)}$$

The unit vector $\mathbf{\Omega}$ is uniquely identified by the polar coordinates, θ and φ, say, of its oriented ending point as $u = e^{i\varphi/2}\cos(\theta/2)$, $v = e^{-i\varphi/2}\sin(\theta/2)$. Wave functions for spin-s are then polynomials of degree $2s$ in u and v, and the scalar product between any two such wave functions can be written as an integral over the unit sphere. The VBS wave function becomes

$$\Psi = \sqrt{2}\prod_i (u_i v_{i+1} - v_i u_{i+1}), \tag{G.4}$$

with

$$|\Psi|^2 = \prod_i (1 - \mathbf{\Omega}_i \cdot \mathbf{\Omega}_{i+1}). \tag{G.5}$$

Restricting Eqs. (G.3) to the subspace of polynomials of degree 2 in u and v, it can be seen that the matrix elements of the spin operators coincide with those of

$$S^+ \mapsto 4uv^*, \tag{G.6a}$$

$$S^- \mapsto 4vu^*, \tag{G.6b}$$

$$S^z \mapsto 2(v^*u - u^*v), \tag{G.6c}$$

$$\mathbf{S}_i \cdot \mathbf{S}_j \mapsto 4\mathbf{\Omega}_i \cdot \mathbf{\Omega}_j. \tag{G.6d}$$

The wave function for a chain of N sites is normalized as

$$\int \prod_i \left(\frac{d^2\mathbf{\Omega}_i}{4\pi}\right) |\Psi|^2 = 1 + \left(-\frac{1}{3}\right)^N, \tag{G.7}$$

the right-hand side tending to unity in the thermodynamic limit ($N \to \infty$). More generally, the spin correlation function can be expressed as

$$\langle \mathbf{S}_0 \cdot \mathbf{S}_r \rangle = 4\langle \mathbf{\Omega}_0 \cdot \mathbf{\Omega}_r \rangle = 4\left(-\frac{1}{3}\right)^r, \tag{G.8}$$

which is alternating in sign, as is typical for an antiferromagnetic state, but is exponentially vanishing with a correlation length $\xi = (\ln 3)^{-1} \simeq 0.9$. It has been proved that also this model does possess a finite gap (Affleck *et al.*, 1987, 1988).

Affleck then considers the following generalization of the Hamiltonian, Eq. (G.2):

$$H = \sum_i [\mathbf{S}_i \cdot \mathbf{S}_{i+1} - \beta(\mathbf{S}_i \cdot \mathbf{S}_{i+1})^2]. \tag{G.9}$$

Equation (G.2) is recovered when $\beta = -\frac{1}{3}$, in which limit the gap is finite. On the other hand, one has a vanishing gap when $\beta = 1$. Affleck therefore surmises that some sort of 'phase transition' may occur at intermediate values of β, and specifically where it changes sign, *i.e.* at $\beta = 0$. To get some further insight, Affleck (1989) considers the ground state for an $s = 1$ chain. This is shown to be the dimerized state, where two valence bonds connect every second pair of sites. It turns out that it is possible to construct an Hamiltonian, analogous to Eq. (G.2), for which the dimerized state is the ground state. This is supportive of the conjecture that dimer states are the only ground states to this class of spin chain Hamiltonians. These two simple valence-bond states, *viz.* the VBS and the dimerized state, suggest two possible phases for an $s = 1$ antiferromagnet: one being characterized by a unique ground state, and another with two ground states and spontaneously broken translational symmetry. In particular, the dimerized phase has dimer long-range order:

$$\langle(\mathbf{S}_i \cdot \mathbf{S}_{i+1})(\mathbf{S}_j \cdot \mathbf{S}_{j+1})\rangle \mapsto \text{const} \cdot (-1)^{i-j} + \text{const}, \tag{G.10}$$

while $\langle\mathbf{S}_i \cdot \mathbf{S}_{i+1}\rangle$ decays exponentially. It has been conjectured (Affleck *et al.*, 1987, 1988) that the case $\beta = 1$ corresponds to the transition point between the VBS and the dimerized phase for a class of spin-chain bilinear-biquadratic Hamiltonians, including nearest neighbour correlations.

In the case of arbitrary β, although the VBS and dimerized phases are not exact ground states, yet they provide some reasonable variational guess for the true ground state. The variational energies in these states turn out to be

$$E_{\text{VBS}} = -\frac{4}{3} - 2\beta, \tag{G.11a}$$

$$E_{\text{dimerized}} = -1 - \frac{8}{3}\beta. \tag{G.11b}$$

As a term of comparison, the Néel (staggered) phase has a variation energy of $E_{\text{Néel}} = -1 - 2\beta$. From Eqs. (G.11), one concludes that the VBS state is variationally favoured for $\beta < \frac{1}{2}$, while the dimerized state is variationally favoured for $\beta > \frac{1}{2}$. This leads to a variational estimate of a critical value $\beta = \frac{1}{2}$ for the transition between the two valence-bond states, the vanishing of the gap at $\beta = 1$ being however a strong hint as for an improvement for such a variational estimate (Affleck *et al.*, 1987, 1988).

G.3 Experimental tests of the Haldane conjecture

Haldane (1983) conjectured that the ground state of the Heisenberg model in 1D should depend on the possible values of the spins involved. Specifically, Haldane (1983) predicted that half-odd-integer spin systems should behave like the spin-$\frac{1}{2}$ chain, whereas models with integer spins, as is the antiferromagnetic Heisenberg model in 1D, should be characterized by a (sizeable) energy gap between the ground state and the first excited state. Moreover, the spin correlation function should decay rapidly (exponentially) as a function of distance.

Actually, the absence of the spin-gap had been proved by Lieb *et al.* (1961) for the spin-$\frac{1}{2}$ Heisenberg model. This theorem was extended to larger spin values, but the gaplessness of the model could be proved only for the half-odd-integer spin systems in 1D (Affleck and Lieb, 1986; Schultz *et al.*, 1964), and more recently also in higher dimensions (Hastings, 2004). In other words, the extended Lieb-Schultz-Mattis theorem (Lieb *et al.*, 1961) could not eliminate the possibility of the Haldane gap.

Although there has been no rigorous proof of Haldane's conjectures for the integer-spin Heisenberg model, there is an antiferromagnetic Hamiltonian describing an $S = 1$ spin chain, which was rigorously proved to have the features of the Haldane system (Affleck and Marston, 1988). There is however strong numerical evidence in support of Haldane's conjecture (Golinelli *et al.*, 1994).

Several quasi-1D antiferromagnetic materials have now studied experimentally, whereof two Ni-based materials provided early examples of spin-1 antiferromagnetic chains for which the Haldane conjecture holds, *viz.* $CsNiCl_3$ and $Ni(C_2H_8N_2)_2NO_2ClO_4$.

The former material, $CsNiCl_3$ is characterized by chains with a strong $Ni^{2+} - Cl^- - Ni^{2+}$ superexchange coupling $J \approx 33$ K. Such chains form in turn a triangular lattice, with a weaker $Ni^{2+} - Cl^- - Cl^- - Ni^{2+}$ superexchange coupling $J'/J \approx 0.018$ (Affleck, 1989). This $CsNiCl_3$ material undergoes two Néel transitions, at $T = 4.84$ K and $T = 4.40$ K. The ground state for an isotropic classical antiferromagnet on a triangular lattice can be described as being formed by three interpenetrating triangular lattices, each of which characterized by a fixed direction of the spin, the spin directions in two such sublattices differing by (multiples of) $2\pi/3$. The spin directions alternate from one plane to another. This phase is observed below the lower critical temperature. As a result of crystal-field anisotropy, only the z-components of the spins turn out to be ordered in the intermediate temperature phase. Any spin order is eventually lost above the upper critical temperature. The main evidence for the occurrence of an Haldane gap here comes from neutron scattering experiments (Buyers *et al.*, 1986; Morra *et al.*, 1988) in the high-temperature phase of $CsNiCl_3$ ($T \simeq 10$ K), with gaps of $\sim 0.4J$ (see Affleck, 1989, for further details).

The second material, $Ni(C_2H_8N_2)_2NO_2ClO_4$, is characterized by a superexchange coupling between Ni^{2+} ions of $J \approx 55$ K, and by an inter-chain coupling $J'/J \simeq 4 \cdot 10^{-4}$. This material does not exhibit antiferromagnetic ordering until below 1.2 K, where again a finite gap is measured, with an average value of $\approx 0.4J$ (Renard *et al.*, 1987).

Several quasi-one-dimensional antiferromagnetic materials have become available and experimentally accessible nowadays, where it is nearly possible to control the desired degree of anisotropy and coupling ratio. These more recent experimental studies, as well as various theoretical and numerical techniques, have confirmed the existence of an Haldane gap, and allowed a thorough study of the low temperature phase diagram of the ordered phase. In particular, this seems to be a symmetry protected topological state, also known as a topological spin liquid (see Wierschem and Sengupta, 2014, and references therein).

As the present review was nearing completion, a very recent paper by Cubitt *et al.* (2015) appeared, where Haldane's conjecture that any integer-spin infinite chain should be gapped is proved to be actually *undecidable,* in general. Specifically, by examining models of quantum spin systems on a two-dimensional lattice with translationally invariant, nearest-neighbour interactions, Cubitt *et al.* (2015) showed that the question whether such systems are characterized by a gapped ground state is equivalent to the outcome of the 'halting problem' of a universal Turing machine (Turing, 1937). As Cubitt *et al.* (2015) explicitly state, 'there cannot exist an algorithm or a computable criterion that solves the spectral gap problem in general.' This rules out the possibility, in principle, to determine whether an arbitrary model is gapped or gapless, and implies that there exist models for which the presence or absence of a spectral gap is independent of the axioms of mathematics. Remarkably, Cubitt *et al.* (2015) relate this form of undecidability to that encountered in Gödel's incompleteness theorem (Gödel, 1931). Of course, the exact and general result by Cubitt *et al.* (2015) does not exclude the possibility of devising specific models for which Haldane's conjecture can be verified, as is in the fact the case of the model studied by Affleck and Marston (1988).

The results of Cubitt *et al.* (2015) also restrict the validity, in general, of the thermodynamic limit. This consists in extrapolating results obtained for models applied to systems of finite size, to systems of infinite size. Numerical techniques to pursue such a limit include finite-size scaling, which have their more rigorous counterpart in the renormalization group approach (Cardy, 1996; Kardar, 2007; Reichl, 2009). However, at least for the families of model Hamiltonians considered by Cubitt *et al.* (2015), the low-energy spectrum may initially look gapless as the system size increases, but will eventually be characterized by a gap of finite magnitude, at a certain threshold lattice size (or, vice versa, a gap will abruptly close). The critical value of the lattice size at which this would happen is likewise uncomputable. Analogous implications would also apply to other undecidable low-temperature properties. This conclusion would suggest the possibility of (possibly quantum) phase transitions not driven by temperature or other external parameters, but by the size of the system. This would be relevant not only for condensed matter systems, but also for many-body models of quantum chromodynamics on a lattice (Rebbi, 1983).

Appendix H

Quantum mechanics and number theory

Ever since the identification of the discrete regular patterns underlying the spectral (Lyman, Balmer, Paschen, ...) series of hydrogen, which led to the foundations of the theory of quanta, number theory has provided quantum mechanics with much inspiration. Remarkable connections have been established between number theory and the discrete structure of the energy levels of the Hamiltonians of certain quantum systems, and it is suggestive that, inversely, implications from quantum theory have been proposed for number theory.

In this context, the well-known, and still challenging, conjecture in number theory is Riemann hypothesis. This states that all non-trivial zeroes of the Riemann function, defined by $\zeta(s) = 0$, are such that $\mathrm{Re}\, s = \frac{1}{2}$ (Titchmarsh and Heath-Brown, 1986; Edwards, 2001; Cartier, 1995). Berry (1986); Berry and Keating (1999) then proposed that the imaginary parts of such zeroes might occur as the eigenvalues of some suitable Hamiltonian. It has been surmised that the statistical distribution of these zeros would possess connections with the random matrix theory and the theory of chaotic systems (Odlyzko, 1987, 1990; Stöckmann, 1999). More recently, Sierra and Townsend (2008); Sierra (2008) have proposed specific Hamiltonians whose spectral structure would be related to the distribution of zeroes of the Riemann function along the 'critical line', $\mathrm{Re}\, s = \frac{1}{2}$. Their proposal is appealing, in that it would be potentially accessible to experiments, especially in lower spatial dimensions (see Schumayer and Hutchinson, 2011, for a recent review).

Yet another connection between physical systems and the Riemann ζ function is that proposed by Dyson (2009), who claimed that the non-trivial zeroes of the ζ function would constitute a one-dimensional quasicrystal, *i.e.* a regular but aperiodic structure, endowed with a pointwise discrete diffraction pattern (Penrose, 1978, 1979; Shechtman *et al.*, 1984).

Below, we shall only briefly review some exact results by Crandall (1996) on the closely related subject of a suitably defined 'quantum ζ function', and its relation to the spectrum of several Hamiltonians, including the possibility of yielding quite accurate numerical estimates of its ground-state eigenvalue.

H.1 Quantum zeta function, and some properties of power potentials

Given a quantum system described by the Hamiltonian H, characterized by a spectrum $\{E\}$ of eigenvalues being discrete and possibly infinite, Crandall (1996) defines the 'quantum zeta function' as a formal sum over such eigenvalues

$$Z(s) = \sum_E \frac{1}{E^s}, \tag{H.1}$$

for $s \in \mathbb{C}$, when such a sum exists. The quantum zeta function is of course reminiscent of the Riemann ζ function

$$\zeta(s) = \sum_{n=1}^{\infty} \frac{1}{n^s}, \tag{H.2}$$

the above sum converging for $\operatorname{Re} s > 1$, from which domain it can then analytically continued elsewhere in the complex plane (Titchmarsh and Heath-Brown, 1986; Edwards, 2001; Cartier, 1995). Similarly, provided the ordered eigenvalues E are discrete, non-vanishing, and diverging sufficiently rapidly, Crandall (1996) concludes that $Z(s)$ should also converge for sufficiently large $\operatorname{Re} s$. When the eigenstates of the Hamiltonian H can be assigned a definite parity, ± 1 say, then it proves useful to introduce also a 'parity zeta function', defined as (Crandall, 1996)

$$Z(s) = \sum_E \frac{\pm 1}{E^s}. \tag{H.3}$$

Crandall (1996) then remarks that it is possible to evaluate explicitly $Z(s)$ and/or $Y(s)$ for given systems and for particular values of s, even in cases when the spectrum $\{E\}$ is not known exactly or completely. Inversely, knowledge of the behaviour of such quantum zeta functions can provide useful information on the spectrum, *e.g.* approximate estimates of its ground-state energy.

A straightforward example, where of course the complete spectrum is known exactly, is that of a quantum harmonic oscillator,

$$H = T + V \equiv -\frac{1}{2}\frac{d^2}{dx^2} + \frac{1}{2}\omega^2 x^2, \tag{H.4}$$

being characterized by the well-known spectrum $E_n = \left(n + \frac{1}{2}\right)\omega$, with n a non-negative integer. In that case, after a trivial change of summation variable, one arrives at

$$Z(s) = \sum_{n=0}^{\infty} \frac{1}{E_n^s} = \frac{2^s - 1}{\omega^s}\zeta(s), \tag{H.5}$$

the above sum converging for $\operatorname{Re} s > 1$, and being related to the analytic continuation of the Riemann ζ function also elsewhere in the complex plane.

A less trivial example is provided by the Schrödinger equation for a one-dimensional power potential:

$$\left(-\frac{1}{2}\frac{d^2}{dx^2} + |x|^\nu\right)\psi = E\psi, \tag{H.6}$$

which essentially reduces to the quantum harmonic oscillator for $\nu = 2$. Apart from the cases $\nu = 1$ and $\nu = 2$, the energy eigenvalues for such a problem are not known in closed form, even though the case $\nu = 4$ (quartic potential) has been extensively studied, both theoretically and numerically. However, assuming the eigenvalues E_n and the corresponding orthonormal eigenfunctions $\psi_n(x)$ be known, the solution to the associated problem for the Green's function in a generic potential $V(x)$,

$$\left(-\frac{1}{2}\frac{d^2}{dx^2} + V(x) - E\right)G = -\delta(x - x_0), \tag{H.7}$$

can be presented as

$$G(x, x_0; E) = \sum_n \frac{\psi_n(x)\psi_n^*(x_0)}{E - E_n}. \tag{H.8}$$

Knowledge of the Green's function at $E = 0$ can then be formally related to special values of the quantum zeta functions Z and Y. In particular, Crandall (1996) shows immediately that

$$Z(1) = -\int dx\, G(x, x, ; 0), \tag{H.9a}$$

the integral being taken over the natural domain of the eigenfunctions. More generally, for positive integer n,

$$Z(n) = -\frac{1}{(n-1)!}\int dx\, G^{(n-1)}(x, x, ; 0), \tag{H.9b}$$

while for even potentials $V(-x) = V(x)$,

$$Y(1) = -\int dx\, G(x, -x; 0). \tag{H.9c}$$

Coming back to the Green's function associated to Eq. (H.6) for a power potential, knowledge of the exact solutions of Eq. (H.6) for $E = 0$ and arbitrary (even non-integer) ν (Abramowitz and Stegun, 1964), allows Crandall (1996) to determine the closed form of the Green's function as

$$G(x, x_0; E) = -\frac{4}{p}\sqrt{|xx_0|}K_{1/p}(\alpha|x|^{p/2}) \\ \times \left(S(x, x_0)I_{1/p}(\alpha|x|^{p/2}) + \frac{\sin(\pi/p)}{\pi}K_{1/p}(\alpha|x|^{p/2}) \right), \tag{H.10}$$

where $p = \nu + 2$, $\alpha = 2\sqrt{2}/p$, and I and K are modified Bessel functions of the given arguments. In Eq. (H.10), $S(x, x_0) = 1$ if x, x_0 have the same sign, and $S(x, x_0) = 0$ otherwise. Integrating either $G(x, x; 0)$ or $G(x, -x; 0)$ over $-\infty < x < \infty$, and making use of Eqs. (H.9), Crandall (1996) finds

$$Y^{(\nu)}(1) = \left(\frac{2}{(\nu+2)^2}\right)^{\nu/(\nu+2)} \frac{\Gamma^2\left(\frac{2}{\nu+2}\right)\Gamma\left(\frac{3}{\nu+2}\right)}{\Gamma\left(\frac{4}{\nu+2}\right)\Gamma\left(1-\frac{1}{\nu+2}\right)}, \tag{H.11a}$$

$$Z^{(\nu)}(1) = \left(1 + \sec\frac{2\pi}{\nu+2}\right)Y^{(\nu)}(1). \tag{H.11b}$$

For ν a positive integer, these results reduce to those of Voros (1980), and in particular, in the case $\nu = 2$ corresponding to the quantum harmonic oscillator, they actually yield the analytic continuation of otherwise divergent sums.

Crandall (1996) also discusses the limiting case $\nu \to \infty$, for which the special values $Z^{(\infty)}(1) = \frac{4}{3}$ and $Y^{(\infty)}(1) = \frac{2}{3}$ are found. These are consistent with the fact that, in that limit, the system reduces to a particle confined in the box $(-1, 1)$, with infinite barriers at such ends, for which the spectrum is of course well-known.

In the case of the quartic potential ($\nu = 4$), Crandall's results agree again with those of Voros (1980), yielding

$$Z^{(4)}(1) = 3Y^{(4)}(1) = \frac{3^{2/3}}{8\pi^2}\Gamma^5\left(\frac{1}{3}\right) \sim 3.635\,003\,644\,88\ldots. \tag{H.12}$$

The standard WKB approximation of the quartic potential yields for the approximate eigenvalues, $E_n \sim W_n$ say, the expression

$$W_n = \left(c\left(n + \frac{1}{2}\right)\right)^{4/3}, \tag{H.13}$$

where

$$c = 2\sqrt{2\pi}\frac{\Gamma\left(\frac{7}{4}\right)}{\Gamma\left(\frac{1}{4}\right)}. \tag{H.14}$$

Crandall (1996) then remarks that, using only excited WKB approximations, and the relation

$$\frac{1}{E_0} \sim Y(1) + \frac{1}{W_1} - \frac{1}{W_2} + \frac{1}{W_3} - \cdots, \tag{H.15}$$

one obtains a quartic ground-state energy E_0 estimate with an error of less than 0.1%.

Bibliography

Abbott, B. P., *et al.* (LIGO Scientific Collaboration and Virgo Collaboration) (2016). Observation of gravitational waves from a binary black hole merger, *Phys. Rev. Lett.* **116**, p. 061102, doi:10.1103/PhysRevLett.116.061102. *(p. 227)*

Abrahams, E., Anderson, P. W., Licciardello, D. C., and Ramakrishnan, T. V. (1979). Scaling theory of localization: Absence of quantum diffusion in two dimensions, *Phys. Rev. Lett.* **42**, 10, pp. 673–676, doi:10.1103/PhysRevLett.42.673. *(p. 138)*

Abramowitz, M. and Stegun, I. A. (eds.) (1964). *Handbook of mathematical functions with formulas, graphs, and mathematical tables* (Dover, New York), ISBN 9780486612720. *(p. 276)*

Affleck, I. (1989). Quantum spin chains and the Haldane gap, *J. Phys.: Condens. Matter* **1**, 19, p. 3047, doi:10.1088/0953-8984/1/19/001. *(p. 265, 266, 267, 269)*

Affleck, I., Kennedy, T., Lieb, E., and Tasaki, H. (1987). Rigorous results on valence-bond ground states in antiferromagnets, *Phys. Rev. Lett.* **59**, 7, pp. 799–802, doi:10.1103/PhysRevLett.59.799. *(p. 265, 267, 268)*

Affleck, I., Kennedy, T., Lieb, E. H., and Tasaki, H. (1988). Valence bond ground states in isotropic quantum antiferromagnets, *Commun. Math. Phys.* **115**, 3, pp. 477–528, doi:10.1007/BF01218021. *(p. 265, 267, 268)*

Affleck, I. and Lieb, E. H. (1986). A proof of part of Haldane's conjecture on spin chains, *Lett. Math. Phys.* **12**, pp. 57–69, doi:10.1007/BF00400304. *(p. 269)*

Affleck, I. and Marston, J. B. (1988). Large-n limit of the Heisenberg-Hubbard model: Implications for high-T_c superconductors, *Phys. Rev. B* **37**, 7, pp. 3774–3777, doi:10.1103/PhysRevB.37.3774. *(p. 269, 270)*

Aizenman, M. (1981). Proof of the triviality of ϕ^4_d field theory and some mean-field features of Ising models for $d < 4$, *Phys. Rev. Lett.* **47**, 1, pp. 1–4, doi:10.1103/PhysRevLett.47.1. *(p. 152)*

Allen, J. F. and Misener, A. D. (1938a). Flow of liquid Helium II, *Nature* **141**,

*Back references: Slanted page numbers at the end of each bibliography entry refer to the page(s) in this book where the reference appears.

p. 75, doi:10.1038/141075a0. *(p. 94)*

Allen, J. F. and Misener, A. D. (1938b). Flow phenomena in liquid Helium II, *Nature* **142**, pp. 643–644, doi:10.1038/142643a0. *(p. 94)*

Alonso, J. A. and March, N. H. (1989). *Electrons in Metals and Alloys* (Academic Press, New York), ISBN 9780120536207. *(p. 119, 120)*

Altland, A. and Simons, B. (2007). *Condensed Matter Field Theory* (Cambridge University Press, Cambridge), ISBN 9780521845083. *(p. 49)*

Altshuler, B. L. (1985). Fluctuations in the extrinsic conductivity of disordered conductors, *Sov. Phys. JETP Lett.* **41**, pp. 648–651, [Pis'ma Zh. Eksp. Teor. Fiz. **41**, 530–533 (1985)]. *(p. 138)*

Altshuler, B. L. and Aronov, A. G. (1985). Electron-electron interaction in disordered conductors, in A. L. Efros and M. Pollak (eds.), *Electron-Electron Interactions in Disordered Systems, Modern problems in condensed matter sciences*, Vol. 10, chap. 1 (Elsevier, Amsterdam), pp. 1–153. *(p. 138)*

Altshuler, B. L., Kravtsov, V. E., and Lerner, I. V. (1988). in T. Ando and H. Fukuyama (eds.), *Anderson localization, Springer Proceedings in Physics*, Vol. 28 (Springer, Heidelberg), p. 300. *(p. 138, 139, 140, 141, 142)*

Alves, N. A., Berg, B. A., and Villanova, R. (1991). Potts models: Density of states and mass gap from Monte Carlo calculations, *Phys. Rev. B* **43**, pp. 5846–5856, doi:10.1103/PhysRevB.43.5846. *(p. 171)*

Alves, N. A. and Hansmann, U. H. E. (2000). Partition function zeros and finite size scaling of helix-coil transitions in a polypeptide, *Phys. Rev. Lett.* **84**, pp. 1836–1839, doi:10.1103/PhysRevLett.84.1836. *(p. 171)*

Amado, R. D. and Greenwood, F. C. (1973). There is no Efimov effect for four or more particles, *Phys. Rev. D* **7**, 8, pp. 2517–2519, doi:10.1103/PhysRevD.7.2517. *(p. 44)*

Ambegaokar, V., Halperin, B. I., and Langer, J. S. (1971). Hopping conductivity in disordered systems, *Phys. Rev. B* **4**, 8, pp. 2612–2620, doi:10.1103/PhysRevB.4.2612. *(p. 142, 143, 144, 145)*

Amovilli, C. and March, N. (2015). Electrostatic potentials at the nucleus for isoelectronic series of light atomic ions using the QMC method in relation to DFT, *J. Math. Chem.* **53**, 8, pp. 1725–1732, doi:10.1007/s10910-015-0514-6. *(p. 26, 27, 28, 29, 30)*

Amovilli, C. and March, N. H. (1995). Slater sum and kinetic energy tensor in some simple inhomogeneous electron liquids, *Phys. Chem. Liq.* **30**, 3, pp. 135–139, doi:10.1080/00319109508031647. *(p. 231)*

Anderson, E. (2012). The problem of time in quantum gravity, in V. R. Frignanni (ed.), *Classical and Quantum Gravity: Theory, Analysis and Applications*, chap. 4 (Nova, New York), preprint `arXiv:1009.2157`. *(p. 219)*

Anderson, M. H., Ensher, J. R., Matthews, M. R., Wieman, C. E., and Cornell, E. A. (1995). Observation of Bose-Einstein condensation in a dilute atomic vapor, *Science* **269**, 5221, pp. 198–201, doi:10.1126/science.269.5221.198. *(p. 94)*

Anderson, P. W. (1958). Absence of diffusion in certain random lattices, *Phys. Rev.* **109**, 5, pp. 1492–1505, doi:10.1103/PhysRev.109.1492. *(p. 135, 136, 249)*

Anderson, P. W. (1963). Plasmons, gauge invariance, and mass, *Phys. Rev.* **130**, 1, pp. 439–442, doi:10.1103/PhysRev.130.439. *(p. 76)*

Anderson, P. W. (1990). "Luttinger-liquid" behavior of the normal metallic state of the 2D Hubbard model, *Phys. Rev. Lett.* **64**, pp. 1839–1841, doi:10.1103/PhysRevLett.64.1839, reprinted in Anderson (1997). *(p. 253, 255)*

Anderson, P. W. (1997). *The Theory of Superconductivity in the High-T_c Cuprates* (Princeton University Press, Princeton NJ), ISBN 9780691043654. *(p. 281)*

Andrei, E. Y., Deville, G., Glattli, D. C., Williams, F. I. B., Paris, E., and Etienne, B. (1988). Observation of a magnetically induced Wigner solid, *Phys. Rev. Lett.* **60**, 26, pp. 2765–2768, doi:10.1103/PhysRevLett.60.2765. *(p. 239)*

Andrei, N. (1980). Diagonalization of the Kondo hamiltonian, *Phys. Rev. Lett.* **45**, 5, pp. 379–382, doi:10.1103/PhysRevLett.45.379. *(p. 248)*

Andrei, N., Furuya, K., and Lowenstein, J. H. (1983). Solution of the Kondo problem, *Rev. Mod. Phys.* **55**, 2, pp. 331–402, doi:10.1103/RevModPhys.55.331. *(p. 248)*

Andronikashvili, E. L. (1946). Direct observation of two types of motion in Helium II, *Zh. Eksp. Teor. Fiz. (Sov. Phys. JETP)* **16**, p. 780. *(p. 94)*

Angilella, G. G. N. (2011). *Esercizi di Metodi matematici della fisica* (Springer, Milano), ISBN 9788847019522. *(p. 205)*

Angilella, G. G. N., Bartalini, S., Cataliotti, F. S., Herrera, I., March, N. H., and Pucci, R. (2006a). The Gross-Pitaevskii equations and beyond for inhomogeneous condensed bosons, in A. V. Ling (ed.), *Trends in Boson Research*, Vol. 1 (Nova Science, New York), ISBN 1-59454-521-9, p. 35, preprint `cond-mat/0410287`. *(p. 94, 110)*

Angilella, G. G. N., Leys, F. E., March, N. H., and Pucci, R. (2003). Phase transitions wholly within the liquid state, *Phys. Chem. Liq.* **41**, pp. 211–226, doi:10.1080/0031910021000061574. *(p. 94)*

Angilella, G. G. N., Leys, F. E., March, N. H., and Pucci, R. (2004). Correlation between characteristic energies in non–s-wave pairing superconductors, *Phys. Lett. A* **322**, pp. 375–383, doi:10.1016/j.physleta.2003.12.066. *(p. 79)*

Angilella, G. G. N., March, N. H., and Pucci, R. (2000). T_c for non–s-wave pairing superconductors correlated with coherence length and effective mass, *Phys. Rev. B* **62**, pp. 13919–13921, doi:10.1103/PhysRevB.62.13919. *(p. 249)*

Angilella, G. G. N., March, N. H., and Pucci, R. (2001). Electron (hole) liquids flowing through both heavy fermion and cuprate materials in superconducting and normal state, *Phys. Chem. Liq.* **39**, p. 405, doi:10.1080/00319100108031673. *(p. 249)*

Angilella, G. G. N., March, N. H., and Pucci, R. (2005). T_c for heavy fermion superconductors linked with other physical properties at zero and applied pressure, *Supercond. Sci. Technol.* **18**, p. 557, doi:10.1088/0953-2048/18/4/028. *(p. 249)*

Angilella, G. G. N., March, N. H., and Pucci, R. (2006b). Anyon liquid fractional statistics distribution function with applications to phonons in a low-dimensional vibrating lattice and to two-dimensional magnets, *Phys. Chem. Liq.* **44**, pp. 193–202, doi:10.1080/00319100500499680. *(p. 57, 72)*

Angilella, G. G. N., March, N. H., Siringo, F., and Pucci, R. (2006c). Gap equation for anyon superfluids, *Phys. Chem. Liq.* **44**, pp. 343–351, doi:10.1080/00319100600740983. *(p. 57, 58, 59, 72)*

Aoyama, T. (2001). Specific heat of the ideal gas obeying the generalized exclusion statistics, *Eur. Phys. J. B* **20**, pp. 123–131, doi:10.1007/BF01313922. *(p. 56, 59)*

Arovas, D., Auerbach, A., and Haldane, F. (1988). Extended Heisenberg models of antiferromagnetism: Analogies to the fractional quantum Hall effect, *Phys. Rev. Lett.* **60**, 6, pp. 531–534, doi:10.1103/PhysRevLett.60.531. *(p. 266)*

Arriola, E. R. and Soler, J. (1999). Asymptotic behaviour for the 3D Schrödinger-Poisson system in the attractive case with positive energy, *Appl. Math. Lett.* **12**, 8, pp. 1–6, doi:10.1016/S0893-9659(99)00113-5. *(p. 225)*

Ashcroft, N. W. (1966). Electron-ion pseudopotentials in metals, *Phys. Lett.* **23**, pp. 48–50, doi:10.1016/0031-9163(66)90251-4. *(p. 128)*

Ashcroft, N. W. and Mermin, N. D. (1976). *Solid State Physics* (Saunders College Publ., Fort Worth), ISBN 9780030839931. *(p. 127, 234)*

Ashtekar, A. (1986). New variables for classical and quantum gravity, *Phys. Rev. Lett.* **57**, 18, pp. 2244–2247, doi:10.1103/PhysRevLett.57.2244. *(p. 221)*

Ashtekar, A. (1987). New Hamiltonian formulation of general relativity, *Phys. Rev. D* **36**, 6, pp. 1587–1602, doi:10.1103/PhysRevD.36.1587. *(p. 221)*

Baake, M., Grimm, U., and Pisani, C. (1995). Partition function zeros for aperiodic systems, *J. Stat. Phys.* **78**, 1, pp. 285–297, doi:10.1007/BF02183349. *(p. 171)*

Baccarelli, I., Delgado-Barrio, G., Gianturco, F. A., Gonzalez-Lezana, T., Miret-Artés, S., and Villarreal, P. (2000). Searching for Efimov states in triatomic systems: The case of $LiHe_2$, *Europhys. Lett.* **50**, 5, pp. 567–573, doi:10.1209/epl/i2000-00307-8. *(p. 42)*

Bagrov, V. G. and Gitman, D. M. (1990). *Exact solutions of relativistic wave equations, Mathematics and its applications (Soviet series)*, Vol. 39 (Kluwer, Dordrecht), ISBN 9780792302155. *(p. 193, 211)*

Bagrov, V. G. and Gitman, D. M. (2014). *The Dirac Equation and its Solutions, de Gruyter Studies in mathematical physics*, Vol. 4 (de Gruyter, Berlin), ISBN 9783110262926. *(p. 193, 211)*

Bahrami, M., Großardt, A., Donadi, S., and Bassi, A. (2014). The Schrödinger-Newton equation and its foundations, *New J. Phys.* **16**, 11, p. 115007, doi:10.1088/1367-2630/16/11/115007. *(p. 223, 224)*

Baltin, R. and March, N. H. (1987). Electron and kinetic energy densities from Dirac's equation for a model semi-infinite inhomogeneous electron gas, *J. Phys. A: Math. Gen.* **20**, 16, pp. 5517–5528, doi:10.1088/0305-4470/20/16/026. *(p. 199)*

Barbiellini, B. (2014). Natural orbital functional theory and pairing correlation effects in electron momentum density, *Low Temp. Phys.* **40**, 4, pp. 318–322, doi:10.1063/1.4869587. *(p. 91)*

Bardeen, J., Cooper, L. N., and Schrieffer, J. R. (1957). Theory of superconductivity, *Phys. Rev.* **108**, 5, pp. 1175–1204, doi:10.1103/PhysRev.108.1175. *(p. 76, 79, 80)*

Bargmann, V., Michel, L., and Telegdi, V. L. (1959). Precession of the polarization of particles moving in a homogeneous electromagnetic field, *Phys. Rev. Lett.* **2**, 10, pp. 435–436, doi:10.1103/PhysRevLett.2.435. *(p. 211, 214)*

Baxter, R. J. (1971). Eight-vertex model in lattice statistics, *Phys. Rev. Lett.* **26**, 14, pp. 832–833, doi:10.1103/PhysRevLett.26.832. *(p. 181)*

Baxter, R. J. (1972a). One-dimensional anisotropic Heisenberg chain, *Ann. Phys.* **70**, 2, pp. 323–337, doi:10.1016/0003-4916(72)90270-9. *(p. 181)*

Baxter, R. J. (1972b). Partition function of the eight-vertex lattice model, *Ann. Phys.* **70**, 1, pp. 193–228, doi:10.1016/0003-4916(72)90335-1. *(p. 181)*

Baxter, R. J. (1973a). Eight-vertex model in lattice statistics and one-dimensional anisotropic Heisenberg chain. I. Some fundamental eigenvectors, *Ann. Phys.* **76**, 1, pp. 1–24, doi:10.1016/0003-4916(73)90439-9. *(p. 181)*

Baxter, R. J. (1973b). Eight-vertex model in lattice statistics and one-dimensional anisotropic Heisenberg chain. II. Equivalence to a generalized ice-type lattice model, *Ann. Phys.* **76**, 1, pp. 25–47, doi:10.1016/0003-4916(73)90440-5. *(p. 181)*

Baxter, R. J. (1973c). Eight-vertex model in lattice statistics and one-dimensional anisotropic Heisenberg chain. III. Eigenvectors of the transfer matrix and Hamiltonian, *Ann. Phys.* **76**, 1, pp. 48–71, doi:10.1016/0003-4916(73) 90441-7. *(p. 181)*

Baxter, R. J. and Kelland, S. B. (1974). Spontaneous polarization of the eight-vertex model, *J. Phys. C: Solid State Phys.* **7**, 22, pp. L403–L406, doi: 10.1088/0022-3719/7/22/003. *(p. 183)*

Bednorz, J. G. and Müller, K. A. (1986). Possible high T_c superconductivity in the Ba–La–Cu–O system, *Z. Phys. B* **64**, pp. 189–193, doi:10.1007/ BF01303701. *(p. 82)*

Beichert, F., Hooley, C. A., Moessner, R., and Oganesyan, V. (2013). Correlations and phase structure of Ising models at complex temperature, Preprint `arXiv:1304.6314`. *(p. 172)*

Belenchia, A., Benincasa, D. M. T., Liberati, S., Marin, F., Marino, F., and Ortolan, A. (2016). Testing quantum gravity induced nonlocality via optomechanical quantum oscillators, *Phys. Rev. Lett.* **116**, p. 161303, doi: 10.1103/PhysRevLett.116.161303. *(p. 228)*

Belinskiĭ, V. A. and Zakharov, V. E. (1978). Integration of the Einstein equations by means of the inverse scattering problem technique and construction of exact soliton solutions, *Sov. Phys. JETP* **48**, 6, pp. 985–994, [Zh. Eksp. Teor. Fiz. **75**, 1955–1971 (1978)]. *(p. 218)*

Bena, I., Droz, M., and Lipowski, A. (2005). Statistical mechanics od equilibrium and nonequilibrium phase transitions: the Yang-Lee formalism, *Int. J. Mod. Phys. B* **19**, 29, pp. 4269–4329, doi:10.1142/S0217979205032759. *(p. 156, 157, 160, 165, 169)*

Berezinskii, V. L. (1971). Destruction of long-range order in one-dimensional and two-dimensional systems having a continuous symmetry group. I. Classical systems, *Sov. Phys. JETP* **32**, pp. 493–500. *(p. 96)*

Berezinskii, V. L. (1972). Destruction of long-range order in one-dimensional and two-dimensional systems having a continuous symmetry group. II. Quantum systems, *Sov. Phys. JETP* **34**, pp. 610–616. *(p. 96)*

Bergmann, P. G. (1949). Non-linear field theories, *Phys. Rev.* **75**, 4, pp. 680–685, doi:10.1103/PhysRev.75.680. *(p. 218)*

Bergmann, P. G. and Brunings, J. H. M. (1949). Non-linear field theories II. Canonical equations and quantization, *Rev. Mod. Phys.* **21**, 3, pp. 480–487, doi:10.1103/RevModPhys.21.480. *(p. 218)*

Bernasconi, J. (1973). Electrical conductivity in disordered systems, *Phys. Rev. B* **7**, 6, pp. 2252–2260, doi:10.1103/PhysRevB.7.2252. *(p. 143)*

Bernstein, D. H., Giladi, E., and Jones, K. R. W. (1998). Eigenstates of the gravitational Schrödinger equation, *Mod. Phys. Lett. A* **13**, 29, pp. 2327–2336, doi:10.1142/S0217732398002473. *(p. 225)*

Berry, M. V. (1986). Spectral zeta functions for Aharonov-Bohm quantum billiards, *J. Phys. A: Math. Gen.* **19**, 12, pp. 2281–2296, doi:10.1088/0305-4470/19/12/015. *(p. 273)*

Berry, M. V. and Keating, J. P. (1999). The Riemann zeros and eigenvalue asymptotics, *SIAM Rev.* **41**, 2, pp. 236–266, doi:10.1137/S0036144598347497. *(p. 273)*

Berti, E. (2016). Viewpoint: The first sounds of merging black holes, *Physics* **9**, p. 17. *(p. 228)*

Binek, C. (1998). Density of zeros on the Lee-Yang circle obtained from magnetization data of a two-dimensional Ising ferromagnet, *Phys. Rev. Lett.* **81**, pp. 5644–5647, doi:10.1103/PhysRevLett.81.5644. *(p. 164, 165)*

Binek, C., Kleemann, W., and Katori, H. A. (2001). Yang-Lee edge singularities determined from experimental high-field magnetization data, *J. Phys.: Cond. Matt.* **13**, 35, pp. L811–L817, doi:10.1088/0953-8984/13/35/103. *(p. 165)*

Biskup, M., Borgs, C., Chayes, J. T., and Kotecký, R. (2004). Partition function zeros at first-order phase transitions: Pirogov-Sinai theory, *J. Stat. Phys.* **116**, 1, pp. 97–155, doi:10.1023/B:JOSS.0000037243.48527.e3. *(p. 171)*

Black, C. T., Ralph, D. C., and Tinkham, M. (1996). Spectroscopy of the superconducting gap in individual nanometer-scale aluminum particles, *Phys. Rev. Lett.* **76**, 4, pp. 688–691, doi:10.1103/PhysRevLett.76.688. *(p. 82)*

Bloch, I., Dalibard, J., and Zwerger, W. (2008). Many-body physics with ultracold gases, *Rev. Mod. Phys.* **80**, 3, pp. 885–964, doi:10.1103/RevModPhys.80.885. *(p. 110)*

Blöchl, P. E., Pruschke, T., and Potthoff, M. (2013). Density-matrix functionals from Green's functions, *Phys. Rev. B* **88**, 20, p. 205139, doi:10.1103/PhysRevB.88.205139. *(p. 90)*

Blythe, R. A. and Evans, M. R. (2002). Lee-Yang zeros and phase transitions in nonequilibrium steady states, *Phys. Rev. Lett.* **89**, p. 080601, doi:10.1103/PhysRevLett.89.080601. *(p. 169)*

Bogoliubov, N. N. (1947). On the theory of superfluidity, *J. Phys. (Moscow)* **11**, pp. 23–32. *(p. 92, 95)*

Bogoliubov, N. N. (1958). On a new method in the theory of superconductivity, *Nuovo Cimento* **7**, pp. 794–805, doi:10.1007/BF02745585. *(p. 80, 81)*

Bonsall, L. and Maradudin, A. A. (1977). Some static and dynamical properties of a two-dimensional Wigner crystal, *Phys. Rev. B* **15**, 4, pp. 1959–1973, doi:10.1103/PhysRevB.15.1959. *(p. 237, 238, 239)*

Booß-Bavnbek, B., Esposito, G., and Lesch, M. (eds.) (2010). *New Paths Towards Quantum Gravity, Lecture Notes in Physics*, Vol. 807 (Springer, Berlin), ISBN 9783642118968. *(p. 219, 296)*

Braaten, E. and Hammer, H. (2001). Three-body recombination into deep bound states in a bose gas with large scattering length, *Phys. Rev. Lett.* **87**, 16, p. 160407, doi:10.1103/PhysRevLett.87.160407. *(p. 42)*

Braaten, E. and Hammer, H.-W. (2006). Universality in few-body systems with large scattering length, *Phys. Rep.* **428**, 5–6, pp. 259–390, doi:10.1016/j.physrep.2006.03.001. *(p. 32, 33, 36, 39)*

Brau, F. (2003). Upper limit on the number of bound states of the spinless Salpeter equation, *Phys. Lett. A* **313**, 5–6, pp. 363–372, doi:10.1016/S0375-9601(03)00809-0. *(p. 215)*

Brau, F. (2005). Upper limit on the critical strength of central potentials in relativistic quantum mechanics, *J. Math. Phys.* **46**, 3, 032305, doi:10.1063/1.1850997. *(p. 215)*

Braun, J. W., Su, Q., and Grobe, R. (1999). Numerical approach to solve the time-dependent Dirac equation, *Phys. Rev. A* **59**, 1, pp. 604–612, doi:10.1103/PhysRevA.59.604. *(p. 209)*

Brinkman, W. F. and Rice, T. M. (1970). Application of Gutzwiller's variational method to the metal-insulator transition, *Phys. Rev. B* **2**, 10, pp. 4302–4304, doi:10.1103/PhysRevB.2.4302. *(p. 132, 133)*

Brown, L. S. and Kibble, T. W. B. (1964). Interaction of intense laser beams with electrons, *Phys. Rev.* **133**, 3A, pp. A705–A719, doi:10.1103/PhysRev.133.A705. *(p. 209)*

Brühl, R., Kalinin, A., Kornilov, O., Toennies, J. P., Hegerfeldt, G. C., and Stoll, M. (2005). Matter wave diffraction from an inclined transmission grating: Searching for the elusive ^{4}He trimer Efimov state, *Phys. Rev. Lett.* **95**, 6, p. 063002, doi:10.1103/PhysRevLett.95.063002. *(p. 42)*

Bugrij, A. I. and Loktev, V. M. (2007). On the theory of Bose-Einstein condensation of quasiparticles: On the possibility of condensation of ferromagnons at high temperatures, *Low Temp. Phys.* **33**, 1, pp. 37–50, doi:10.1063/1.2409633. *(p. 97)*

Bugrij, A. I. and Loktev, V. M. (2013). On the theory of spatially inhomogeneous Bose-Einstein condensation of magnons in yttrium iron garnet, *Low Temp. Phys.* **39**, 12, pp. 1037–1047, doi:10.1063/1.4843356. *(p. 97)*

Buhmann, H., Joss, W., von Klitzing, K., Kukushkin, I. V., Plaut, A. S., Martinez, G., Ploog, K., and Timofeev, V. B. (1991). Novel magneto-optical behavior in the Wigner-solid regime, *Phys. Rev. Lett.* **66**, 7, pp. 926–929, doi:10.1103/PhysRevLett.66.926. *(p. 237)*

Buijse, M. A. and Baerends, E. J. (2002). An approximate exchange-correlation hole density as a functional of the natural orbitals, *Mol. Phys.* **100**, 4, pp. 401–421, doi:10.1080/00268970110070243. *(p. 8)*

Bulgac, A. (2002). Dilute quantum droplets, *Phys. Rev. Lett.* **89**, 5, p. 050402, doi:10.1103/PhysRevLett.89.050402. *(p. 42)*

Bunkov, Y. M. and Volovik, G. E. (2010). Magnon Bose-Einstein condensation and spin superfluidity, *J. Phys.: Condens. Matter* **22**, 16, p. 164210, doi: 10.1088/0953-8984/22/16/164210. *(p. 97)*

Burioni, R., Cassi, D., and Donetti, L. (1999). Lee-Yang zeros and the Ising model on the Sierpinski gasket, *J. Phys. A: Math. Gen.* **32**, 27, pp. 5017–5028, doi:10.1088/0305-4470/32/27/303. *(p. 171)*

Buyers, W., Morra, R., Armstrong, R., Hogan, M., Gerlach, P., and Hirakawa, K. (1986). Experimental evidence for the Haldane gap in a spin-1 nearly isotropic, antiferromagnetic chain, *Phys. Rev. Lett.* **56**, 4, pp. 371–374, doi: 10.1103/PhysRevLett.56.371. *(p. 269)*

Calabrese, P., Essler, F. H. L., and Fagotti, M. (2011). Quantum quench in the transverse-field Ising chain, *Phys. Rev. Lett.* **106**, p. 227203, doi:10.1103/PhysRevLett.106.227203. *(p. 171)*

Calabrese, P. and Mintchev, M. (2007). Correlation functions of one-dimensional anyonic fluids, *Phys. Rev. B* **75**, p. 233104, doi:10.1103/PhysRevB.75.233104. *(p. 46)*

Camino, F. E., Zhou, W., and Goldman, V. J. (2005). Realization of a Laughlin quasiparticle interferometer: Observation of fractional statistics, *Phys. Rev. B* **72**, p. 075342, doi:10.1103/PhysRevB.72.075342. *(p. 46)*

Canright, G. S. and Johnson, M. D. (1994). Fractional statistics: α to β, *J. Phys. A: Math. Gen.* **27**, pp. 3579–3598, doi:10.1088/0305-4470/27/11/009. *(p. 47, 55)*

Capuzzi, P., March, N. H., and Tosi, M. P. (2005a). Differential equation for the ground-state density of artificial two-electron atoms with harmonic confinement, *J. Phys. A: Math. Gen.* **38**, pp. L439–L442, doi: 10.1088/0305-4470/38/24/L01. *(p. 19, 20, 21)*

Capuzzi, P., March, N. H., and Tosi, M. P. (2005b). Wigner bosonic molecules with repulsive interactions and harmonic confinement, *Phys. Lett. A* **339**, pp. 207–211, doi:10.1016/j.physleta.2005.02.078. *(p. 21)*

Cardy, J. (1996). *Scaling and Renormalization in Statistical Physics* (Cambridge University Press, Cambridge), ISBN 9780521499590. *(p. 147, 177, 178, 270)*

Carmelo, J. and Baeriswyl, D. (1988a). Ground-state properties and elementary excitations of the one-dimensional Hubbard model, *Int. J. Mod. Phys. B* **2**, 05, pp. 1013–1019, doi:10.1142/S0217979288000846. *(p. 255)*

Carmelo, J. and Baeriswyl, D. (1988b). Solution of the one-dimensional Hubbard model for arbitrary electron density and large U, *Phys. Rev. B* **37**, 13, pp. 7541–7548, doi:10.1103/PhysRevB.37.7541. *(p. 255)*

Carmelo, J. and Ovchinnikov, A. A. (1991). Generalization of the Landau liquid concept: example of the Luttinger liquids, *J Phys.: Cond. Matter* **3**, 6, pp. 757–766, doi:10.1088/0953-8984/3/6/012, reprinted in Mattis (1993). *(p. 253, 256)*

Cartier, P. (1995). An introduction to zeta functions, in M. Waldschmidt, P. Moussa, J.-M. Luck, and C. Itzykson (eds.), *From number theory to physics*, chap. 1 (Springer, Berlin), ISBN 9783642080975, pp. 1–63. *(p. 273, 274)*

Castelvecchi, D. (2015). Is string theory science? *Sci. Am.* `http://www.scientificamerican.com/article/is-string-theory-science/`. *(p. 219)*

Castorina, P., Cea, P., Nardulli, G., and Paiano, G. (1984). Partial-wave analysis of a relativistic Coulomb problem, *Phys. Rev. D* **29**, 11, pp. 2660–2663, doi:10.1103/PhysRevD.29.2660. *(p. 215)*

Castro Neto, A. H., Guinea, F., Peres, N. M. R., Novoselov, K. S., and Geim, A. K. (2009). The electronic properties of graphene, *Rev. Mod. Phys.* **81**, 1, pp. 109–162, doi:10.1103/RevModPhys.81.109. *(p. 211)*

Cazalilla, M. A., Citro, R., Giamarchi, T., Orignac, E., and Rigol, M. (2011). One dimensional bosons: From condensed matter systems to ultracold gases, *Rev. Mod. Phys.* **83**, 4, pp. 1405–1466, doi:10.1103/RevModPhys.83.1405. *(p. 110)*

Ceperley, D. M. (1995). Path integrals in the theory of condensed helium, *Rev. Mod. Phys.* **67**, 2, pp. 279–355, doi:10.1103/RevModPhys.67.279. *(p. 92)*

Ceperley, D. M. and Alder, B. J. (1980). Ground state of the electron gas by a stochastic method, *Phys. Rev. Lett.* **45**, pp. 566–569, doi:10.1103/PhysRevLett.45.566. *(p. 127, 237)*

Chapman, R. G. and March, N. H. (1988). Magnetic susceptibility of expanded fluid alkali metals, *Phys. Rev. B* **38**, 1, pp. 792–794, doi:10.1103/PhysRevB.38.792, reprinted in March and Angilella (2009). *(p. 132, 133)*

Chen, D., White, M., Borries, C., and DeMarco, B. (2011). Quantum quench of an atomic Mott insulator, *Phys. Rev. Lett.* **106**, p. 235304, doi:10.1103/PhysRevLett.106.235304. *(p. 170)*

Chihara, J. (1973). in S. Takeuchi (ed.), *The properties of liquid metals. Proceedings of the Second International Conference held at Tokyo, Japan, September 1972.* (Taylor and Francis, London), pp. 137–141. *(p. 131)*

Chin, C., Grimm, R., Julienne, P., and Tiesinga, E. (2010). Feshbach resonances in ultracold gases, *Rev. Mod. Phys.* **82**, 2, pp. 1225–1286, doi:10.1103/RevModPhys.82.1225. *(p. 43)*

Clark, S. (2016). Quantum bounce, *New Scientist* **229**, 3054, pp. 32–35, doi:10.1016/S0262-4079(16)30033-1. *(p. 226)*

Coleman, A. J. (1963). Structure of fermion density matrices, *Rev. Mod. Phys.* **35**, 3, pp. 668–686, doi:10.1103/RevModPhys.35.668. *(p. 4, 87, 88, 89, 100)*

Coleman, A. J. (1964). Electron pairs in the quasichemical-equilibrium and Bardeen-Cooper-Schrieffer theories, *Phys. Rev. Lett.* **13**, 13, pp. 406–407, doi:10.1103/PhysRevLett.13.406. *(p. 90)*

Coleman, A. J. (1997). The AGP model for fermion systems, *Int. J. Quantum Chem.* **63**, 1, pp. 23–30, doi:10.1002/(SICI)1097-461X(1997)63:1<23::AID-QUA5>3.0.CO;2-4. *(p. 88)*

Coleman, A. J. and Yukalov, V. I. (1991). Order indices and mid-range order, *Mod. Phys. Lett. B* **5**, 24n25, pp. 1679–1686, doi:10.1142/S0217984991002021. *(p. 89)*

Coleman, A. J. and Yukalov, V. I. (1992). Order indices for boson density matrices, *Nuovo Cimento B* **107**, 5, pp. 535–552, doi:10.1007/BF02723631.

(p. 89)

Coleman, A. J. and Yukalov, V. I. (1993). Order indices and ordering in macroscopic systems, *Nuovo Cimento B* **108**, 12, pp. 1377–1397, doi: 10.1007/BF02755191. *(p. 89)*

Coleman, A. J. and Yukalov, V. I. (1996). Relation between microscopic and macroscopic characteristics of statistical systems, *Int. J. Mod. Phys. B* **10**, 25, pp. 3505–3515, doi:10.1142/S0217979296001872. *(p. 88, 89, 90, 91, 100, 102)*

Coleman, A. J. and Yukalov, V. I. (2000). *Reduced Density Matrices: Coulson's Challenge*, *Lecture Notes in Chemistry*, Vol. 72 (Springer, New York), ISBN 9783540671480. *(p. 88, 100)*

Cooper, N. R. and Simon, S. H. (2015). Signatures of fractional exclusion statistics in the spectroscopy of quantum Hall droplets, *Phys. Rev. Lett.* **114**, p. 106802, doi:10.1103/PhysRevLett.114.106802. *(p. 72)*

Cordero, N. A., March, N. H., and Alonso, J. A. (2013). Semiempirical fine-tuning for Hartree-Fock ionization potentials of atomic ions with non-integral atomic number, *Phys. Lett. A* **377**, 41, pp. 2955–2958, doi: 10.1016/j.physleta.2013.09.014. *(p. 28)*

Côté, R. and MacDonald, A. H. (1990). Phonons as collective modes: The case of a two-dimensional Wigner crystal in a strong magnetic field, *Phys. Rev. Lett.* **65**, 21, pp. 2662–2665, doi:10.1103/PhysRevLett.65.2662. *(p. 240)*

Côté, R. and MacDonald, A. H. (1991). Collective modes of the two-dimensional Wigner crystal in a strong magnetic field, *Phys. Rev. B* **44**, 16, pp. 8759–8773, doi:10.1103/PhysRevB.44.8759. *(p. 240)*

Coulson, C. A. (1960). Present state of molecular structure calculations, *Rev. Mod. Phys.* **32**, 2, pp. 170–177, doi:10.1103/RevModPhys.32.170. *(p. 3, 87)*

Crandall, R., Whitnell, R., and Bettega, R. (1984). Exactly soluble two-electron atomic model, *Am. J. Phys.* **52**, pp. 438–442, doi:10.1119/1.13650. *(p. 18, 20)*

Crandall, R. E. (1996). On the quantum zeta function, *J. Phys. A: Math. Gen.* **29**, 21, pp. 6795–6816, doi:10.1088/0305-4470/29/21/014. *(p. 274, 275, 276, 277)*

Cubitt, T. S., Perez-Garcia, D., and Wolf, M. M. (2015). Undecidability of the spectral gap, *Nature* **528**, 7581, pp. 207–211, doi:10.1038/nature16059, article. *(p. 270)*

Cusack, S., March, N. H., Parrinello, M., and Tosi, M. P. (1976). Electron-electron pair correlation function in solid and molten nearly-free electron metals, *J. Phys. F: Metal Phys.* **6**, 5, pp. 749–766, doi:10.1088/0305-4608/6/5/017, Reprinted in March and Angilella (2009). *(p. 129, 130, 131)*

Dancz, J., Edwards, S. F., and .March, N. H. (1973). The electronic structure of a disordered network, *J. Phys. C: Solid State Phys.* **6**, 5, pp. 873–879, doi:10.1088/0022-3719/6/5/016. *(p. 138)*

Dasnières de Veigy, A. and Ouvry, S. (1994). Equation of state of an anyon gas in a strong magnetic field, *Phys. Rev. Lett.* **72**, pp. 600–603, doi:10.1103/PhysRevLett.72.600. *(p. 56)*

Dasnières de Veigy, A. and Ouvry, S. (1995). One-dimensional statistical mechanics for identical particles: the Calogero and anyon cases, *Mod. Phys. Lett. B* **9**, pp. 271–284, doi:10.1142/S0217984995000267. *(p. 56)*

Daubechies, I. (1983). An uncertainty principle for fermions with generalized kinetic energy, *Commun. Math. Phys.* **90**, 4, pp. 511–520, doi:10.1007/BF01216182. *(p. 215)*

de Wijs, G. A., Pastore, G., Selloni, A., and van der Lugt, W. (1995). Electron-ion correlation in liquid metals from first principles: Liquid Mg and liquid Bi, *Phys. Rev. Lett.* **75**, 24, pp. 4480–4483, doi:10.1103/PhysRevLett.75.4480. *(p. 127, 128, 129, 132)*

De Witt, C. M. and Wesley, W. G. (1971). Quantum falling charges, *Gen. Rel. Grav.* **2**, 3, pp. 235–245, doi:10.1007/BF00763768. *(p. 221)*

Della Selva, A., Magnin, J., and Masperi, L. (1996). Bargmann-Michel-Telegdi equation and one-particle relativistic approach, *Nuovo Cimento B (Series 11)* **111**, 7, pp. 855–862, doi:10.1007/BF02749016. *(p. 212)*

Delves, L. M. (1960). Tertiary and general-order collisions (II), *Nucl. Phys.* **20**, pp. 275–308, doi:10.1016/0029-5582(60)90174-7. *(p. 33)*

Delves, R. T., Joyce, G. S., and Zucker, I. J. (1997). Exact results in the theory of intermediate statistics, *Proc. R. Soc. London A* **453**, 1961, pp. 1177–1194, doi:10.1098/rspa.1997.0065. *(p. 62)*

Demidov, V. E., Dzyapko, O., Demokritov, S. O., Melkov, G. A., and Slavin, A. N. (2007). Thermalization of a parametrically driven magnon gas leading to Bose-Einstein condensation, *Phys. Rev. Lett.* **99**, 3, p. 037205, doi:10.1103/PhysRevLett.99.037205. *(p. 97)*

Demidov, V. E., Dzyapko, O., Demokritov, S. O., Melkov, G. A., and Slavin, A. N. (2008). Observation of spontaneous coherence in Bose-Einstein condensate of magnons, *Phys. Rev. Lett.* **100**, 4, p. 047205, doi:10.1103/PhysRevLett.100.047205. *(p. 97)*

Demokritov, S. O., Demidov, V. E., Dzyapko, O., Melkov, G. A., Serga, A. A., Hillebrands, B., and Slavin, A. N. (2006). Bose-Einstein condensation of quasi-equilibrium magnons at room temperature under pumping, *Nature* **443**, 7110, pp. 430–433, doi:10.1038/nature05117. *(p. 97, 98)*

Deser, S., Jackiw, R., and Templeton, S. (1982). Topologically massive gauge theories, *Ann. Phys.* **140**, 2, pp. 372–411, doi:10.1016/0003-4916(82)90164-6. *(p. 50)*

Deser, S. and van Nieuwenhuizen, P. (1974a). One-loop divergences of quantized Einstein-Maxwell fields, *Phys. Rev. D* **10**, 2, pp. 401–410, doi:10.1103/PhysRevD.10.401. *(p. 219)*

Deser, S. and van Nieuwenhuizen, P. (1974b). Nonrenormalizability of the quantized Dirac-Einstein system, *Phys. Rev. D* **10**, 2, pp. 411–420, doi:10.1103/PhysRevD.10.411. *(p. 219)*

DeWitt, B. S. (1967a). Quantum theory of gravity. I. The canonical theory, *Phys. Rev.* **160**, 5, pp. 1113–1148, doi:10.1103/PhysRev.160.1113. *(p. 218)*

DeWitt, B. S. (1967b). Quantum theory of gravity. II. The manifestly covariant theory, *Phys. Rev.* **162**, 5, pp. 1195–1239, doi:10.1103/PhysRev.162.1195. *(p. 218)*

DeWitt, B. S. (1967c). Quantum theory of gravity. III. Applications of the covariant theory, *Phys. Rev.* **162**, 5, pp. 1239–1256, doi:10.1103/PhysRev.162.1239. *(p. 218)*

DeWitt, B. S. (2009). Quantum gravity: yesterday and today, *Gen. Rel. Grav.* **41**, 2, pp. 413–419, doi:10.1007/s10714-008-0719-0, [Erratum: *ibid.*, p. 671]; preprint arXiv:0805.2935. *(p. 218)*

Diósi, L. (1984). Gravitation and quantum-mechanical localization of macro-objects, *Phys. Lett. A* **105**, 4–5, pp. 199–202, doi:10.1016/0375-9601(84)90397-9. *(p. 222, 223)*

Dirac, P. A. M. (1928). The quantum theory of the electron, *Proc. Roy. Soc. A (London): Math., Phys. Eng. Sci.* **117**, 778, pp. 610–624, doi:10.1098/rspa.1928.0023. *(p. 194)*

Dirac, P. A. M. (1958). The theory of gravitation in Hamiltonian form, *Proc. R. Soc. Lond. A* **246**, 1246, pp. 333–343, doi:10.1098/rspa.1958.0142. *(p. 218)*

Dirac, P. A. M. (1959). Fixation of coordinates in the Hamiltonian theory of gravitation, *Phys. Rev.* **114**, 3, pp. 924–930, doi:10.1103/PhysRev.114.924. *(p. 218)*

Drummond, N. D., Radnai, Z., Trail, J. R., Towler, M. D., and Needs, R. J. (2004). Diffusion quantum Monte Carlo study of three-dimensional Wigner crystals, *Phys. Rev. B* **69**, p. 085116, doi:10.1103/PhysRevB.69.085116. *(p. 236)*

Dukelsky, J., Pittel, S., and Sierra, G. (2004). *Colloquium:* Exactly solvable Richardson-Gaudin models for many-body quantum systems, *Rev. Mod. Phys.* **76**, 3, pp. 643–662, doi:10.1103/RevModPhys.76.643. *(p. 86)*

Dyson, F. (2009). Birds and frogs, *Notices of the AMS* **56**, 2, pp. 212–223. *(p. 273)*

Dziarmaga, J. (2005). Dynamics of a quantum phase transition: Exact solution of the quantum Ising model, *Phys. Rev. Lett.* **95**, p. 245701, doi:10.1103/PhysRevLett.95.245701. *(p. 171)*

Dzyapko, O., Demidov, V. E., Buchmeier, M., Stockhoff, T., Schmitz, G., Melkov, G. A., and Demokritov, S. O. (2009). Excitation of two spatially separated Bose-Einstein condensates of magnons, *Phys. Rev. B* **80**, 6, p. 060401, doi:10.1103/PhysRevB.80.060401. *(p. 100)*

Dzyapko, O., Demidov, V. E., Demokritov, S. O., Melkov, G. A., and Slavin, A. N. (2007a). Direct observation of Bose-Einstein condensation in a parametrically driven gas of magnons, *New J. Phys.* **9**, 3, p. 64, doi:10.1088/1367-2630/9/3/064. *(p. 97)*

Dzyapko, O., Demidov, V. E., Demokritov, S. O., Melkov, G. A., and Slavin, A. N. (2007b). Quasiequilibrium gas of magnons with a nonzero chemical potential: A way to Bose-Einstein condensation, *J. Appl. Phys.* **101**, 9, 09C103, doi:10.1063/1.2693891. *(p. 100)*

Edwards, H. M. (2001). *Riemann's zeta function* (Dover, New York), ISBN 9780486417400. *(p. 273, 274)*

Efimov, V. (1970). Energy levels arising from resonant two-body forces in a three-body system, *Phys. Lett. B* **33**, 8, pp. 563–564, doi:10.1016/0370-2693(70)90349-7. *(p. 31, 42)*

Efimov, V. N. (1971). Weakly-bound states of three resonantly-interacting particles, *Sov. J. Nucl. Phys.* **12**, pp. 589–595, [Yad. Fiz. **12** 1080 (1970)]. *(p. 31, 39, 41, 42)*

Egelstaff, P. A., March, N. H., and McGill, N. C. (1974). Electron correlation functions in liquids from scattering data, *Can. J. Phys.* **52**, 17, pp. 1651–1659, doi:10.1139/p74-216, Reprinted in March and Angilella (2009). *(p. 127, 128, 132)*

Einstein, A. (1916). Näherungsweise Integration der Feldgleichungen der Gravitation, *Sitzungsber. K. Preuss. Akad. Wiss.: Phys. Math. Kl.* **1916 (part 1)**, pp. 688–696. *(p. 227)*

Einstein, A. (1924). Quantentheorie des einatomigen idealen Gases, *Sitzungsber. K. Preuss. Akad. Wiss.: Phys. Math. Kl.* , pp. 261–267. *(p. 91)*

Einstein, A. (1925). Quantentheorie des einatomigen idealen Gases. Zweite Abhandlung, *Sitzungsber. K. Preuss. Akad. Wiss.: Phys. Math. Kl.* , pp. 3–14. *(p. 91)*

Ellis, G. and Silk, J. (2014). Defend the integrity of physics, *Nature* **516**, pp. 321–323. *(p. 219)*

Englert, F. and Brout, R. (1964). Broken symmetry and the mass of gauge vector mesons, *Phys. Rev. Lett.* **13**, 9, pp. 321–323, doi:10.1103/PhysRevLett.13.321. *(p. 76)*

Enz, C. P. (1992). *A Course on Many-Body Theory Applied to Solid-State Physics* (World Scientific, Singapore), ISBN 9789971503369. *(p. 58)*

Ewald, P. P. (1921). Die Berechnung optischer und elektrostatischer Gitterpotentiale, *Annalen der Physik* **369**, 3, pp. 253–287, doi:10.1002/andp.19213690304. *(p. 237)*

Faddeev, L. D. (1961). Scattering theory for a three particle system, *Sov. Phys. JETP* **12**, pp. 1014–1019, [Zh. Eksp. Teor. Fiz. **39** 1459 (1960)]. *(p. 36)*

Faddeev, L. D. and Takhtajan, L. A. (1981). What is the spin of a spin wave? *Phys. Lett. A* **85**, pp. 375–377, doi:10.1016/0375-9601(81)90335-2. *(p. 255)*

Fano, U. (1935). Sullo spettro di assorbimento dei gas nobili presso il limite dello spettro d'arco, *Nuovo Cimento* **12**, pp. 154–161, doi:10.1007/BF02958288. *(p. 42)*

Fano, U. (1961). Effects of configuration interaction on intensities and phase shifts, *Phys. Rev.* **124**, 6, pp. 1866–1878, doi:10.1103/PhysRev.124.1866. *(p. 42)*

Fazekas, P. and Müller-Hartmann, E. (1991). Magnetic and non-magnetic ground states of the Kondo lattice, *Z. Phys. B* **85**, 2, pp. 285–300, doi:10.1007/BF01313231. *(p. 250, 251)*

Fedorov, D. V. and Jensen, A. S. (1993). Efimov effect in coordinate space Faddeev equations, *Phys. Rev. Lett.* **71**, 25, pp. 4103–4106, doi:10.1103/PhysRevLett.71.4103. *(p. 39)*

Ferconi, M. and Tosi, M. P. (1991a). Phonon dispersion curves in high-temperature solids from liquid structure factors, *Europhys. Lett.* **14**, 8, pp. 797–802, doi:10.1209/0295-5075/14/8/013. *(p. 240)*

Ferconi, M. and Tosi, M. P. (1991b). Density functional approach to phonon dispersion relations and elastic constants of high-temperature crystals, *J.*

Phys.: Condens. Matter **3**, 50, pp. 9943–9964, doi:10.1088/0953-8984/3/50/001. *(p. 240)*

Ferconi, M. and Vignale, G. (1992). Phonons of the 2D Wigner crystal in strong magnetic field from density functional theory, *Europhys. Lett.* **20**, 5, pp. 457–462, doi:10.1209/0295-5075/20/5/013. *(p. 240)*

Ferlaino, F. and Grimm, R. (2010). Trend: Forty years of Efimov physics: How a bizarre prediction turned into a hot topic, *Physics* **3**, p. 9, doi:10.1103/Physics.3.9. *(p. 41, 44)*

Ferlaino, F., Zenesini, A., Berninger, M., Huang, B., Nägerl, H.-C., and Grimm, R. (2011). Efimov resonances in ultracold quantum gases, *Few-Body Systems* **51**, 2–4, pp. 113–133, doi:10.1007/s00601-011-0260-7. *(p. 41)*

Fermi, E. (1927). Un metodo statistico per la determinazione di alcune proprietà dell'atomo, *Rendiconti dell'Accademia Nazionale dei Lincei* **6**, p. 602. *(p. 13)*

Fermi, E. (1928). Eine statistische Methode zur Bestimmung einiger Eigenschaften des Atoms und ihre Anwendung auf die Theorie des periodischen Systems der Elemente, *Z. Physik* **48**, pp. 73–79, doi:10.1007/BF01351576. *(p. 13)*

Ferraz, A., March, N. H., and Suzuki, M. (1978). Melting curve of Wigner electron crystal, *Phys. Chem. Liq.* **8**, 3, pp. 153–156, doi:10.1080/00319107808084749. *(p. 240)*

Ferraz, A., March, N. H., and Suzuki, M. (1979). Approximate melting curve of Wigner electron crystal, *Phys. Chem. Liq.* **9**, 1, pp. 59–65, doi:10.1080/00319107908084767. *(p. 240)*

Feshbach, H. (1958). A unified theory of nuclear reactions, *Ann. Phys. (N.Y.)* **5**, p. 357, doi:10.1016/0003-4916(58)90007-1. *(p. 42)*

Feshbach, H. (1962). A unified theory of nuclear reactions. II, *Ann. Phys. (N.Y.)* **19**, p. 287, doi:10.1016/0003-4916(62)90221-X. *(p. 42)*

Fetter, A. L. and Walecka, J. D. (2004). *Quantum Theory of Many-Particle Systems* (Dover, New York), doi:9780486428277. *(p. 45, 80, 233, 234, 235)*

Feynman, R. (1953). Atomic theory of the λ transition in helium, *Phys. Rev.* **91**, 6, pp. 1291–1301, doi:10.1103/PhysRev.91.1291. *(p. 109)*

Feynman, R. (1954). Atomic theory of the two-fluid model of liquid helium, *Phys. Rev.* **94**, 2, pp. 262–277, doi:10.1103/PhysRev.94.262. *(p. 109)*

Feynman, R. P. (1949). The theory of positrons, *Phys. Rev.* **76**, 6, pp. 749–759, doi:10.1103/PhysRev.76.749. *(p. 202)*

Feynman, R. P. (1963). The quantum theory of gravitation, *Acta Physica Polonica* **24**, pp. 697–722. *(p. 218, 221)*

Feynman, R. P. and Hibbs, A. R. (1965). *Quantum Mechanics and Path Integrals* (McGraw-Hill, New York), ISBN 9780070206502. *(p. 229)*

Feynman, R. P., Moringo, F. B., and Wagner, W. G. (1999). *Feynman lectures on gravitation* (Penguin, New York), ISBN 9780140284508. *(p. 219, 220, 221)*

Fierz, M. (1939). Über die relativistische Theorie kräftefreier Teilchen mit beliebigem Spin, *Helv. Phys. Acta* **12**, pp. 3–37, doi:10.5169/seals-110930. *(p. 45, 217, 221)*

Fierz, M. and Pauli, W. (1939). On relativistic wave equations for particles of arbitrary spin in an electromagnetic field, *Proc. R. Soc. (London) A* **173**, pp. 211–232, doi:10.1098/rspa.1939.0140. *(p. 217, 221, 314)*

Fisher, M. E. (1965). The nature of critical points, in W. E. Brittin (ed.), *Lectures in theoretical physics*, Vol. 7c: Statistical physics, weak interactions, field theory (University of Colorado Press, Boulder), pp. 1–159. *(p. 156, 165, 166, 167)*

Fisher, M. E. (1967). The theory of equilibrium critical phenomena, *Rep. Prog. Phys.* **30**, 2, p. 615, [**31**, 418 (1968)]. *(p. 156, 165)*

Fisher, M. E. (1978). Yang-Lee edge singularity and ϕ^3 field theory, *Phys. Rev. Lett.* **40**, pp. 1610–1613, doi:10.1103/PhysRevLett.40.1610. *(p. 171)*

Fisher, M. P. A., Weichman, P. B., Grinstein, G., and Fisher, D. S. (1989). Boson localization and the superfluid-insulator transition, *Phys. Rev. B* **40**, 1, pp. 546–570, doi:10.1103/PhysRevB.40.546. *(p. 97, 135)*

Fock, V. (1937). Die Eigenzeit in der klassischen und in der Quantenmechanik, *Physik. Z. Sowjetunion* **12**, p. 404. *(p. 204, 213)*

Foldy, L. L. (1951). A note on atomic binding energies, *Phys. Rev.* **83**, 2, pp. 397–399, doi:10.1103/PhysRev.83.397. *(p. 24)*

Foldy, L. L. and Wouthuysen, S. A. (1950). On the Dirac theory of spin 1/2 particles and its non-relativistic limit, *Phys. Rev.* **78**, 1, pp. 29–36, doi: 10.1103/PhysRev.78.29. *(p. 202)*

Forte, S. (1992). Quantum mechanics and field theory with fractional statistics, *Rev. Mod. Phys.* **64**, pp. 193–236, doi:10.1103/RevModPhys.64.193. *(p. 46, 64, 65)*

Fradkin, E., Kivelson, S. A., and Tranquada, J. M. (2015). Colloquium: Theory of intertwined orders in high temperature superconductors, *Rev. Mod. Phys.* **87**, 2, pp. 457–482, doi:10.1103/RevModPhys.87.457. *(p. 251)*

Frank, R. L., Lieb, E. H., Seiringer, R., and Siedentop, H. (2007). Müller's exchange-correlation energy in density-matrix-functional theory, *Phys. Rev. A* **76**, p. 052517, doi:10.1103/PhysRevA.76.052517. *(p. 8)*

Frau, M., Lerda, A., and Sciuto, S. (1996). Anyons and deformed Lie algebra, in L. Castellani and J. Wess (eds.), *Proceedings of the CXXVII International School of Physics "E. Fermi" on "Quantum Groups and their Applications in Physics"* (IOS Press, Amsterdam), ISBN 9789051992472, pp. 215–244. *(p. 47, 48, 49, 51)*

Frau, M., R.-Monteiro, M. A., and Sciuto, S. (1994). q-deformed classical Lie algebras and their anyonic realization, *J. Phys. A: Math. Gen.* **27**, 3, pp. 801–816, doi:10.1088/0305-4470/27/3/022. *(p. 47)*

Friedel, J. (1952). The distribution of electrons round impurities in monovalent metals, *Phil. Mag.* **43**, 337, pp. 153–189, doi:10.1080/14786440208561086. *(p. 116, 117)*

Frölich, H. (1954). On the theory of superconductivity: the one-dimensional case, *Proc. R. Soc. A* **223**, pp. 296–305, doi:10.1098/rspa.1954.0116. *(p. 76)*

Fukuyama, H. (1975). Two-dimensional Wigner crystal under magnetic field, *Solid State Commun.* **17**, 10, pp. 1323–1326, doi:10.1016/0038-1098(75)90696-1. *(p. 239)*

Fulde, P. (1999). Solids with weak and strong electron correlations, in N. H. March (ed.), *Electron correlation in the solid state*, chap. 2 (World Scientific, Singapore), pp. 47–102. *(p. 248)*

Fulde, P. and Ferrell, R. A. (1964). Superconductivity in a strong spin-exchange field, *Phys. Rev.* **135**, pp. A550–A563, doi:10.1103/PhysRev.135.A550. *(p. 78)*

Gál, T. and March, N. H. (2009). Density functional for the ground-state energy of artificial two-electron atoms with harmonic confinement, *J. Phys. B: At. Mol. Opt. Phys.* **42**, p. 025001, doi:10.1088/0953-4075/42/2/025001. *(p. 19, 20, 21)*

Gambacurta, D. and Lacroix, D. (2012). Thermodynamical properties of small superconductors with a fixed number of particles, *Phys. Rev. C* **85**, 4, p. 044321, doi:10.1103/PhysRevC.85.044321. *(p. 86)*

Gaspari, G. D. and Gyorffy, B. L. (1972). Electron-phonon interactions, d resonances, and superconductivity in transition metals, *Phys. Rev. Lett.* **28**, 13, pp. 801–805, doi:10.1103/PhysRevLett.28.801. *(p. 120)*

Gaudin, M. (1967). Un systeme a une dimension de fermions en interaction, *Phys. Lett. A* **24**, 1, pp. 55–56, doi:10.1016/0375-9601(67)90193-4. *(p. 243)*

Gavoret, J. and Nozières, P. (1964). Structure of the perturbation expansion for the Bose liquid at zero temperature, *Ann. Phys. (NY)* **28**, pp. 349–399, doi:10.1016/0003-4916(64)90200-3. *(p. 93, 95)*

Geldof, D., Van Alsenoy, C., and March, N. H. (2013). The central role of the von Weizsäcker kinetic energy density in determining the energy functional of a model two-electron atom with harmonic confinement and inverse square interparticle repulsion, *J. Math. Chem.* **51**, 6, pp. 1561–1568, doi:10.1007/s10910-013-0164-5. *(p. 21, 22)*

Gell-Mann, M. and Brueckner, K. A. (1957). Correlation energy of an electron gas at high density, *Phys. Rev.* **106**, pp. 364–368, doi:10.1103/PhysRev.106.364. *(p. 235)*

Gerjuoy, E. (1965). Momentum transfer cross-section theorem, *J. Math. Phys.* **6**, 6, pp. 993–996, doi:10.1063/1.1704359. *(p. 120)*

Gerlach, W. and Stern, O. (1922a). Das magnetische Moment des Silberatoms, *Z. Phys.* **9**, 1, pp. 353–355, doi:10.1007/BF01326984. *(p. 200)*

Gerlach, W. and Stern, O. (1922b). Der experimentelle Nachweis der Richtungsquantelung im Magnetfeld, *Z. Phys.* **9**, 1, pp. 349–352, doi:10.1007/BF01326983. *(p. 200)*

Gersch, H. A. and Knollman, G. C. (1963). Quantum cell model for bosons, *Phys. Rev.* **129**, 2, pp. 959–967, doi:10.1103/PhysRev.129.959. *(p. 97)*

Ghassib, H. B. and Chester, G. V. (1984). ^{4}He n-mers and Bose-Einstein condensation in He II, *J. Chem. Phys.* **81**, 1, pp. 585–586, doi:10.1063/1.447350. *(p. 42)*

Ghulghazaryan, R. G., Sargsyan, K. G., and Ananikian, N. S. (2007). Partition function zeros of the one-dimensional Blume-Capel model in transfer matrix formalism, *Phys. Rev. E* **76**, p. 021104, doi:10.1103/PhysRevE.76.021104. *(p. 171)*

Giamarchi, T. and Schulz, H. J. (1988). Anderson localization and interactions in one-dimensional metals, *Phys. Rev. B* **37**, 1, pp. 325–340, doi:10.1103/PhysRevB.37.325. *(p. 97, 135)*

Gilbert, T. L. (1975). Hohenberg-Kohn theorem for nonlocal external potentials, *Phys. Rev. B* **12**, pp. 2111–2120, doi:10.1103/PhysRevB.12.2111. *(p. 6, 7, 8)*

Ginzburg, V. L. (1960). *Soviet Solid State* **2**, p. 61. *(p. 82)*

Ginzburg, V. L. and Landau, L. D. (1950). On the theory of superconductivity, *Zh. Eksp. Teor. Fiz. (Sov. Phys. JETP)* **20**, pp. 1064–1082. *(p. 75, 93)*

Girvin, S. M. and MacDonald, A. H. (1987). Off-diagonal long-range order, oblique confinement, and the fractional quantum Hall effect, *Phys. Rev. Lett.* **58**, 12, pp. 1252–1255, doi:10.1103/PhysRevLett.58.1252. *(p. 251)*

Giuliani, G. and Vignale, G. (2005). *Quantum Theory of the Electron Liquid* (Cambridge University Press, Cambridge), ISBN 9780521821124. *(p. 64, 65, 66, 71, 233, 235, 236, 237)*

Glyde, H. R., Azuah, R. T., and Stirling, W. G. (2000). Condensate, momentum distribution, and final-state effects in liquid ^{4}He, *Phys. Rev. B* **62**, 21, pp. 14337–14349, doi:10.1103/PhysRevB.62.14337. *(p. 92)*

Gödel, K. (1931). Über formal unentscheidbare Sätze der Principia Mathematica und verwandter Systeme I, *Monatsh. Math. Phys.* **38**, pp. 173–198, doi:10.1007/BF01700692. *(p. 270)*

Goedecker, S. and Umrigar, C. J. (1998). Natural orbital functional for the many-electron problem, *Phys. Rev. Lett.* **81**, pp. 866–869, doi:10.1103/PhysRevLett.81.866. *(p. 8)*

Goldman, V. J., Shayegan, M., and Tsui, D. C. (1988). Evidence for the fractional quantum Hall state at $\nu = \frac{1}{7}$, *Phys. Rev. Lett.* **61**, pp. 881–884, doi:10.1103/PhysRevLett.61.881. *(p. 239)*

Goldstein, G. and Andrei, N. (2013). Equilibration and generalized GGE in the Lieb Liniger gas, Preprint `arXiv:1309.3471`. *(p. 112)*

Goldstein, G. and Andrei, N. (2014a). Failure of the local generalized Gibbs ensemble for integrable models with bound states, *Phys. Rev. A* **90**, 4, p. 043625, doi:10.1103/PhysRevA.90.043625. *(p. 112)*

Goldstein, G. and Andrei, N. (2014b). How to experimentally detect a GGE? universal spectroscopic signatures of the GGE in the Tonks gas, Preprint `arXiv:1405.6365`. *(p. 112)*

Goldstein, G. and Andrei, N. (2014c). Quench between a Mott insulator and a Lieb-Liniger liquid, *Phys. Rev. A* **90**, 4, p. 043626, doi:10.1103/PhysRevA.90.043626. *(p. 112)*

Goldstein, G. and Andrei, N. (2014d). Strong eigenstate thermalization hypothesis, Preprint `arXiv:1408.3589`. *(p. 112)*

Goldstein, G. and Andrei, N. (2015). Equilibration and generalized Gibbs ensemble for hard wall boundary conditions, *Phys. Rev. B* **92**, 15, p. 155103, doi:10.1103/PhysRevB.92.155103, preprint `arXiv:1505.02585`. *(p. 112)*

Golinelli, O., Jolicoeur, T., and Lacaze, R. (1994). Finite-lattice extrapolations for a Haldane-gap antiferromagnet, *Phys. Rev. B* **50**, 5, pp. 3037–3044, doi:10.1103/PhysRevB.50.3037. *(p. 269)*

Gorter, C. J. (1949). The two fluid model for helium II, *Nuovo Cimento* **6**, Suppl.

2, pp. 245–250, doi:10.1007/BF02780988. *(p. 313)*

Gracia-Bondía, J. M. (2010). Notes on "quantum gravity" and noncommutative geometry, in Booß-Bavnbek, Bernhelm, G. Esposito, and M. Lesch (eds.), *New Paths Towards Quantum Gravity, Lecture Notes in Physics*, Vol. 807 (Springer, Berlin), ISBN 9783642118968, pp. 3–58, doi:10.1007/978-3-642-11897-5_1, in Booß-Bavnbek *et al.* (2010). *(p. 219)*

Gradshteyn, I. S. and Ryzhik, I. M. (1994). *Table of Integrals, Series, and Products*, 5th edn. (Academic Press, Boston), ISBN 9780122947551. *(p. 17, 58, 66, 67, 70, 151, 167, 196, 215, 237, 262)*

Green, H. S. (1952). *Molecular theory of fluids* (North-Holland, Amsterdam). *(p. 125)*

Green, M., Schwarz, J., and Witten, E. (2012a). *Superstring theory. Vol. 1: Introduction* (Cambridge University Press, Cambridge), ISBN 9781107029118. *(p. 219)*

Green, M., Schwarz, J., and Witten, E. (2012b). *Superstring theory. Vol. 2: Loop amplitudes, anomalies and phenomenology* (Cambridge University Press, Cambridge), ISBN 978-1107029132. *(p. 219)*

Greenberg, O. W. (1991). Particles with small violations of Fermi or Bose statistics, *Phys. Rev. D* **43**, pp. 4111–4120, doi:10.1103/PhysRevD.43.4111. *(p. 66)*

Greenfield, A. J., Wellendorf, J., and Wiser, N. (1971). X-ray determination of the static structure factor of liquid Na and K, *Phys. Rev. A* **4**, 4, pp. 1607–1616, doi:10.1103/PhysRevA.4.1607. *(p. 125)*

Greiner, M., Mandel, O., Esslinger, T., Hänsch, T. W., and Bloch, I. (2002a). Quantum phase transition from a superfluid to a Mott insulator in a gas of ultracold atoms, *Nature* **415**, 6867, pp. 39–44, doi:10.1038/415039a. *(p. 170)*

Greiner, M., Mandel, O., Hänsch, T. W., and Bloch, I. (2002b). Collapse and revival of the matter wave field of a Bose-Einstein condensate, *Nature* **419**, 6902, pp. 51–54, doi:10.1038/nature00968. *(p. 170)*

Greschner, S. and Santos, L. (2015). Anyon hubbard model in one-dimensional optical lattices, *Phys. Rev. Lett.* **115**, p. 053002, doi:10.1103/PhysRevLett.115.053002. *(p. 72)*

Grimes, C. C. and Adams, G. (1979). Evidence for a liquid-to-crystal phase transition in a classical, two-dimensional sheet of electrons, *Phys. Rev. Lett.* **42**, pp. 795–798, doi:10.1103/PhysRevLett.42.795. *(p. 237)*

Grisenti, R. E., Schöllkopf, W., Toennies, J. P., Hegerfeldt, G. C., Köhler, T., and Stoll, M. (2000). Determination of the bond length and binding energy of the helium dimer by diffraction from a transmission grating, *Phys. Rev. Lett.* **85**, 11, pp. 2284–2287, doi:10.1103/PhysRevLett.85.2284. *(p. 42)*

Gritsenko, O., Pernal, K., and Baerends, E. J. (2005). An improved density matrix functional by physically motivated repulsive corrections, *J. Chem. Phys.* **122**, 20, 204102, doi:10.1063/1.1906203. *(p. 9)*

Gross, N., Shotan, Z., Kokkelmans, S., and Khaykovich, L. (2009). Observation of universality in ultracold ^{7}Li three-body recombination, *Phys. Rev. Lett.* **103**, 16, p. 163202, doi:10.1103/PhysRevLett.103.163202. *(p. 44)*

Grossmann, S. and Lehmann, V. (1969). Phase transitions and the distribution of temperature zeros of the partition function. II. Applications and examples, *Z. Phys.* **218**, 5, pp. 449–459, doi:10.1007/BF01392424. *(p. 165, 169)*

Grossmann, S. and Rosenhauer, W. (1967). Temperature dependence near phase transitions in classical and quant. mech. canonical statistics, *Z. Phys.* **207**, 2, pp. 138–152, doi:10.1007/BF01326224. *(p. 165, 169)*

Grossmann, S. and Rosenhauer, W. (1969). Phase transitions and the distribution of temperature zeros of the partition function. I. General relations, *Z. Phys.* **218**, 5, pp. 437–448, doi:10.1007/BF01392423. *(p. 165, 169)*

Guan, X.-W., Batchelor, M. T., and Lee, C. (2013). Fermi gases in one dimension: From Bethe ansatz to experiments, *Rev. Mod. Phys.* **85**, pp. 1633–1691, doi: 10.1103/RevModPhys.85.1633. *(p. 72)*

Gubser, S. S., Klebanov, I. R., and Polyakov, A. M. (1998). Gauge theory correlators from non-critical string theory, *Phys. Lett. B* **428**, 1–132, pp. 105–114, doi:10.1016/S0370-2693(98)00377-3. *(p. 174, 219)*

Gulácsi, M. (2006). The Kondo lattice model, *Phil. Mag.* **86**, 13-14, pp. 1907–1946, doi:10.1080/14786430500355045. *(p. 251)*

Gürsey, F. (1950). Classical statistical mechanics of a rectilinear assembly, *Math. Proc. Camb. Phil. Soc.* **46**, 01, pp. 182–194, doi:10.1017/S0305004100025603. *(p. 173)*

Gutierrez, T. D. (2004). Intensity interferometry with anyons, *Phys. Rev. A* **69**, p. 063614, doi:10.1103/PhysRevA.69.063614. *(p. 64)*

Gutzwiller, M. (1963). Effect of correlation on the ferromagnetism of transition metals, *Phys. Rev. Lett.* **10**, 5, pp. 159–162, doi:10.1103/PhysRevLett.10.159. *(p. 241)*

Gutzwiller, M. C. (1962). Fermion ensembles of maximum entropy, *Phys. Rev.* **125**, 5, pp. 1455–1460, doi:10.1103/PhysRev.125.1455. *(p. 90)*

Ha, Z. N. C. (1996). *Quantum many-body systems in one dimension* (World Scientific, Singapore), ISBN 9789810222758. *(p. 46, 65)*

Haggard, H. M. and Rovelli, C. (2015). Quantum-gravity effects outside the horizon spark black to white hole tunneling, *Phys. Rev. D* **92**, p. 104020, doi: 10.1103/PhysRevD.92.104020. *(p. 226, 227)*

Haldane, F. D. M. (1981). 'Luttinger liquid theory' of one-dimensional quantum fluids. I. Properties of the Luttinger model and their extension to the general 1D interacting spinless Fermi gas, *J. Phys. C: Solid State Phys.* **14**, 19, pp. 2585–2610, doi:10.1088/0022-3719/14/19/010. *(p. 253)*

Haldane, F. D. M. (1983). Nonlinear field theory of large-spin Heisenberg antiferromagnets: Semiclassically quantized solitons of the one-dimensional easy-axis Néel state, *Phys. Rev. Lett.* **50**, 15, pp. 1153–1156, doi:10.1103/PhysRevLett.50.1153. *(p. 268)*

Haldane, F. D. M. (1991). "Fractional statistics" in arbitrary dimensions: A generalization of the Pauli principle, *Phys. Rev. Lett.* **67**, pp. 937–940, doi: 10.1103/PhysRevLett.67.937. *(p. 46, 47, 54, 63)*

Hall, R. L., Lucha, W., and Schöberl, F. F. (2003). Discrete spectra of semirelativistic Hamiltonians, *Int. J. Mod. Phys. A* **18**, 15, pp. 2657–2680, doi: 10.1142/S0217751X0301406X. *(p. 215)*

Halperin, B. I. (1984). Statistics of quasiparticles and the hierarchy of fractional quantized Hall states, *Phys. Rev. Lett.* **52**, pp. 1583–1586, doi:10.1103/PhysRevLett.52.1583, [**52**, 2390(E) (1984)]. *(p. 46)*

Hamidian, M. H., Edkins, S. D., Joo, S. H., Kostin, A., Eisaki, H., Uchida, S., Lawler, M. J., Kim, E., Mackenzie, A. P., Fujita, K., Lee, J., and Davis, J. C. S. (2016). Detection of a Cooper-pair density wave in $Bi_2Sr_2CaCu_2O_{8+x}$, *Nature* **532**, 7599, pp. 343–347, doi:10.1038/nature17411. *(p. 78)*

Hammer, H., Nogga, A., and Schwenk, A. (2013). *Colloquium*: Three-body forces: From cold atoms to nuclei, *Rev. Mod. Phys.* **85**, 1, pp. 197–217, doi:10.1103/RevModPhys.85.197. *(p. 41)*

Hardy, L. (2007). Towards quantum gravity: a framework for probabilistic theories with non-fixed causal structure, *J. Phys. A: Math. Theor.* **40**, 12, pp. 3081–3100, doi:10.1088/1751-8113/40/12/S12. *(p. 217)*

Harrison, N. (2001). *A numerical study of the Schrödinger-Newton equations*, Thesis submitted for the degree of D. Phil., University of Oxford, UK. *(p. 226)*

Harrison, R., Moroz, I., and Tod, K. P. (2003). A numerical study of the Schrödinger-Newton equations, *Nonlinearity* **16**, 1, pp. 101–122, doi:10.1088/0951-7715/16/1/307. *(p. 226)*

Hastings, M. B. (2004). Lieb-Schultz-Mattis in higher dimensions, *Phys. Rev. B* **69**, 10, p. 104431, doi:10.1103/PhysRevB.69.104431. *(p. 269)*

Hedin, L. and Lundqvist, S. (1970). Effects of electron-electron and electron-phonon interactions on the one-electron states of solids, in F. Seitz, D. Turnbull, and H. Ehrenreich (eds.), *Solid State Physics*, *Solid State Physics*, Vol. 23 (Academic Press), pp. 1–181, doi:10.1016/S0081-1947(08)60615-3. *(p. 125)*

Hew, W. K., Thomas, K. J., Pepper, M., Farrer, I., Anderson, D., Jones, G. A. C., and Ritchie, D. A. (2009). Incipient formation of an electron lattice in a weakly confined quantum wire, *Phys. Rev. Lett.* **102**, p. 056804, doi:10.1103/PhysRevLett.102.056804. *(p. 238)*

Hewson, A. C. (1993). *The Kondo Problem to Heavy Fermions* (Cambridge University Press, Cambridge), ISBN 9780511470752, doi:10.1017/CBO9780511470752. *(p. 248)*

Heyl, M., Polkovnikov, A., and Kehrein, S. (2013). Dynamical quantum phase transitions in the transverse-field Ising model, *Phys. Rev. Lett.* **110**, p. 135704, doi:10.1103/PhysRevLett.110.135704. *(p. 170, 171)*

Higgs, P. W. (1964). Broken symmetries and the masses of gauge bosons, *Phys. Rev. Lett.* **13**, 16, pp. 508–509, doi:10.1103/PhysRevLett.13.508. *(p. 76)*

Hohenberg, P. C. (1967). Existence of long-range order in one and two dimensions, *Phys. Rev.* **158**, 2, pp. 383–386, doi:10.1103/PhysRev.158.383. *(p. 96)*

Hohenberg, P. C. and Kohn, W. (1964). Inhomogeneous electron gas, *Phys. Rev.* **136**, pp. B864–B871, doi:10.1103/PhysRev.136.B864. *(p. 6, 7, 14, 26)*

Hohenberg, P. C. and Martin, P. C. (1965). Microscopic theory of superfluid helium, *Ann. Phys. (N.Y.)* **34**, 2, pp. 291–359, doi:10.1016/0003-4916(65)90280-0. *(p. 95)*

Holas, A., Howard, I. A., and March, N. H. (2003). Wave functions and low-order density matrices for a class of two-electron 'artificial atoms' embracing

Hookean and Moshinsky models, *Phys. Lett. A* **310**, pp. 451–456, doi:10.1016/S0375-9601(03)00408-0. *(p. 14, 16, 17, 18)*

Holas, A. and March, N. H. (1995). Exact exchange-correlation potential and approximate exchange potential in terms of density matrices, *Phys. Rev. A* **51**, pp. 2040–2048, doi:10.1103/PhysRevA.51.2040, reprinted in March and Angilella (2009). *(p. 2, 9, 10, 230)*

Holas, A. and March, N. H. (1999). Field dependence of the energy of a molecule in a magnetic field, *Phys. Rev. A* **60**, 4, pp. 2853–2866, doi:10.1103/PhysRevA.60.2853. *(p. 18)*

Holas, A. and March, N. H. (2006). Generalization to higher dimensions of one-dimensional differential equation for Slater sum of an inhomogeneous electron liquid for a given local potential, *Phys. Chem. Liq.* **44**, 4, pp. 465–475, doi:10.1080/00319100500412287. *(p. 230, 231)*

Holas, A., March, N. H., and Rubio, A. (2005). Differential virial theorem in relation to a sum rule for the exchange-correlation force in density-functional theory, *J. Chem. Phys.* **123**, p. 194104, doi:10.1063/1.2114848, reprinted in March and Angilella (2009). *(p. 11)*

Holzmann, M. and Baym, G. (2007). Condensate superfluidity and infrared structure of the single-particle Green's function: The Josephson relation, *Phys. Rev. B* **76**, 9, p. 092502, doi:10.1103/PhysRevB.76.092502. *(p. 95)*

Horinouchi, Y. and Ueda, M. (2015). Onset of a limit cycle and universal three-body parameter in Efimov physics, *Phys. Rev. Lett.* **114**, 2, p. 025301, doi:10.1103/PhysRevLett.114.025301. *(p. 32)*

Horowitz, G. T. (2000). Quantum gravity at the turn of the millennium, in V. G. Gurzadyan, R. T. Jantzen, and R. Ruffini (eds.), *Proceedings of the 9th Marcel Grossmann Meeting (MGIX MM): on recent developments in theoretical and experimental general relativity, gravitation, and relativistic field theories* (World Scientific, Singapore), pp. 55–67, preprint `arXiv:gr-qc/0011089`. *(p. 219)*

Hu, Y., Stirling, S. D., and Wu, Y.-S. (2014). Emergent exclusion statistics of quasiparticles in two-dimensional topological phases, *Phys. Rev. B* **89**, p. 115133, doi:10.1103/PhysRevB.89.115133. *(p. 72)*

Huang, K. (1948). Quantum mechanical calculation of the heat of solution and residual resistance of gold in silver, *Proc. Phys. Soc.* **60**, 2, pp. 161–174, doi:10.1088/0959-5309/60/2/305. *(p. 119, 120, 121)*

Huang, W.-H. (1995). Statistical interparticle potential between two anyons, *Phys. Rev. B* **52**, pp. 15090–15092, doi:10.1103/PhysRevB.52.15090. *(p. 65)*

Hubbard, J. (1963). Electron correlations in narrow energy bands, *Proc. R. Soc. Lond. A* **276**, 1365, pp. 238–257, doi:10.1098/rspa.1963.0204. *(p. 241)*

Hubbard, J. (1964a). Electron correlations in narrow energy bands. II. The degenerate band case, *Proc. R. Soc. Lond. A* **277**, 1369, pp. 237–259, doi:10.1098/rspa.1964.0019. *(p. 241)*

Hubbard, J. (1964b). Electron correlations in narrow energy bands. III. An improved solution, *Proc. R. Soc. Lond. A* **281**, 1386, pp. 401–419, doi:10.1098/rspa.1964.0190. *(p. 241)*

Hubbard, J. (1965). Electron correlations in narrow energy bands. IV. The atomic representation, *Proc. R. Soc. Lond. A* **285**, 1403, pp. 542–560, doi:10.1098/rspa.1965.0124. *(p. 241)*

Hubbard, J. (1967a). Electron correlations in narrow energy bands. V. A perturbation expansion about the atomic limit, *Proc. R. Soc. Lond. A* **296**, 1444, pp. 82–99, doi:10.1098/rspa.1967.0007. *(p. 241)*

Hubbard, J. (1967b). Electron correlations in narrow energy bands. VI. The connexion with many-body perturbation theory, *Proc. R. Soc. Lond. A* **296**, 1444, pp. 100–112, doi:10.1098/rspa.1967.0008. *(p. 241)*

Hubbard, J. and Schofield, P. (1972). Wilson theory of a liquid-vapour critical point, *Phys. Lett. A* **40**, 3, p. 245, doi:10.1016/0375-9601(72)90675-5. *(p. 177, 179, 191)*

Ichimaru, S. (1982). Strongly coupled plasmas: high-density classical plasmas and degenerate electron liquids, *Rev. Mod. Phys.* **54**, pp. 1017–1059, doi:10.1103/RevModPhys.54.1017. *(p. 125)*

Iguchi, K. (1997). Quantum statistical mechanics of an ideal gas with fractional exclusion statistics in arbitrary dimensions, *Phys. Rev. Lett.* **78**, pp. 3233–3236, doi:10.1103/PhysRevLett.78.3233. *(p. 46, 59, 60, 61)*

Iguchi, K. (1998). Generalization of a Fermi liquid to a liquid with fractional exclusion statistics in arbitrary dimensions: Theory of a Haldane liquid, *Phys. Rev. Lett.* **80**, pp. 1698–1701, doi:10.1103/PhysRevLett.80.1698. *(p. 59, 63)*

Iguchi, K. and Sutherland, B. (2000). Interacting particles as neither bosons nor fermions: A microscopic origin for fractional statistics, *Phys. Rev. Lett.* **85**, pp. 2781–2784, doi:10.1103/PhysRevLett.85.2781. *(p. 59, 64)*

Illner, R., Kavian, O., and Lange, H. (1998). Stationary solutions of quasi-linear Schrödinger-Poisson systems, *J. Diff. Equations* **145**, 1, pp. 1–16, doi:10.1006/jdeq.1997.3405. *(p. 225)*

Illner, R., Zweifel, P. F., and Lange, H. (1994). Global existence, uniqueness and asymptotic behaviour of solutions of the Wigner-Poisson and Schrödinger-Poisson systems, *Math. Meth. Appl. Sci.* **17**, 5, pp. 349–376, doi:10.1002/mma.1670170504. *(p. 225)*

Imry, Y. (1986). Physics of microscopic systems, in G. Grinstein and G. Mazenko (eds.), *Directions in Condensed Matter Physics* (World Scientific, Singapore), ISBN 9789971978426, pp. 101–163. *(p. 139)*

Inguscio, M. (2003). How to freeze out collisions, *Science* **300**, pp. 1671–1673, doi:10.1126/science.1085845. *(p. 94, 110)*

Ioffe, A. F. and Regel, A. R. (1960). Non-crystalline, amorphous and liquid electronic semiconductors, *Prog. Semicond.* **4**, pp. 237–291. *(p. 135)*

Iqbal, N., Liu, H., and Mezei, M. (2011). Lectures on holographic non-Fermi liquids and quantum phase transitions, in M. Dine, T. Banks, and S. Sachdev (eds.), *String Theory and Its Applications. TASI 2010: From meV to the Planck Scale* (World Scientific, Singapore), pp. 707–815, Proceedings of the 2010 Theoretical Advanced Study Institute in Elementary Particle Physics. Boulder, Colorado, 1-25 June 2010. Also available as preprint `arXiv:1110.3814`. *(p. 147, 148, 174, 175, 176, 177, 178, 180)*

Isihara, A. (1991). *Condensed matter physics* (Oxford University Press, Oxford), ISBN 9780486151649. *(p. 247)*

Ising, E. (1925). Beitrag zur Theorie des Ferromagnetismus, *Z. Phys.* **31**, 1, pp. 253–258, doi:10.1007/BF02980577. *(p. 147, 148)*

Itzykson, C., Pearson, R. B., and Zuber, J. B. (1983). Distribution of zeros in Ising and gauge models, *Nucl. Phys. B* **220**, 4, pp. 415–433, doi:10.1016/0550-3213(83)90499-6. *(p. 167, 168)*

Itzykson, C. and Zuber, J. (1980). *Quantum Field Theory* (McGraw-Hill, New York), ISBN 0070320713. *(p. 199, 201, 202, 204, 209)*

Iyer, D. and Andrei, N. (2012). Quench dynamics of the interacting bose gas in one dimension, *Phys. Rev. Lett.* **109**, 11, p. 115304, doi:10.1103/PhysRevLett.109.115304. *(p. 110, 111, 112)*

Iyer, D., Guan, H., and Andrei, N. (2013). Exact formalism for the quench dynamics of integrable models, *Phys. Rev. A* **87**, 5, p. 053628, doi:10.1103/PhysRevA.87.053628. *(p. 110, 111, 112)*

Jackiw, R. and Rebbi, C. (1976). Solitons with fermion number $\frac{1}{2}$, *Phys. Rev. D* **13**, 12, pp. 3398–3409, doi:10.1103/PhysRevD.13.3398. *(p. 264)*

Jackiw, R. and Schrieffer, J. R. (1981). Solitons with fermion number $\frac{1}{2}$ in condensed matter and relativistic field theories, *Nucl. Phys. B* **190** **[FS3]**, 2, pp. 253–265, doi:10.1016/0550-3213(81)90557-5. *(p. 257, 259)*

Jacobson, T. and Smolin, L. (1988). Nonperturbative quantum geometries, *Nucl. Phys. B* **299**, 2, pp. 295–345, doi:10.1016/0550-3213(88)90286-6. *(p. 221)*

Janak, J. F. (1974). Simplification of total-energy and pressure calculations in solids, *Phys. Rev. B* **9**, p. 3985, doi:10.1103/PhysRevB.9.3985. *(p. 19)*

Janke, W., Johnston, D., and Kenna, R. (2005). Critical exponents from general distributions of zeroes, *Computer Phys. Commun.* **169**, 1–3, pp. 457–461, doi:10.1016/j.cpc.2005.03.101, proceedings of the Europhysics Conference on Computational Physics 2004 (CCP 2004). *(p. 169)*

Janke, W. and Kenna, R. (2001). The strength of first and second order phase transitions from partition function zeroes, *J. Stat. Phys.* **102**, 5, pp. 1211–1227, doi:10.1023/A:1004836227767. *(p. 167, 168)*

Jensen, A. S., Riisager, K., Fedorov, D. V., and Garrido, E. (2004). Structure and reactions of quantum halos, *Rev. Mod. Phys.* **76**, 1, pp. 215–261, doi:10.1103/RevModPhys.76.215. *(p. 41)*

Johnson, J., Krinsky, S., and McCoy, B. (1973). Vertical-arrow correlation length in the eight-vertex model and the low-lying excitations of the $X-Y-Z$ hamiltonian, *Phys. Rev. A* **8**, 5, pp. 2526–2547, doi:10.1103/PhysRevA.8.2526. *(p. 183)*

Johnson, M. D. and March, N. H. (1963). Long-range oscillatory interaction between ions in liquid metals, *Phys. Lett.* **3**, pp. 313–314, doi:10.1016/0031-9163(63)90170-7, reprinted in March and Angilella (2009). *(p. 125)*

Jones, P. H., Maragò, O. M., and Volpe, G. (2015). *Optical Tweezers: Principles and Applications* (Cambridge University Press, Cambridge), ISBN 9781107051164. *(p. 224)*

Jones, W. and March, N. H. (1986a). *Theoretical Solid-State Physics. Perfect Lattices in Equilibrium*, Vol. 1 (Dover, New York), ISBN 9780486650159.

(p. 115, 253)

Jones, W. and March, N. H. (1986b). *Theoretical Solid-State Physics. Non-equilibrium and Disorder*, Vol. 2 (Dover, New York), ISBN 9780486650166. *(p. 80, 116, 117, 136, 138)*

Jordan, P. and Wigner, E. (1928). Über das Paulische Äquivalenzverbot, *Z. Phys.* **47**, 9, pp. 631–651, doi:10.1007/BF01331938. *(p. 48, 185)*

Josephson, B. D. (1966). Relation between the superfluid density and order parameter for superfluid He near T_c, *Phys. Lett.* **21**, 6, pp. 608–609, doi: 10.1016/0031-9163(66)90088-6. *(p. 95)*

Joyce, G. S., Sarkar, S., Spałek, J., and Byczuk, K. (1996). Thermodynamic properties of particles with intermediate statistics, *Phys. Rev. B* **53**, p. 990. *(p. 46, 56, 57, 58, 59, 61, 62)*

Kadanoff, L. and Wegner, F. (1971). Some critical properties of the eight-vertex model, *Phys. Rev. B* **4**, 11, pp. 3989–3993, doi:10.1103/PhysRevB.4.3989. *(p. 184)*

Kamerlingh Onnes, H. (1911a). Further experiments with liquid helium. C. On the change of electric resistance of pure metals at very low temperatures, etc. IV. The resistance of pure mercury at helium temperatures. *Comm. Phys. Lab. Univ. Leiden* **120b**. *(p. 75)*

Kamerlingh Onnes, H. (1911b). Further experiments with liquid helium. D. On the change of electric resistance of pure metals at very low temperatures, etc. V. The disappearance of the resistance of mercury. *Comm. Phys. Lab. Univ. Leiden* **122b**. *(p. 75)*

Kamerlingh Onnes, H. (1911c). Further experiments with liquid helium. G. On the electrical resistance of pure metals, etc. VI. On the sudden change in the rate at which the resistance of mercury disappears. *Comm. Phys. Lab. Univ. Leiden* **124c**. *(p. 75)*

Kapitza, P. (1938). Viscosity of liquid Helium below the λ-point, *Nature* **141**, p. 74, doi:10.1038/141074a0. *(p. 94)*

Kar, R., Misra, A., March, N. H., and Klein, D. J. (2012). Bose-Einstein condensation of magnons in ferromagnetic thin films, *Phase Trans.* **85**, 9, pp. 831–839, doi:10.1080/01411594.2012.682060. *(p. 98, 99, 100)*

Kardar, M. (2007). *Statistical Physics of Fields* (Cambridge University Press, Cambridge), ISBN 9780521873413. *(p. 147, 152, 178, 179, 182, 270)*

Károlyházy, F. (1966). Gravitation and quantum mechanics of macroscopic objects, *Nuovo Cim. A* **42**, 2, pp. 390–402, doi:10.1007/BF02717926. *(p. 223)*

Kasuya, T. (1956). A theory of metallic ferro- and antiferromagnetism on Zener's model, *Prog. Theor. Phys.* **16**, 1, pp. 45–57, doi:10.1143/PTP.16.45. *(p. 247)*

Katsnelson, M. I. (2006). Zitterbewegung, chirality, and minimal conductivity in graphene, *Eur. Phys. J. B* **51**, pp. 157–160, doi:10.1140/epjb/e2006-00203-1. *(p. 211)*

Katsnelson, M. I. (2007). Graphene: Carbon in two dimensions, *Materials Today* **10**, pp. 20–27, doi:10.1016/S1369-7021(06)71788-6. *(p. 211)*

Katsnelson, M. I. (2012). *Graphene: Carbon in Two Dimensions* (Cambridge University Press, Cambridge), ISBN 9780521195409. *(p. 211)*

Katsnelson, M. I. and Novoselov, K. S. (2007). Graphene: new bridge between condensed matter physics and quantum electrodynamics, *Solid State Commun.* **143**, pp. 3–13, doi:10.1016/j.ssc.2007.02.043. *(p. 211)*

Katsnelson, M. I., Novoselov, K. S., and Geim, A. K. (2006). Chiral tunnelling and the Klein paradox in graphene, *Nature Phys.* **2**, pp. 620–625, doi:10.1038/nphys384. *(p. 211)*

Katsura, S. (1954). On the phase transition, *J. Chem. Phys.* **22**, 7, pp. 1277–1278, doi:10.1063/1.1740378. *(p. 162)*

Katsura, S. (1955). Phase transition of Husimi-Temperley model of imperfect gas, *Prog. Theor. Phys.* **13**, 6, pp. 571–586, doi:10.1143/PTP.13.571. *(p. 162)*

Kaufman, B. (1949). Crystal statistics. II. Partition function evaluated by spinor analysis, *Phys. Rev.* **76**, 8, pp. 1232–1243, doi:10.1103/PhysRev.76.1232. *(p. 148, 151, 166)*

Kaufman, B. and Onsager, L. (1949). Crystal statistics. III. Short-range order in a binary Ising lattice, *Phys. Rev.* **76**, 8, pp. 1244–1252, doi:10.1103/PhysRev.76.1244. *(p. 148, 151, 166)*

Keilmann, T., Lanzmich, S., McCulloch, I., and Roncaglia, M. (2011). Statistically induced phase transitions and anyons in 1D optical lattices, *Nat. Commun.* **2**, p. 361, doi:10.1038/ncomms1353. *(p. 72)*

Keller, J. (2002). *Theory of the electron: A theory of matter from START*, *Fundamental Theories of Physics*, Vol. 115 (Kluwer, Dordrecht). *(p. 211)*

Keller, J. and Oziewicz, Z. (eds.) (1997). *The theory of the electron* (Universidad Nacional Autónoma de Mexico (UNAM), Mexico City). *(p. 211, 212)*

Kemeny, G. (1965a). A model of the insulator-metal transition. Part I. Two-body correlations, *Ann. Phys.* **32**, 1, pp. 69–99, doi:10.1016/0003-4916(65)90060-6. *(p. 241)*

Kemeny, G. (1965b). A model of the insulator-metal transition. Part II. Quasi-chemical equilibrium theory, *Ann. Phys.* **32**, 3, pp. 404–415, doi:10.1016/0003-4916(65)90140-5. *(p. 241)*

Kenna, R., Johnston, D. A., and Janke, W. (2006a). Scaling relations for logarithmic corrections, *Phys. Rev. Lett.* **96**, p. 115701, doi:10.1103/PhysRevLett.96.115701. *(p. 169)*

Kenna, R., Johnston, D. A., and Janke, W. (2006b). Self-consistent scaling theory for logarithmic-correction exponents, *Phys. Rev. Lett.* **97**, p. 155702, doi:10.1103/PhysRevLett.97.155702, [Erratum *ibid.* **97**, 169901 (2006)]. *(p. 169)*

Khare, A. (2005). *Fractional statistics and quantum theory*, 2nd edn. (World Scientific, Singapore), ISBN 9789812561602. *(p. 47, 54, 64)*

Kim, E.-A., Lawler, M., Vishveshwara, S., and Fradkin, E. (2005). Signatures of fractional statistics in noise experiments in quantum Hall fluids, *Phys. Rev. Lett.* **95**, p. 176402, doi:10.1103/PhysRevLett.95.176402. *(p. 46)*

Kim, S.-Y. and Creswick, R. J. (1998). Fisher zeros of the **Q**-state Potts model in the complex temperature plane for nonzero external magnetic field, *Phys. Rev. E* **58**, pp. 7006–7012, doi:10.1103/PhysRevE.58.7006. *(p. 171)*

Kleinert, H. (2016). *Particles and quantum fields* (World Scientific, Singapore), ISBN 9789814740890. *(p. 197, 199, 200, 201, 202, 204, 209)*

Knoop, S., Ferlaino, F., Mark, M., Berninger, M., Schöbel, H., Nägerl, H., and Grimm, R. (2009). Observation of an Efimov-like trimer resonance in ultracold atom-dimer scattering, *Nat. Phys.* **5**, 3, pp. 227–230, doi: 10.1038/nphys1203. *(p. 44)*

Kondo, J. (1964). Resistance minimum in dilute magnetic alloys, *Prog. Theor. Phys.* **32**, 1, pp. 37–49, doi:10.1143/PTP.32.37. *(p. 247)*

Kortman, P. J. and Griffiths, R. B. (1971). Density of zeros on the Lee-Yang circle for two ising ferromagnets, *Phys. Rev. Lett.* **27**, pp. 1439–1442, doi: 10.1103/PhysRevLett.27.1439. *(p. 162)*

Kosterlitz, J. M. and Thouless, D. J. (1973). Ordering, metastability and phase transitions in two-dimensional systems, *J. Phys. C: Solid State Phys.* **6**, 7, pp. 1181–1203, doi:10.1088/0022-3719/6/7/010. *(p. 96)*

Kowalski, J. M. (1972). Differential equation for the partition function of the Ising chain, *Phys. Lett. A* **38**, 6, pp. 421–422, doi:10.1016/0375-9601(72)90234-4. *(p. 154, 155)*

Kraemer, T., Mark, M., Waldburger, P., Danzl, J. G., Chin, C., Engeser, B., Lange, A. D., Pilch, K., Jaakkola, A., Nägerl, H., and Grimm, R. (2006). Evidence for Efimov quantum states in an ultracold gas of caesium atoms, *Nature* **440**, 7082, pp. 315–318, doi:10.1038/nature04626. *(p. 43)*

Kravchenko, S. V., Kravchenko, G. V., Furneaux, J. E., Pudalov, V. M., and D'Iorio, M. (1994). Possible metal-insulator transition at $B = 0$ in two dimensions, *Phys. Rev. B* **50**, pp. 8039–8042, doi:10.1103/PhysRevB.50.8039. *(p. 237)*

Kravchenko, S. V., Simonian, D., Sarachik, M. P., Mason, W., and Furneaux, J. E. (1996). Electric field scaling at a $B = 0$ metal-insulator transition in two dimensions, *Phys. Rev. Lett.* **77**, pp. 4938–4941, doi:10.1103/PhysRevLett.77.4938. *(p. 237)*

Kravtsov, V. E. and Lerner, I. V. (1984). Instability of the scaling theory of 2-d localization, *Solid State Commun.* **52**, 6, pp. 593–598, doi:10.1016/0038-1098(84)90885-8. *(p. 141)*

Kummer, H. (1967). N-representability problem for reduced density matrices, *J. Math. Phys.* **8**, 10, pp. 2063–2081, doi:10.1063/1.1705122. *(p. 4)*

Kurtze, D. A. (1983). The Yang-Lee edge singularity in one-dimensional Ising and N-vector models, *J. Stat. Phys.* **30**, 1, pp. 15–35, doi:10.1007/BF01010866. *(p. 171)*

Kuznetsov, A. and Mikheev, N. (2013). Solutions of the Dirac equation in an external electromagnetic field, in *Electroweak Processes in External Active Media, Springer Tracts in Modern Physics*, Vol. 252 (Springer, Berlin), ISBN 978-3-642-36225-5, pp. 13–24, doi:10.1007/978-3-642-36226-2_2. *(p. 211)*

Lacroix, C. and Cyrot, M. (1979). Phase diagram of the Kondo lattice, *Phys. Rev. B* **20**, 5, pp. 1969–1976, doi:10.1103/PhysRevB.20.1969. *(p. 250)*

Landau, L. D. (1941). The theory of superfluidity of helium II, *J. Phys. (Moscow)* **5**, pp. 71–90. *(p. 94)*

Lange, H., Toomire, B., and Zweifel, P. F. (1995). An overview of Schrödinger-Poisson problems, *Rep. Math. Phys.* **36**, 2–3, pp. 331–345, doi:10.1016/0034-4877(96)83629-9, Proceedings of the 26th Symposium on Mathematical Physics. *(p. 222)*

Langer, J. S. and Ambegaokar, V. (1961). Friedel sum rule for a system of interacting electrons, *Phys. Rev.* **121**, 4, pp. 1090–1092, doi:10.1103/PhysRev.121.1090. *(p. 117, 118)*

Larkin, A. and Varlamov, A. A. (2002). Fluctuation phenomena in superconductors, in K. Bennemann and J. B. Ketterson (eds.), *Handbook on Superconductivity: Conventional and Unconventional Superconductors*, chap. 10 (Springer Verlag, Berlin), pp. 369–458, doi:10.1007/978-3-540-73253-2_10. *(p. 82)*

Larkin, A. I. and Ovchinnikov, Y. N. (1964). Inhomogeneous state of superconductors, *Zh. Eksp. Teor. Fiz.* **47**, p. 1136, [Sov. Phys. JETP **20**, 762 (1965)]. *(p. 78)*

Larkin, A. I. and Varlamov, A. A. (2005). *Theory of Fluctuations in Superconductors* (Oxford University Press, Oxford), ISBN 9780191523700. *(p. 82)*

Läuchli, A. M. and Kollath, C. (2008). Spreading of correlations and entanglement after a quench in the one-dimensional Bose-Hubbard model, *J. Stat. Mech.* **2008**, 05, p. P05018, doi:10.1088/1742-5468/2008/05/P05018. *(p. 170)*

Laughlin, R. B. (1983a). Anomalous quantum Hall effect: An incompressible quantum fluid with fractionally charged excitations, *Phys. Rev. Lett.* **50**, 18, pp. 1395–1398, doi:10.1103/PhysRevLett.50.1395. *(p. 46, 48, 239)*

Laughlin, R. B. (1983b). Quantized motion of three two-dimensional electrons in a strong magnetic field, *Phys. Rev. B* **27**, 6, pp. 3383–3389, doi:10.1103/PhysRevB.27.3383. *(p. 46, 239)*

Laughlin, R. B. (1988a). The relationship between high-temperature superconductivity and the fractional quantum Hall effect, *Science* **242**, 4878, pp. 525–533, doi:10.1126/science.242.4878.525. *(p. 46)*

Laughlin, R. B. (1988b). Superconducting ground state of noninteracting particles obeying fractional statistics, *Phys. Rev. Lett.* **60**, pp. 2677–2680, doi:10.1103/PhysRevLett.60.2677, [**61**, 379(E) (1988)]. *(p. 46)*

Layzer, D. (1959). On a screening theory of atomic spectra, *Ann. Phys.* **8**, p. 271, doi:10.1016/0003-4916(59)90023-5. *(p. 28)*

Lea, M. J., March, N. H., and Sung, W. (1991). Thermodynamics of melting of a two-dimensional Wigner electron crystal, *J. Phys.: Condens. Matter* **3**, 23, pp. 4301–4306, doi:10.1088/0953-8984/3/23/020, reprinted in March and Angilella (2009). *(p. 46, 239)*

Lea, M. J., March, N. H., and Sung, W. (1992). Melting of Wigner electron crystals; phenomenology and anyon magnetism, *J. Phys.: Condens. Matter* **4**, 23, pp. 5263–5272, doi:10.1088/0953-8984/4/23/003. *(p. 46)*

Lee, J. (2013a). Exact partition function zeros of the Wako-Saitô-Muñoz-Eaton protein model, *Phys. Rev. Lett.* **110**, p. 248101, doi:10.1103/PhysRevLett.110.248101. *(p. 171)*

Lee, J. (2013b). Exact partition function zeros of the Wako-Saitô-Muñoz-Eaton β hairpin model, *Phys. Rev. E* **88**, p. 022710, doi:10.1103/PhysRevE.88.022710. *(p. 171)*

Lee, P. A. and Ramakrishnan, T. V. (1985). Disordered electronic systems, *Rev. Mod. Phys.* **57**, 2, pp. 287–337, doi:10.1103/RevModPhys.57.287. *(p. 135,*

138)

Lee, P. A. and Stone, A. D. (1985). Universal conductance fluctuations in metals, *Phys. Rev. Lett.* **55**, 15, pp. 1622–1625, doi:10.1103/PhysRevLett.55.1622. *(p. 138)*

Lee, T. D. and Yang, C. N. (1952). Statistical theory of equations of state and phase transitions. II. Lattice gas and Ising model, *Phys. Rev.* **87**, pp. 410–419, doi:10.1103/PhysRev.87.410. *(p. 156, 161, 162, 169)*

Leggett, A. J. (1975). A theoretical description of the new phases of liquid ^{3}He, *Rev. Mod. Phys.* **47**, pp. 331–414, doi:10.1103/RevModPhys.47.331. *(p. 80)*

Leggett, A. J. (1999). Superfluidity, *Rev. Mod. Phys.* **71**, 2, pp. S318–S323, doi:10.1103/RevModPhys.71.S318. *(p. 93)*

Leggett, A. J. (2001). Bose-Einstein condensation in the alkali gases: Some fundamental concepts, *Rev. Mod. Phys.* **73**, 2, pp. 307–356, doi:10.1103/RevModPhys.73.307. *(p. 94)*

Leinaas, J. M. and Myrheim, J. (1977). On the theory of identical particles, *Nuovo Cimento B* **37**, 1, pp. 1–23, doi:10.1007/BF02727953. *(p. 46, 47)*

Lerda, A. (1992). *Anyons: Quantum mechanics of particles with fractional statistics* (Springer, Berlin), ISBN 9783540561057. *(p. 47)*

Lerda, A. and Sciuto, S. (1993). Anyons and quantum groups, *Nucl. Phys. B* **401**, 3, pp. 613–643, doi:10.1016/0550-3213(93)90316-H. *(p. 47)*

Levanyuk, A. P. (1959). Contribution to the theory of light scattering near the second-order phase-transition points, *Sov. Phys. JETP* **9**, 3, pp. 571–576, [Zh. Eksp. Theor. Fyz. **36**, 810–818 (1959)]. *(p. 82)*

Levy, M. (1979). Universal variational functionals of electron densities, first-order density matrices, and natural spin-orbitals and solution of the *v*-representability problem, *Proc. Nat. Acad. Sci.* **76**, 12, pp. 6062–6065. *(p. 6)*

Lewis, G. N. (1916). The atom and the molecule, *J. Am. Chem. Soc.* **38**, pp. 762–785, doi:10.1021/ja02261a002. *(p. 87)*

Leys, F. E. (2003). *Inhomogeneous electron liquid theory applied to metallic phases and nanostructures*, Doctoral thesis, University of Antwerp. *(p. 153)*

Leys, F. E. and March, N. H. (2003). Electron-electron correlations in liquid *s-p* metals, *J. Phys. A: Math. Gen.* **36**, 22, pp. 5893–5898, doi:10.1088/0305-4470/36/22/309. *(p. 128, 129, 131)*

Leys, F. E., March, N. H., Angilella, G. G. N., and Zhang, M. (2002). Similarity and contrasts between thermodynamic properties at the critical point of liquid alkali metals and of electron-hole droplets, *Phys. Rev. B* **66**, p. 073314, doi:10.1103/PhysRevB.66.073314. *(p. 153)*

Leys, F. E., March, N. H., and Lamoen, D. (2004). Relativistic virial relations for both homogeneous and spatially varying electron liquids, *Phys. Chem. Liq.* **42**, 4, pp. 423–431, doi:10.1080/00319100410001697873. *(p. 199)*

Li, W., Liu, Z., Wu, Y.-S., and Chen, Y. (2014). Exotic fractional topological states in a two-dimensional organometallic material, *Phys. Rev. B* **89**, p. 125411, doi:10.1103/PhysRevB.89.125411. *(p. 73)*

Liao, Y., Rittner, A. S. C., Paprotta, T., Li, W., Partridge, G. B., Hulet, R. G., Baur, S. K., and Mueller, E. J. (2010). Spin-imbalance in a one-dimensional

Fermi gas, *Nature* **467**, 7315, pp. 567–569, doi:10.1038/nature09393. *(p. 78)*

Lieb, E. (1963). Exact analysis of an interacting Bose gas. II. The excitation spectrum, *Phys. Rev.* **130**, 4, pp. 1616–1624, doi:10.1103/PhysRev.130.1616. *(p. 103, 106, 107, 108, 109, 110)*

Lieb, E. and Liniger, W. (1963). Exact analysis of an interacting Bose gas. I. The general solution and the ground state, *Phys. Rev.* **130**, 4, pp. 1605–1616, doi:10.1103/PhysRev.130.1605. *(p. 103, 106, 107, 110)*

Lieb, E. and Mattis, D. (1962). Theory of ferromagnetism and the ordering of electronic energy levels, *Phys. Rev.* **125**, 1, pp. 164–172, doi:10.1103/PhysRev.125.164. *(p. 241)*

Lieb, E., Schultz, T., and Mattis, D. (1961). Two soluble models of an antiferromagnetic chain, *Ann. Phys.* **16**, 3, pp. 407–466, doi:10.1016/0003-4916(61)90115-4. *(p. 189, 269)*

Lieb, E. and Wu, F. (1968a). Absence of Mott transition in an exact solution of the short-range, one-band model in one dimension, *Phys. Rev. Lett.* **20**, 25, p. 1445, doi:10.1103/PhysRevLett.20.1445. *(p. 241, 242, 244, 245, 254, 256)*

Lieb, E. and Wu, F. (1968b). Erratum: Absence of Mott transition in an exact solution of the short-range, one-band model in one dimension, *Phys. Rev. Lett.* **21**, 3, pp. 192–192, doi:10.1103/PhysRevLett.21.192.2. *(p. 241, 254)*

Lieb, E. H. (1977). Existence and uniqueness of the minimizing solution of Choquard's nonlinear equation, *Studies in Applied Mathematics* **57**, 2, pp. 93–105, doi:10.1002/sapm197757293. *(p. 225)*

Lieb, E. H. and Narnhofer, H. (1975). The thermodynamic limit for jellium, *J. Stat. Phy.* **12**, 4, pp. 291–310, doi:10.1007/BF01012066. *(p. 236)*

Lieb, E. H. and Simon, B. (1977). The Thomas-Fermi theory of atoms, molecules and solids, *Adv. Math.* **23**, 1, pp. 22–116, doi:10.1016/0001-8708(77)90108-6. *(p. 14)*

Lim, T. K., Duffy, S. K., and Damer, W. C. (1977). Efimov state in the ^{4}He trimer, *Phys. Rev. Lett.* **38**, 7, pp. 341–343, doi:10.1103/PhysRevLett.38.341. *(p. 41)*

Limacher, P. A., Ayers, P. W., Johnson, P. A., De Baerdemacker, S., Van Neck, D., and Bultinck, P. (2013). A new mean-field method suitable for strongly correlated electrons: Computationally facile antisymmetric products of nonorthogonal geminals, *J. Chem. Theory Comp.* **9**, 3, pp. 1394–1401, doi:10.1021/ct300902c. *(p. 88)*

Liu, W. and Andrei, N. (2014). Quench dynamics of the anisotropic Heisenberg model, *Phys. Rev. Lett.* **112**, 25, p. 257204, doi:10.1103/PhysRevLett.112.257204. *(p. 112)*

Liu, Y., Christandl, M., and Verstraete, F. (2007). Quantum computational complexity of the N-representability problem: QMA complete, *Phys. Rev. Lett.* **98**, p. 110503, doi:10.1103/PhysRevLett.98.110503. *(p. 6)*

Longhi, S. and Della Valle, G. (2012). Anyonic bloch oscillations, *Phys. Rev. B* **85**, p. 165144, doi:10.1103/PhysRevB.85.165144. *(p. 72)*

Löwdin, P. O. (1955). Quantum theory of many-particle systems. I. Physical interpretations by means of density matrices, natural spin-orbitals, and convergence problems in the method of configurational interaction, *Phys. Rev.* **97**, pp. 1474–1489, doi:10.1103/PhysRev.97.1474. *(p. 4, 5, 88, 89)*

Lu, W. T. and Wu, F. Y. (2001). Density of the Fisher zeroes for the Ising model, *J. Stat. Phys.* **102**, 3, pp. 953–970, doi:10.1023/A:1004863322373. *(p. 166)*

Lucha, W. and Scöberl, F. F. (1999). Semirelativistic treatment of bound states, *Int. J. Mod. Phys. A* **14**, pp. 2309–2333, doi:10.1142/S0217751X99001160. *(p. 216)*

Luther, A. and Peschel, I. (1974). Single-particle states, Kohn anomaly, and pairing fluctuations in one dimension, *Phys. Rev. B* **9**, 7, pp. 2911–2919, doi: 10.1103/PhysRevB.9.2911. *(p. 182, 187)*

Luther, A. and Peschel, I. (1975). Calculation of critical exponents in two dimensions from quantum field theory in one dimension, *Phys. Rev. B* **12**, 9, pp. 3908–3917, doi:10.1103/PhysRevB.12.3908. *(p. 182)*

Luttinger, J. M. (1960). Fermi surface and some simple equilibrium properties of a system of interacting fermions, *Phys. Rev.* **119**, 4, pp. 1153–1163, doi: 10.1103/PhysRev.119.1153. *(p. 118)*

Luttinger, J. M. (1963). An exactly soluble model of a many-fermion system, *J. Math. Phys.* **4**, 9, pp. 1154–1162, doi:10.1063/1.1704046. *(p. 182)*

Luttinger, J. M. and Ward, J. C. (1960). Ground-state energy of a many-fermion system. II, *Phys. Rev.* **118**, 5, pp. 1417–1427, doi:10.1103/PhysRev.118.1417. *(p. 118)*

Lynton, E. A. (1964). *Superconductivity* (Methuen, London). *(p. 94)*

Macek, J. (1968). Properties of autoionizing states of He, *J. Phys. B: Atom. Mol. Phys.* **1**, 5, pp. 831–843, doi:10.1088/0022-3700/1/5/309. *(p. 36)*

Macek, J. (1986). Loosely bound states of three particles, *Z. Phys. D: At. Mol. Clus.* **3**, 1, pp. 31–37, doi:10.1007/BF01442345. *(p. 39)*

Macke, W. (1950). Über die Wechselwirkungen im Fermi-Gas Polarisationserscheinungen, Correlationsenergie, Elektronenkondensation, *Z. Naturforsch. A* **5**, pp. 192–208. *(p. 235)*

Magro, W. R. and Ceperley, D. M. (1994). Ground-state properties of the two-dimensional Bose Coulomb liquid, *Phys. Rev. Lett.* **73**, 6, pp. 826–829, doi:10.1103/PhysRevLett.73.826. *(p. 96)*

Majumdar, C. K. and Ghosh, D. K. (1969). On next-nearest-neighbor interaction in linear chain. I, *J. Math. Phys.* **10**, 8, pp. 1388–1398, doi: 10.1063/1.1664978. *(p. 265)*

Maldacena, J. (1999). The large-N limit of superconformal field theories and supergravity, *Int. J. Theor. Phys.* **38**, 4, pp. 1113–1133, doi:10.1023/A: 1026654312961. *(p. 174, 219)*

March, N. H. (1975a). Exact resistivity formula for finite-range spherical potential of arbitrary strength, *Phil. Mag.* **32**, 2, pp. 497–500, doi:10.1080/14786437508219971. *(p. 120)*

March, N. H. (1975b). *Self-Consistent Fields in Atoms* (Pergamon Press, Oxford). *(p. 194)*

March, N. H. (1983a). Bound and Efimov states for $^4He_2$ and $^4He_3$ and Bose-Einstein condensation, *J. Chem. Phys.* **79**, 1, pp. 529–530, doi:10.1063/1.

445510. *(p. 42)*

March, N. H. (1983b). Origins — the Thomas-Fermi theory, in S. Lundqvist and N. H. March (eds.), *Theory of the Inhomogeneous Electron Gas*, chap. 1 (Springer, New York), p. 1, doi:10.1007/978-1-4899-0415-7_1. *(p. 14)*

March, N. H. (1984). Reply to "^{4}He n-mers and Bose-Einstein condensation in He II", *J. Chem. Phys.* **81**, 1, pp. 587–587, doi:10.1063/1.447351. *(p. 42)*

March, N. H. (1988). Comment on "Melting of the Wigner crystal at finite temperature", *Phys. Rev. A* **37**, pp. 4526–4526, doi:10.1103/PhysRevA.37.4526. *(p. 240)*

March, N. H. (1990). *Liquid metals. Concepts and theory* (Cambridge University Press, Cambridge), ISBN 9780080032290. *(p. 119, 120, 121, 123, 125)*

March, N. H. (1993). Melting of a magnetically induced wigner electron solid and anyon properties, *J. Phys.: Condens. Matter* **5**, 34B, pp. B149–B156, doi:10.1088/0953-8984/5/34B/018. *(p. 46, 56)*

March, N. H. (1996). *Electron correlation in molecules and condensed phases* (Plenum Press, New York), ISBN 9780306448447. *(p. 240)*

March, N. H. (1997). Chemical collision model and statistical distribution function for an anyon liquid, *Phys. Chem. Liq.* **34**, pp. 61–64, doi:10.1080/00319109708035914. *(p. 46, 56)*

March, N. H. (2002). First-order density matrix as a functional of the ground-state electron density for harmonic confinement of two electrons which also interact harmonically, *Phys. Lett. A* **306**, pp. 63–65, doi:10.1016/S0375-9601(02)01598-0. *(p. 14)*

March, N. H. (2012). Superfluidity with and without a condensate in two and three dimensions, *Phys. Chem. Liq.* **50**, pp. 819–821, doi:10.1080/00319104.2012.728305. *(p. 96)*

March, N. H. (2014a). Inclusion of an applied magnetic field of arbitrary strength in the Ising model, *Phys. Lett. A* **378**, 30–31, pp. 2295–2296, doi:10.1016/j.physleta.2014.05.040. *(p. 155, 156)*

March, N. H. (2014b). Similarities and contrasts between critical point behavior of heavy fluid alkalis and d-dimensional Ising model, *Phys. Lett. A* **378**, 3, pp. 254–256, doi:10.1016/j.physleta.2013.10.030. *(p. 153, 154)*

March, N. H. (2015). Toward a final theory of critical exponents in terms of dimensionality d plus universality class n, *Phys. Lett. A* **379**, 9, pp. 820–822, doi:10.1016/j.physleta.2014.11.043. *(p. 174, 178, 181, 182)*

March, N. H. and Angilella, G. G. N. (eds.) (2009). *Many-body Theory of Molecules, Clusters, and Condensed Phases* (World Scientific, Singapore), ISBN 9789814271776. *(p. vii, 287, 288, 291, 299, 301, 305, 310, 315)*

March, N. H. and Deb, B. M. (eds.) (1987). *The single particle density in physics and chemistry* (Academic Press, New York), ISBN 9780124705180. *(p. 25, 26)*

March, N. H., Gidopoulos, N., Theophilou, A. K., Lea, M. J., and Sung, W. (1993). The statistical distribution function for an anyon liquid, *Phys. Chem. Liq.* **26**, pp. 135–141, doi:10.1080/00319109308030827. *(p. 46, 56)*

March, N. H. and Murray, A. M. (1960a). Electronic wave functions around a vacancy in a metal, *Proc. Roy. Soc. Lond. A* **256**, pp. 400–415, doi:10.

1098/rspa.1960.0115, reprinted in March and Angilella (2009). *(p. 113)*

March, N. H. and Murray, A. M. (1960b). Relation between Dirac and canonical density matrices, with applications to imperfections in metals, *Phys. Rev.* **120**, pp. 830–836, doi:10.1103/PhysRev.120.830, reprinted in March and Angilella (2009). *(p. 121, 130, 229, 230)*

March, N. H. and Murray, A. M. (1961). Self-consistent perturbation treatment of impurities and imperfections in metals, *Proc. Roy. Soc. Lond. A* **261**, pp. 119–133, doi:10.1098/rspa.1961.0065, reprinted in March and Angilella (2009). *(p. 113, 114, 118, 119)*

March, N. H. and Murray, A. M. (1962). Self-consistent perturbation treatment of impurities and imperfections in metals. II. Second-order perturbation corrections, *Proc. R. Soc. Lond. A* **266**, 1327, pp. 559–567, doi:10.1098/rspa.1962.0078. *(p. 118, 119)*

March, N. H., Suzuki, M., and Parrinello, M. (1979). Phenomenological theory of first- and second-order metal-insulator transitions at absolute zero, *Phys. Rev. B* **19**, 4, pp. 2027–2029, doi:10.1103/PhysRevB.19.2027. *(p. 132)*

March, N. H. and Tosi, M. P. (1973). Quantum theory of pure liquid metals as two-component systems, *Ann. Phys. (NY)* **81**, 2, pp. 414–437, doi:10.1016/0003-4916(73)90164-4, Reprinted in March and Angilella (2009). *(p. 131)*

March, N. H. and Tosi, M. P. (1985). Melting criteria for classical and quantal Wigner crystals, *Phys. Chem. Liq.* **14**, 4, pp. 303–306, doi:10.1080/00319108508080993. *(p. 240)*

March, N. H. and Tosi, M. P. (2002). *Introduction to Liquid State Physics* (World Scientific, Singapore), ISBN 9789812778482. *(p. 93, 94, 123, 131, 233)*

March, N. H., Tosi, M. P., and Chapman, R. G. (1988). Liquid-vapour coexistence curves from a model equation of state, *Phy. Chem. Liq.* **18**, 3, pp. 195–205, doi:10.1080/00319108808078594. *(p. 153)*

March, N. H., Young, W. H., and Sampanthar, S. (1995). *The Many-Body Problem in Quantum Mechanics* (Dover, New York), ISBN 9780486687544. *(p. 77, 79, 80, 233)*

Mari, A., De Palma, G., and Giovannetti, V. (2016). Experiments testing macroscopic quantum superpositions must be slow, *Sci. Rep.* **6**, p. 22777, doi:10.1038/srep22777. *(p. 228)*

Marques, M. A. L. and Lathiotakis, N. N. (2008). Empirical functionals for reduced-density-matrix-functional theory, *Phys. Rev. A* **77**, p. 032509, doi:10.1103/PhysRevA.77.032509. *(p. 9)*

Mastellone, A., Falci, G., and Fazio, R. (1998). Small superconducting grain in the canonical ensemble, *Phys. Rev. Lett.* **80**, 20, pp. 4542–4545, doi:10.1103/PhysRevLett.80.4542. *(p. 86)*

Mattis, D. C. (1974). New wave-operator identity applied to the study of persistent currents in 1D, *J. Math. Phys.* **15**, 5, pp. 609–612, doi:10.1063/1.1666693. *(p. 187)*

Mattis, D. C. (1993). *The Many-Body Problem. An Encyclopedia of Exactly Solved Models in One Dimension* (World Scientific, Singapore), ISBN 9789810209759. *(p. vii, 286)*

Mattuck, R. D. (1976). *A guide to Feynman diagrams in the many-body problem*

(Dover, New York), ISBN 9780486670478. *(p. 31)*

Matveev, V. and Shrock, R. (1995). Complex-temperature properties of the 2D Ising model with beta $H = \pm i\pi/2$, *J. Phys. A: Math. Gen.* **28**, 17, pp. 4859–4882, doi:10.1088/0305-4470/28/17/018. *(p. 169, 172)*

Matveev, V. and Shrock, R. (1996). Complex-temperature properties of the two-dimensional Ising model for nonzero magnetic field, *Phys. Rev. E* **53**, pp. 254–267, doi:10.1103/PhysRevE.53.254. *(p. 169, 172)*

Mazziotti, D. A. (1998). Contracted Schrödinger equation: Determining quantum energies and two-particle density matrices without wave functions, *Phys. Rev. A* **57**, pp. 4219–4234, doi:10.1103/PhysRevA.57.4219. *(p. 4)*

Mazziotti, D. A. (2000). Geminal functional theory: A synthesis of density and density matrix methods, *J. Chem. Phys.* **112**, 23, pp. 10125–10130, doi: 10.1063/1.481653. *(p. 88)*

Mazziotti, D. A. and Erdahl, R. M. (2001). Uncertainty relations and reduced density matrices: Mapping many-body quantum mechanics onto four particles, *Phys. Rev. A* **63**, p. 042113, doi:10.1103/PhysRevA.63.042113. *(p. 4)*

McCoy, B. (1968). Spin correlation functions of the $X - Y$ model, *Phys. Rev.* **173**, 2, pp. 531–541, doi:10.1103/PhysRev.173.531. *(p. 184, 189)*

Meissner, W. and Ochsenfeld, R. (1933). Ein neuer Effekt bei Eintritt der Supraleitfähigkeit, *Naturwiss.* **21**, 44, pp. 787–788, doi:10.1007/BF01504252. *(p. 75)*

Mermin, N. D. and Wagner, H. (1966). Absence of ferromagnetism or antiferromagnetism in one- or two-dimensional isotropic Heisenberg models, *Phys. Rev. Lett.* **17**, pp. 1133–1136, doi:10.1103/PhysRevLett.17.1133, [Erratum *ibid.* **17**, 1307 (1966)]. *(p. 96)*

Meyer, J. S. and Matveev, K. A. (2009). Wigner crystal physics in quantum wires, *J. Phys.: Condens. Matter* **21**, 2, p. 023203, doi:10.1088/0953-8984/21/2/023203. *(p. 238)*

Mezio, A., Lobos, A. M., Dobry, A. O., and Gazza, C. J. (2015). Haldane phase in one-dimensional topological Kondo insulators, *Phys. Rev. B* **92**, 20, p. 205128, doi:10.1103/PhysRevB.92.205128. *(p. 251)*

Milne, E. A. (1927). The total energy of binding of a heavy atom, *Math. Proc. Cambr. Phil. Soc.* **23**, pp. 794–799, doi:10.1017/S0305004100015589. *(p. 14, 24)*

Minami, Y., Araki, K., Dao, T. D., Nagao, T., Kitajima, M., Takeda, J., and Katayama, I. (2015). Terahertz-induced acceleration of massive Dirac electrons in semimetal bismuth, *Sci. Rep.* **5**, p. 15870, doi:10.1038/srep15870. *(p. 209)*

Minguzzi, A., March, N. H., and Tosi, M. P. (2003). Superfluidity with and without a condensate in interacting Bose fluids, *Phys. Chem. Liq.* **41**, p. 323, doi:10.1080/0031910031000117300. *(p. 91, 92, 93, 95, 96, 97)*

Misner, C. W. (1957). Feynman quantization of general relativity, *Rev. Mod. Phys.* **29**, 3, pp. 497–509, doi:10.1103/RevModPhys.29.497. *(p. 218)*

Modugno, G. (2009). Universal few-body binding, *Science* **326**, 5960, pp. 1640–1641, doi:10.1126/science.1184235. *(p. 44)*

Møller, C. (1962). Les théories relativistes de la gravitation, in A. L. M. A. Tonnelat (ed.), *Colloques Internationaux CNRS*, Vol. 91 (CNRS, Paris), pp. 1–96. *(p. 223)*

Moreva, E., Brida, G., Gramegna, M., Giovannetti, V., Maccone, L., and Genovese, M. (2014). Time from quantum entanglement: An experimental illustration, *Phys. Rev. A* **89**, 5, p. 052122, doi:10.1103/PhysRevA.89.052122. *(p. 219)*

Moroni, S., Senatore, G., and Fantoni, S. (1997). Momentum distribution of liquid helium, *Phys. Rev. B* **55**, 2, pp. 1040–1049, doi:10.1103/PhysRevB.55.1040. *(p. 92)*

Moroz, I. M., Penrose, R., and Tod, P. (1998). Spherically-symmetric solutions of the Schrödinger-Newton equations, *Class. Quantum Grav.* **15**, 9, pp. 2733–2742, doi:10.1088/0264-9381/15/9/019. *(p. 225, 226)*

Morra, R., Buyers, W., Armstrong, R., and Hirakawa, K. (1988). Spin dynamics and the Haldane gap in the spin-1 quasi-one-dimensional antiferromagnet $CsNiCl_3$, *Phys. Rev. B* **38**, 1, pp. 543–555, doi:10.1103/PhysRevB.38.543. *(p. 269)*

Morsch, O. and Oberthaler, M. (2006). Dynamics of Bose-Einstein condensates in optical lattices, *Rev. Mod. Phys.* **78**, 1, pp. 179–215, doi:10.1103/RevModPhys.78.179. *(p. 103)*

Moshinsky, M. (1952). How good is the Hartree-Fock approximation, *Am. J. Phys.* **36**, pp. 52–53, doi:10.1119/1.1974410. *(p. 14)*

Mott, N. F. (1936). The electrical resistance of dilute solid solutions, *Math. Proc. Camb. Phil. Soc.* **32**, 02, pp. 281–290, doi:10.1017/S0305004100001845. *(p. 120)*

Mott, N. F. (1949). The basis of the electron theory of metals, with special reference to the transition metals, *Proc. Phys. Soc. A* **62**, 7, pp. 416–421, doi:10.1088/0370-1298/62/7/303. *(p. 135)*

Mott, N. F. (1968). Metal-insulator transition, *Rev. Mod. Phys.* **40**, 4, pp. 677–683, doi:10.1103/RevModPhys.40.677. *(p. 135)*

Mott, N. F. (1969a). Charge transport in non-crystalline semiconductors, in O. Madelung (ed.), *Festkörperprobleme 9, Advances in Solid State Physics*, Vol. 9 (Springer, Berlin and Heidelberg), ISBN 978-3-540-75325-4, pp. 22–45, doi:10.1007/BFb0109150. *(p. 142)*

Mott, N. F. (1969b). Conduction in non-crystalline materials. III. Localized states in a pseudogap and near extremities of conduction and valence bands, *Phil. Mag.* **19**, 160, pp. 835–852, doi:10.1080/14786436908216338. *(p. 142, 143, 145)*

Mülken, O., Borrmann, P., Harting, J., and Stamerjohanns, H. (2001). Classification of phase transitions of finite Bose-Einstein condensates in power-law traps by Fisher zeros, *Phys. Rev. A* **64**, p. 013611, doi:10.1103/PhysRevA.64.013611. *(p. 165, 170)*

Müller, A. M. K. (1984). Explicit approximate relation between reduced two- and one-particle density matrices, *Phys. Lett. A* **105**, 9, pp. 446–452, doi:10.1016/0375-9601(84)91034-X. *(p. 8)*

Murguía, G., Raya, A., Sánchez, A., and Reyes, E. (2010). The electron propagator in external electromagnetic fields in low dimensions, *Am. J. Phys.* **78**,

7, pp. 700–707, doi:10.1119/1.3311656. *(p. 211)*

Murthy, M. V. N. and Shankar, R. (1994). Haldane exclusion statistics and second virial coefficient, *Phys. Rev. Lett.* **72**, 23, pp. 3629–3633, doi: 10.1103/PhysRevLett.72.3629. *(p. 47, 55, 64)*

Nagara, H., Nagata, Y., and Nakamura, T. (1987). Melting of the Wigner crystal at finite temperature, *Phys. Rev. A* **36**, pp. 1859–1873, doi:10.1103/PhysRevA.36.1859. *(p. 240)*

Nayak, C. and Wilczek, F. (1994). Exclusion statistics: Low-temperature properties, fluctuations, duality, and applications, *Phys. Rev. Lett.* **73**, pp. 2740–2743, doi:10.1103/PhysRevLett.73.2740. *(p. 47, 55, 63)*

Nielsen, E., Fedorov, D. V., Jensen, A. S., and Garrido, E. (2001). The three-body problem with short-range interactions, *Phys. Rep.* **347**, 5, pp. 373–459, doi: 10.1016/S0370-1573(00)00107-1. *(p. 33, 39)*

Novoselov, K. S., Geim, A. K., Morozov, S. V., Jiang, D., Katsnelson, M. I., Grigorieva, I. V., Dubonos, S. V., and Firsov, A. A. (2005a). Two-dimensional gas of massless Dirac fermions in graphene, *Nature* **438**, pp. 197–200, doi: 10.1038/nature04233. *(p. 211)*

Novoselov, K. S., Geim, A. K., Morozov, S. V., Jiang, D., Zhang, Y., Dubonos, S. V., Grigorieva, I. V., and Firsov, A. A. (2004). Electric field effect in atomically thin carbon films, *Science* **306**, pp. 666–669, doi:10.1126/science.1102896. *(p. 211)*

Novoselov, K. S., Jiang, D., Schedin, F., Booth, T. J., Khotkevich, V. V., Morozov, S. V., and Geim, A. K. (2005b). Two-dimensional atomic crystals, *Proc. Nat. Acad. Sci.* **102**, pp. 10451–10453, doi:10.1073/pnas.0502848102. *(p. 211)*

Odlyzko, A. M. (1987). On the distribution of spacings between zeros of the zeta function, *Math. Comp.* **48**, pp. 273–308, doi:10.1090/S0025-5718-1987-0866115-0. *(p. 273)*

Odlyzko, A. M. (1990). Primes, quantum chaos, and computers, in *Number Theory: Proceedings of a Symposium, 4 May 1989* (National Research Council, Washington, DC), pp. 35–46. *(p. 273)*

Onsager, L. (1944). Crystal statistics. I. A two-dimensional model with an order-disorder transition, *Phys. Rev.* **65**, 3-4, pp. 117–149, doi:10.1103/PhysRev.65.117. *(p. 147, 148, 151, 165, 166, 177, 178)*

Onsager, L. (1949a). Remark in the discussion following a paper by Gorter (1949) on the two-fluid model of liquid helium, *Nuovo Cimento* **6**, Suppl. 2, pp. 249–250. *(p. 93)*

Onsager, L. (1949b). Statistical hydromechanics, *Nuovo Cimento* **6**, Suppl. 2, pp. 279–287, doi:10.1007/BF02780991. *(p. 93)*

Orbach, R. (1958). Linear antiferromagnetic chain with anisotropic coupling, *Phys. Rev.* **112**, 2, pp. 309–316, doi:10.1103/PhysRev.112.309. *(p. 104)*

Ornstein, L. S. and Zernike, F. (1914). Accidental deviations of density and opalescence at the critical point of a single substance, *Proc. Acad. Sci. (Amsterdam)* **17**, pp. 793–806. *(p. 123)*

Page, D. N. and Wootters, W. K. (1983). Evolution without evolution: Dynamics described by stationary observables, *Phys. Rev. D* **27**, 12, pp. 2885–2892, doi:10.1103/PhysRevD.27.2885. *(p. 219)*

Parr, R. G. and Yang, W. (1989). *Density Functional Theory of Atoms and Molecules* (Oxford University Press, Oxford), ISBN 9780195092769. *(p. 14)*

Parrinello, M. and March, N. H. (1976). Thermodynamics of Wigner crystallization, *J. Phys. C: Solid State Phys.* **9**, 6, pp. L147–L150, doi:10.1088/0022-3719/9/6/003. *(p. 240)*

Pauli, W. (1927). Zur Quantenmechanik des magnetischen Elektrons, *Z. Phys.* **43**, 9–10, pp. 601–623, doi:10.1007/BF01397326. *(p. 201)*

Pauli, W. (1940). The connection between spin and statistics, *Phys. Rev.* **58**, pp. 716–722, doi:10.1103/PhysRev.58.716. *(p. 45)*

Pauli, W. and Fierz, M. (1939). Über relativistische Feldgleichungen von Teilchen mit beliebigem Spin im elektromagnetischen Feld, *Helv. Phys. Acta* **12**, pp. 297–300, in Bericht über die Tagung der Schweizerischen Physikalischen Gesellschaft; an expanded version then appeared as Fierz and Pauli (1939). *(p. 217, 221)*

Peierls, R. E. (1955). *Quantum Theory of Solids* (Oxford University Press, Oxford), ISBN 9780191516481. *(p. 262)*

Pelissetto, A. and Vicari, E. (2002). Critical phenomena and renormalization-group theory, *Phys. Rep.* **368**, 6, pp. 549–727, doi:10.1016/S0370-1573(02)00219-3. *(p. 153)*

Pellegrino, F. M. D., Angilella, G. G. N., March, N. H., and Pucci, R. (2007). Statistical correlations in an ideal gas of particles obeying fractional exclusion statistics, *Phys. Rev. E* **76**, p. 061123, doi:10.1103/PhysRevE.76.061123. *(p. 60, 64, 65, 66, 67, 68, 71)*

Pellegrino, F. M. D., Angilella, G. G. N., March, N. H., and Pucci, R. (2008). Statistical correlations of an anyon liquid at low temperature, *Phys. Chem. Liq.* **46**, pp. 342–348, doi:10.1080/00319100801972708. *(p. 64, 69, 71)*

Peng, X., Zhou, H., Wei, B.-B., Cui, J., Du, J., and Liu, R.-B. (2015). Experimental observation of lee-yang zeros, *Phys. Rev. Lett.* **114**, p. 010601, doi:10.1103/PhysRevLett.114.010601. *(p. 172)*

Penrose, O. (1951). CXXXVI. On the quantum mechanics of helium II, *Phil. Mag. (Series 7)* **42**, 335, pp. 1373–1377, doi:10.1080/14786445108560954. *(p. 92)*

Penrose, O. and Onsager, L. (1956). Bose-Einstein condensation and liquid helium, *Phys. Rev.* **104**, 3, pp. 576–584, doi:10.1103/PhysRev.104.576. *(p. 92)*

Penrose, R. (1978). Pentaplexity. A class of non-periodic tilings of the plane, *Eureka* **39**, pp. 16–32. *(p. 274)*

Penrose, R. (1979). Pentaplexity a class of non-periodic tilings of the plane, *Math. Intelligencer* **2**, 1, pp. 32–37, doi:10.1007/BF03024384. *(p. 274)*

Penrose, R. (1996). On gravity's role in quantum state reduction, *Gen. Rel. Gravit.* **28**, 5, pp. 581–600, doi:10.1007/BF02105068. *(p. 222)*

Penrose, R. (1998). Quantum computation, entanglement and state reduction, *Phil. Trans. R. Soc. Lond. A* **356**, 1743, pp. 1927–1939, doi:10.1098/rsta.1998.0256. *(p. 222)*

Penrose, R. (2000). Gravitational collapse of the wavefunction: an experimentally testable proposal, in V. G. Gurzadyan, R. T. Jantzen, and R. Ruffini (eds.),

Proceedings of the 9th Marcel Grossmann Meeting (MGIX MM): on recent developments in theoretical and experimental general relativity, gravitation, and relativistic field theories (World Scientific, Singapore), pp. 3–6. *(p. 224)*

Penrose, R. (2014a). On the gravitization of quantum mechanics 1: Quantum state reduction, *Found. Phys.* **44**, 5, pp. 557–575, doi:10.1007/s10701-013-9770-0. *(p. 217, 224)*

Penrose, R. (2014b). On the gravitization of quantum mechanics 2: Conformal cyclic cosmology, *Found. Phys.* **44**, 8, pp. 873–890, doi:10.1007/s10701-013-9763-z. *(p. 217, 224)*

Peres, A. (1962). On Cauchy's problem in general relativity. II, *Nuovo Cimento* **26**, 1, pp. 53–62, doi:10.1007/BF02754342. *(p. 218)*

Peres, A. (1968). Canonical quantization of gravitational field, *Phys. Rev.* **171**, 5, pp. 1335–1344, doi:10.1103/PhysRev.171.1335. *(p. 219)*

Perrot, F. and March, N. H. (1990). Pair potentials for liquid sodium near freezing from electron theory and from inversion of the measured structure factor, *Phys. Rev. A* **41**, 8, pp. 4521–4523, doi:10.1103/PhysRevA.41.4521, reprinted in March and Angilella (2009). *(p. 125, 126, 127)*

Pirani, F. A. E. (1965). Introduction to gravitational radiation theory, in A. Trautman, F. A. E. Pirani, and H. Bondi (eds.), *Lectures on General Relativity, Brandeis Summer Institute of Theoretical Physics*, Vol. 1 (1964) (Prentice-Hall, Englewood, NJ), pp. 249–273. *(p. 227, 228)*

Piraux, B., L'Huillier, A., and Rzążewski, K. (eds.) (1993). *Super-Intense Laser-Atom Physics, NATO Advanced Study Institute, Series B: Physics*, Vol. 316 (Plenum, New York), doi:9780306445873. *(p. 209)*

Piris, M. (2006). A new approach for the two-electron cumulant in natural orbital functional theory, *Int. J. Quantum Chem.* **106**, 5, pp. 1093–1104, doi:10.1002/qua.20858. *(p. 9)*

Piris, M., Lopez, X., Ruipérez, F., Matxain, J. M., and Ugalde, J. M. (2011). A natural orbital functional for multiconfigurational states, *J. Chem. Phys.* **134**, 16, 164102, doi:10.1063/1.3582792. *(p. 9)*

Polchinski, J. (1998a). *String Theory. Vol. 1: An Introduction to the Bosonic String* (Cambridge University Press, Cambridge), ISBN 0521633036. *(p. 219)*

Polchinski, J. (1998b). *String Theory. Vol. 2: Superstring Theory and Beyond* (Cambridge University Press, Cambridge), ISBN 0521633044. *(p. 219)*

Politzer, P., Ma, Y., Jalbout, A. F., and Murray, J. S. (2005). Atomic energies from electrostatic potentials at nuclei: direct evaluation, *Mol. Phys.* **103**, 15-16, pp. 2105–2108, doi:10.1080/00268970500084067. *(p. 26, 28)*

Politzer, P. and Parr, R. G. (1974). Some new energy formulas for atoms and molecules, *J. Chem. Phys.* **61**, 10, pp. 4258–4262, doi:10.1063/1.1681726. *(p. 25)*

Pollack, S. E., Dries, D., and Hulet, R. G. (2009). Universality in three- and four-body bound states of ultracold atoms, *Science* **326**, 5960, pp. 1683–1685, doi:10.1126/science.1182840. *(p. 44)*

Rafanelli, K. and Schiller, R. (1964). Classical motions of spin-$\frac{1}{2}$ particles, *Phys. Rev.* **135**, 1B, pp. B279–B281, doi:10.1103/PhysRev.135.B279. *(p. 211,*

212, 213, 214)

Rajagopal, A. K. (1995). von Neumann entropy associated with the Haldane exclusion statistics, *Phys. Rev. Lett.* **74**, pp. 1048–1051, doi:10.1103/PhysRevLett.74.1048, [Erratum: *ibid.*, **75**, 2452 (1995)]. *(p. 55)*

Randeria, M. and Campuzano, J. (1999). High-T_c superconductors: New insights from angle resolved photoemission, in G. Iadonisi, J. R. Schrieffer, and M. L. Chiofalo (eds.), *Models and phenomenology for conventional and high-temperature superconductivity*, Proceedings of the CXXXVI International School of Physics "E. Fermi", Varenna (Italy), 1997 (IOS, Amsterdam), ISBN 9781614992219, pp. 115–139. *(p. 175)*

Rapisarda, F. and Senatore, G. (1996). Diffusion Monte Carlo study of electrons in two-dimensional layers, *Austral. J. Phys.* **49**, pp. 161–182, doi:10.1071/PH960161. *(p. 236, 237)*

Reatto, L., Levesque, D., and Weis, J. J. (1986). Iterative predictor-corrector method for extraction of the pair interaction from structural data for dense classical liquids, *Phys. Rev. A* **33**, 5, pp. 3451–3465, doi:10.1103/PhysRevA.33.3451. *(p. 125, 126)*

Rebbi, C. (ed.) (1983). *Lattice gauge theories and Monte Carlo simulations* (World Scientific, Singapore), ISBN 9789971950712. *(p. 271)*

Reichl, L. E. (2009). *A modern course in statistical physics* (Wiley, Weinheim), ISBN 9783527407828. *(p. 45, 101, 147, 149, 151, 155, 156, 160, 270)*

Reisenberger, M. P. and Rovelli, C. (1997). "Sum over surfaces" form of loop quantum gravity, *Phys. Rev. D* **56**, 6, pp. 3490–3508, doi:10.1103/PhysRevD.56.3490. *(p. 219)*

Renard, J. P., Verdaguer, M., Regnault, L. P., Erkelens, W. A. C., Rossat-Mignod, J., and Stirling, W. G. (1987). Presumption for a quantum energy gap in the quasi-one-dimensional $S = 1$ Heisenberg antiferromagnet $Ni(C_2H_8N_2)_2NO_2ClO_4$, *Europhys. Lett.* **3**, 8, p. 945, doi:10.1209/0295-5075/3/8/013. *(p. 270)*

Rice, M. J. (1979). Charged π-phase kinks in lightly doped polyacetylene, *Phys. Lett. A* **71**, 1, pp. 152–154, doi:10.1016/0375-9601(79)90905-8. *(p. 257)*

Rice, T. M., Ueda, K., Ott, H. R., and Rudigier, H. (1985). Normal-state properties of heavy-electron systems, *Phys. Rev. B* **31**, 1, pp. 594–596, doi:10.1103/PhysRevB.31.594. *(p. 133)*

Richardson, R. W. (1963). A restricted class of exact eigenstates of the pairing-force Hamiltonian, *Phys. Lett.* **3**, 6, pp. 277–279, doi:10.1016/0031-9163(63)90259-2. *(p. 82)*

Rickayzen, G. (1965). *Theory of Superconductivity* (Interscience Publ./J. Wiley & Sons, New York), ISBN 9780470720509. *(p. 80, 91, 93, 94)*

Ritus, V. I. (1970). The mass operator and exact electron Green's function in intense field, *Pis'ma Zh. Eksp. Teor. Fiz.* **12**, pp. 416–1418, (in Russian). *(p. 210)*

Ritus, V. I. (1972). Radiative corrections in quantum electrodynamics with intense field and their analytical properties, *Ann. Phys.* **69**, 2, pp. 555–582, doi:10.1016/0003-4916(72)90191-1. *(p. 210)*

Ritus, V. I. (1974). On diagonality of the electron mass operator in the constant field, *Pis'ma Zh. Eksp. Teor. Fiz.* **20**, pp. 135–138, (in Russian). *(p. 210)*

Ritus, V. I. (1978). The eigenfunction method and the mass operator in quantum electrodynamics of the constant field, *Pis'ma Zh. Eksp. Teor. Fiz.* **75**, pp. 1560–1583, (in Russian). *(p. 210)*

Ritus, V. I. (1979). Quantum effects of the interaction of elementary particles with an intense electromagnetic field, in *Quantum electrodynamics of phenomena in an intense field, Proceedings of the P. N. Lebedev Institute*, Vol. 111 (Nauka, Moscow), pp. 5–151, (in Russian). *(p. 210)*

Robertshaw, O. and Tod, P. (2006). Lie point symmetries and an approximate solution for the Schrödinger-Newton equations, *Nonlinearity* **19**, 7, pp. 1507–1514, doi:10.1088/0951-7715/19/7/002. *(p. 225)*

Rohr, D. R., Pernal, K., Gritsenko, O. V., and Baerends, E. J. (2008). A density matrix functional with occupation number driven treatment of dynamical and nondynamical correlation, *J. Chem. Phys.* **129**, 16, 164105, doi:10.1063/1.2998201. *(p. 9)*

Rohr, D. R., Toulouse, J., and Pernal, K. (2010). Combining density-functional theory and density-matrix-functional theory, *Phys. Rev. A* **82**, p. 052502, doi:10.1103/PhysRevA.82.052502. *(p. 9)*

Rosenfeld, L. (1930a). Zur Quantelung der Wellenfelder, *Ann. Phys.* **397**, 1, pp. 113–152, doi:10.1002/andp.19303970107. *(p. 217)*

Rosenfeld, L. (1930b). Über die Gravitationswirkungen des Lichtes, *Z. Phys.* **65**, 9-10, pp. 589–599, doi:10.1007/BF01391161. *(p. 217)*

Rosenfeld, L. (1963). On quantization of fields, *Nucl. Phys.* **40**, pp. 353–356, doi:10.1016/0029-5582(63)90279-7. *(p. 223)*

Rousseau, J. S., Stoddart, J. C., and March, N. H. (1972). The electrical resistivity of liquid metals, *J. Phys. C: Solid State Phys.* **5**, 14, pp. L175–L177, doi:10.1088/0022-3719/5/14/001. *(p. 121)*

Rovelli, C. (2000). Notes for a brief history of quantum gravity, in V. G. Gurzadyan, R. T. Jantzen, and R. Ruffini (eds.), *Proceedings of the 9th Marcel Grossmann Meeting (MGIX MM): on recent developments in theoretical and experimental general relativity, gravitation, and relativistic field theories* (World Scientific, Singapore), pp. 742–768, preprint `arXiv:gr-qc/0006061`. *(p. 217, 218)*

Rovelli, C. (2004). *Quantum Gravity* (Cambridge University Press, Cambridge), ISBN 9780521837330. *(p. 219, 226)*

Rowlinson, J. S. (1986). Physics of liquids: Are diameters rectilinear? *Nature* **319**, 6052, pp. 362–362, doi:10.1038/319362a0. *(p. 154)*

Ruffini, R. and Bonazzola, S. (1969). Systems of self-gravitating particles in general relativity and the concept of an equation of state, *Phys. Rev.* **187**, 5, pp. 1767–1783, doi:10.1103/PhysRev.187.1767. *(p. 222)*

Rushbrooke, G. S. (1960). On the hyper-chain approximation in the theory of classical fluids, *Physica* **26**, 4, pp. 259–265, doi:0.1016/0031-8914(60)90020-3. *(p. 125)*

Sachdev, S. (2011a). Condensed matter and AdS/CFT, in E. Papantonopoulos (ed.), *From Gravity to Thermal Gauge Theories: The AdS/CFT Correspondence, Lect. Notes in Phys.*, Vol. 828 (Springer, Berlin), ISBN 978-

3-642-04863-0, pp. 273–311, doi:10.1007/978-3-642-04864-7_9, preprint `arXiv:1002.2947`. *(p. 147, 174)*

Sachdev, S. (2011b). *Quantum Phase Transitions*, 2nd edn. (Cambridge University Press, Cambdridge), ISBN 9780521514682. *(p. 170, 245)*

Sachdev, S. (2012). What can gauge-gravity duality teach us about condensed matter physics? *Ann. Rev. Cond. Matter Phys.* **3**, 1, pp. 9–33, doi:10.1146/annurev-conmatphys-020911-125141, preprint `arXiv:1108.1197`. *(p. 147, 174)*

Salpeter, E. E. (1952). Mass corrections to the fine structure of hydrogen-like atoms, *Phys. Rev.* **87**, 2, pp. 328–343, doi:10.1103/PhysRev.87.328. *(p. 215)*

Salpeter, E. E. and Bethe, H. A. (1951). A relativistic equation for bound-state problems, *Phys. Rev.* **84**, 6, pp. 1232–1242, doi:10.1103/PhysRev.84.1232. *(p. 215)*

Sawada, K. (1957). Correlation energy of an electron gas at high density, *Phys. Rev.* **106**, 2, pp. 372–383, doi:10.1103/PhysRev.106.372. *(p. 235)*

Schiller, R. (1962). Quasi-classical theory of the nonspinning electron, *Phys. Rev.* **125**, 3, pp. 1100–1108, doi:10.1103/PhysRev.125.1100. *(p. 213)*

Schonfeld, J. F. (1981). A mass term for three-dimensional gauge fields, *Nucl. Phys. B* **185**, 1, pp. 157–171, doi:10.1016/0550-3213(81)90369-2. *(p. 50)*

Schrieffer, J. R. and Wolff, P. A. (1966). Relation between the Anderson and Kondo hamiltonians, *Phys. Rev.* **149**, 2, pp. 491–492, doi:10.1103/PhysRev.149.491. *(p. 249)*

Schultz, T., Mattis, D., and Lieb, E. (1964). Two-dimensional Ising model as a soluble problem of many fermions, *Rev. Mod. Phys.* **36**, 3, pp. 856–871, doi:10.1103/RevModPhys.36.856. *(p. 182, 269)*

Schumayer, D. and Hutchinson, D. A. W. (2011). *Colloquium*: Physics of the Riemann hypothesis, *Rev. Mod. Phys.* **83**, pp. 307–330, doi:10.1103/RevModPhys.83.307. *(p. 273)*

Schwinger, J. (1947). A variational principle for scattering problems, *Phys. Rev.* **72**, p. 742. *(p. 32)*

Schwinger, J. (1951). On gauge invariance and vacuum polarization, *Phys. Rev.* **82**, 5, pp. 664–679, doi:10.1103/PhysRev.82.664. *(p. 204)*

Schwinger, J. (1954a). The theory of quantized fields. V, *Phys. Rev.* **93**, 3, pp. 615–628, doi:10.1103/PhysRev.93.615. *(p. 204)*

Schwinger, J. (1954b). The theory of quantized fields. VI, *Phys. Rev.* **94**, 5, pp. 1362–1384, doi:10.1103/PhysRev.94.1362. *(p. 204)*

SCM-LIQ (2015). Database of the Research Group for Structural Characterization of Materials at the Institute for Advanced Materials Processing, Tohoku, Japan. URL: `http://res.tagen.tohoku.ac.jp/~waseda/scm/`. *(p. 129, 130)*

Sen, D., Stone, M., and Vishveshwara, S. (2008). Quasiparticle propagation in quantum Hall systems, *Phys. Rev. B* **77**, p. 115442, doi:10.1103/PhysRevB.77.115442. *(p. 65)*

Sharma, S., Dewhurst, J. K., Lathiotakis, N. N., and Gross, E. K. U. (2008). Reduced density matrix functional for many-electron systems, *Phys. Rev.*

B **78**, p. 201103, doi:10.1103/PhysRevB.78.201103. *(p. 9)*

Shechtman, D., Blech, I., Gratias, D., and Cahn, J. W. (1984). Metallic phase with long-range orientational order and no translational symmetry, *Phys. Rev. Lett.* **53**, pp. 1951–1953, doi:10.1103/PhysRevLett.53.1951. *(p. 274)*

Si, T. and Yu, Y. (2008). Anyonic loops in three-dimensional spin liquid and chiral spin liquid, *Nucl. Phys. B* **803**, 3, pp. 428–449, doi:10.1016/j.nuclphysb.2008.06.009. *(p. 46)*

Siegel, W. (1979). Unextended superfields in extended supersymmetry, *Nucl. Phys. B* **156**, 1, pp. 135–143, doi:10.1016/0550-3213(79)90498-X. *(p. 50)*

Sierra, G. (2008). A quantum mechanical model of the Riemann zeros, *New J. Phys.* **10**, 3, p. 033016, doi:10.1088/1367-2630/10/3/033016. *(p. 273)*

Sierra, G. and Townsend, P. K. (2008). Landau levels and Riemann zeros, *Phys. Rev. Lett.* **101**, p. 110201, doi:10.1103/PhysRevLett.101.110201. *(p. 273)*

Simon, H. and Baake, M. (1997). Lee-Yang zeros in the scaling region of a two-dimensional quasiperiodic Ising model, *J. Phys. A: Math. Gen.* **30**, 15, pp. 5319–5328, doi:10.1088/0305-4470/30/15/018. *(p. 171)*

Simon, H., Baake, M., and Grimm, U. (1995). Lee-Yang zeros for substitutional systems, in C. Janot and R. Mosseri (eds.), *Proceedings of the 5th International Conference on Quasicrystals* (World Scientific, Singapore), pp. 100–103, preprint `arXiv:cond-mat/9902052`. *(p. 171)*

Slater, J. C. (1934a). Electronic energy bands in metals, *Phys. Rev.* **45**, pp. 794–801, doi:10.1103/PhysRev.45.794. *(p. 64, 65)*

Slater, J. C. (1934b). The electronic structure of metals, *Rev. Mod. Phys.* **6**, pp. 209–280, doi:10.1103/RevModPhys.6.209. *(p. 64)*

Slater, J. C. (1972). Hellmann-Feynman and virial theorems in the $X\ \alpha$ method, *J. Chem. Phys.* **57**, pp. 2389–2396, doi:10.1063/1.1678599. *(p. 19)*

Smith, D. W. (1966). N-representability problem for fermion density matrices. II. The first-order density matrix with N even, *Phys. Rev.* **147**, 4, pp. 896–898, doi:10.1103/PhysRev.147.896. *(p. 88)*

Speliotopoulos, A. D. (1997). Turning bosons into fermions: exclusion statistics, fractional statistics and the simple harmonic oscillator, *J. Phys. A: Math. Gen.* **30**, 17, pp. 6177–6184, doi:10.1088/0305-4470/30/17/024. *(p. 47, 55)*

Squire, R. H. and March, N. H. (2014). Are there Efimov trimers in hexafluorobenzene rather than in benzene vapor itself? *Phys. Lett. A* **378**, 41, pp. 3018–3020, doi:10.1016/j.physleta.2014.08.027. *(p. 42)*

Steglich, F., Aarts, J., Bredl, C. D., Lieke, W., Meschede, D., Franz, W., and Schäfer, H. (1979). Superconductivity in the presence of strong Pauli paramagnetism: $CeCu_2Si_2$, *Phys. Rev. Lett.* **43**, 25, pp. 1892–1896, doi:10.1103/PhysRevLett.43.1892. *(p. 248)*

Stewart, G. R. (1984). Heavy-fermion systems, *Rev. Mod. Phys.* **56**, 4, pp. 755–787, doi:10.1103/RevModPhys.56.755. *(p. 248)*

Stöckmann, H. (1999). *Quantum Chaos. An Introduction* (Cambridge University Press, Cambridge), ISBN 9780521592840. *(p. 273)*

Stormer, H. L., Tsui, D. C., and Gossard, A. C. (1999). The fractional quantum Hall effect, *Rev. Mod. Phys.* **71**, pp. S298–S305, doi:10.1103/RevModPhys.71.S298. *(p. 46, 48)*

Stueckelberg, E. C. G. (1941). Un nouveau modèle de l'électron ponctuel en théorie classique, *Helv. Phys. Acta* **14**, doi:10.5169/seals-111170. *(p. 202)*

Su, G. and Suzuki, M. (1998). Quantum statistical mechanics of ideal gas obeying fractional exclusion statistics: a systematic study, *Eur. Phys. J. B* **5**, pp. 577–582, doi:10.1007/s100510050481. *(p. 59, 64)*

Su, W., Schrieffer, J., and Heeger, A. (1979). Solitons in polyacetylene, *Phys. Rev. Lett.* **42**, 25, pp. 1698–1701, doi:10.1103/PhysRevLett.42.1698. *(p. 257, 261, 262, 263)*

Su, W., Schrieffer, J., and Heeger, A. (1980). Soliton excitations in polyacetylene, *Phys. Rev. B* **22**, 4, pp. 2099–2111, doi:10.1103/PhysRevB.22.2099. *(p. 257, 261, 262, 263)*

Su, W., Schrieffer, J., and Heeger, A. (1983). Erratum: Soliton excitations in polyacetylene, *Phys. Rev. B* **28**, 2, pp. 1138–1138, doi:10.1103/PhysRevB. 28.1138. *(p. 257, 261, 262, 263)*

Sun, K., Liu, W. V., Hemmerich, A., and Das Sarma, S. (2011). Topological semimetal in a fermionic optical lattice, *Nature Phys.* **8**, 1, pp. 67–70, doi: 10.1038/nphys2134. *(p. x)*

Sutherland, B. (1970). Two-dimensional hydrogen bonded crystals without the ice rule, *J. Math. Phys.* **11**, 11, pp. 3183–3186, doi:10.1063/1.1665111. *(p. 182, 184)*

Sutherland, B. (1971a). Exact results for a quantum many-body problem in one dimension, *Phys. Rev. A* **4**, pp. 2019–2021, doi:10.1103/PhysRevA.4.2019. *(p. 54)*

Sutherland, B. (1971b). Quantum many-body problem in one dimension: Ground state, *J. Math. Phys.* **12**, pp. 246–250, doi:10.1063/1.1665584. *(p. 54, 64)*

Sutherland, B. (1971c). Quantum many-body problem in one dimension: Thermodynamics, *J. Math. Phys.* **12**, pp. 251–256, doi:10.1063/1.1665585. *(p. 54)*

Sutherland, B. (1972). Exact results for a quantum many-body problem in one dimension. II, *Phys. Rev. A* **5**, pp. 1372–1376, doi:10.1103/PhysRevA.5. 1372. *(p. 54)*

Sutherland, B. (1997). Microscopic theory of exclusion statistics, *Phys. Rev. B* **56**, pp. 4422–4431, doi:10.1103/PhysRevB.56.4422. *(p. 54)*

Suzuki, M. (1967a). A theory on the critical behaviour of ferromagnets, *Prog. Theor. Phys.* **38**, 1, pp. 289–290, doi:10.1143/PTP.38.289. *(p. 162)*

Suzuki, M. (1967b). A theory of the second order phase transitions in spin systems. II: complex magnetic field, *Prog. Theor. Phys.* **38**, 6, pp. 1225–1242, doi:10.1143/PTP.38.1225. *(p. 162, 164)*

't Hooft, G. (1973). An algorithm for the poles at dimension four in the dimensional regularization procedure, *Nucl. Phys. B* **62**, pp. 444–460, doi: 10.1016/0550-3213(73)90263-0. *(p. 219)*

Talbot, E. and Griffin, A. (1983). High- and low-frequency behaviour of response functions in a Bose liquid: One-loop approximation, *Ann. Phys. (N.Y.)* **151**, 1, pp. 71–98, doi:10.1016/0003-4916(83)90315-9. *(p. 95)*

Tang, G., Eggert, S., and Pelster, A. (2015). Ground-state properties of anyons in a one-dimensional lattice, *New J. Phys.* **17**, 12, p. 123016, doi:10.1088/ 1367-2630/17/12/123016. *(p. 72)*

Taylor, E. (2008). Josephson relation for the superfluid density in the BCS-BEC crossover, *Phys. Rev. B* **77**, 14, p. 144521, doi:10.1103/PhysRevB.77.144521. *(p. 95)*

ter Haar, D. (1961). Theory and applications of the density matrix, *Rep. Prog. Phys.* **24**, 1, pp. 304–362, doi:10.1088/0034-4885/24/1/307. *(p. 2)*

Thirring, W. E. (1958). A soluble relativistic field theory, *Ann. Phys.* **3**, 1, pp. 91–112, doi:10.1016/0003-4916(58)90015-0. *(p. 182)*

Thomas, L. H. (1926). The calculation of atomic fields, *Math. Proc. Cambridge Phil. Soc.* **23**, pp. 542–548, doi:10.1017/S0305004100011683. *(p. 13)*

Thomas, L. H. (1935). The interaction between a neutron and a proton and the structure of H^3, *Phys. Rev.* **47**, 12, pp. 903–909, doi:10.1103/PhysRev.47.903. *(p. 33)*

Thouless, D. J. (1974). Electrons in disordered systems and the theory of localization, *Phys. Rep.* **13**, 3, pp. 93–142, doi:10.1016/0370-1573(74)90029-5. *(p. 135)*

Thouless, D. J. (1998). *Topological Quantum Numbers in Nonrelativistic Physics* (World Scientific, Singapore), ISBN 9789810229009. *(p. 76)*

Tisza, L. (1938). Transport phenomena in Helium II, *Nature* **141**, p. 913, doi:10.1038/141913a0. *(p. 94)*

Titchmarsh, E. C. and Heath-Brown, D. R. (eds.) (1986). *The theory of the Riemann zeta function* (Oxford University Press, Oxford), ISBN 9780198533696. *(p. 273, 274)*

Tod, K. P. (2001). The ground state energy of the Schrödinger-Newton equation, *Phys. Lett. A* **280**, 4, pp. 173–176, doi:10.1016/S0375-9601(01)00059-7. *(p. 225)*

Tonks, L. (1936). The complete equation of state of one, two and three-dimensional gases of hard elastic spheres, *Phys. Rev.* **50**, 10, pp. 955–963, doi:10.1103/PhysRev.50.955. *(p. 172)*

Tosi, M. P. and March, N. H. (1973). Small-angle scattering from liquid metals and alloys and electronic correlation functions, *Nuovo Cimento B* **15**, 2, pp. 308–319, doi:10.1007/BF02894788. *(p. 128)*

Tozzini, V. and Tosi, M. P. (1993). Vibrational and elastic properties of the Wigner electron lattice near melting, *Europhys. Lett.* **23**, 6, pp. 433–438, doi:10.1209/0295-5075/23/6/009. *(p. 240)*

Trotzky, S., Chen, Y., Flesch, A., McCulloch, I. P., Schöllwock, U., Eisert, J., and Bloch, I. (2012). Probing the relaxation towards equilibrium in an isolated strongly correlated one-dimensional Bose gas, *Nat. Phys.* **8**, 4, pp. 325–330, doi:10.1038/nphys2232. *(p. 170)*

Trovato, M. (2014). Quantum maximum entropy principle and quantum statistics in extended thermodynamics, *Acta Applicandae Mathematicae* **132**, 1, pp. 605–619, doi:10.1007/s10440-014-9934-8. *(p. 54)*

Trovato, M. and Reggiani, L. (2013). Quantum maximum entropy principle for fractional exclusion statistics, *Phys. Rev. Lett.* **110**, p. 020404, doi:10.1103/PhysRevLett.110.020404. *(p. 54)*

Tsui, D. C., Stormer, H. L., and Gossard, A. C. (1982). Two-dimensional magneto-transport in the extreme quantum limit, *Phys. Rev. Lett.* **48**, pp. 1559–1562,

doi:10.1103/PhysRevLett.48.1559. *(p. 239)*
Tsunetsugu, H., Sigrist, M., and Ueda, K. (1997). The ground-state phase diagram of the one-dimensional Kondo lattice model, *Rev. Mod. Phys.* **69**, 3, pp. 809–864, doi:10.1103/RevModPhys.69.809. *(p. 250, 251)*
Tsvelik, A. M. and Yevtushenko, O. M. (2015). Quantum phase transition and protected ideal transport in a Kondo chain, *Phys. Rev. Lett.* **115**, 21, p. 216402, doi:10.1103/PhysRevLett.115.216402. *(p. 251)*
Tung, S.-K., Jiménez-García, K., Johansen, J., Parker, C. V., and Chin, C. (2014). Geometric scaling of efimov states in a ^{6}Li-^{133}Cs mixture, *Phys. Rev. Lett.* **113**, 24, p. 240402, doi:10.1103/PhysRevLett.113.240402. *(p. 44)*
Turing, A. M. (1937). On computable numbers, with an application to the Entscheidungsproblem, *Proc. Lond. Math. Soc.* **s2-42**, 1, pp. 230–265, doi: 10.1112/plms/s2-42.1.230. *(p. 270)*
Uhlenbeck, G. E. and Goudsmit, S. (1925). Ersetzung der Hypothese vom unmechanischen Zwang durch eine Forderung bezüglich des inneren Verhaltens jedes einzelnen Elektrons, *Naturwiss.* **13**, 47, pp. 953–954, doi: 10.1007/BF01558878. *(p. 200)*
Uhlenbeck, G. E. and Goudsmit, S. (1926). Spinning electrons and the structure of spectra, *Nature* **117**, pp. 264–265, doi:10.1038/117264a0. *(p. 201)*
Uhlenbeck, G. E. and Gropper, L. (1932). The equation of state of a non-ideal Einstein-Bose or Fermi-Dirac gas, *Phys. Rev.* **41**, pp. 79–90, doi:10.1103/PhysRev.41.79. *(p. 65)*
Valatin, J. G. (1958). Comments on the theory of superconductivity, *Nuovo Cimento* **7**, pp. 843–857, doi:10.1007/BF02745589. *(p. 80, 81)*
Valatin, J. G. (1961). Generalized Hartree-Fock method, *Phys. Rev.* **122**, 4, pp. 1012–1020, doi:10.1103/PhysRev.122.1012. *(p. 90)*
Van Tiggelen, B. (1999). Localization of waves, in J. Fouque (ed.), *Diffuse Waves in Complex Media (course at Les Houches, 1998)*, *NATO Science Series*, Vol. 531 (Springer Netherlands), ISBN 978-0-7923-5680-6, pp. 1–60, doi: 10.1007/978-94-011-4572-5_1. *(p. 135)*
Varma, C. M., Littlewood, P. B., Schmitt-Rink, S., Abrahams, E., and Ruckenstein, A. E. (1989). Phenomenology of the normal state of Cu-O high-temperature superconductors, *Phys. Rev. Lett.* **63**, 18, pp. 1996–1999, doi: 10.1103/PhysRevLett.63.1996. *(p. 175)*
Varma, C. M. and Yafet, Y. (1976). Magnetic susceptibility of mixed-valence rare-earth compounds, *Phys. Rev. B* **13**, 7, pp. 2950–2954, doi:10.1103/PhysRevB.13.2950. *(p. 249)*
Varró, S. (2013). New exact solutions of the Dirac equation of a charged particle interacting with an electromagnetic plane wave in a medium, *Laser Phys. Lett.* **10**, 9, p. 095301, doi:10.1088/1612-2011/10/9/095301. *(p. 209)*
Volkov, D. M. (1935). Über eine Klasse von Lösungen der Diracschen Gleichung, *Z. Phys.* **94**, pp. 250–260, doi:10.1007/BF01331022. *(p. 211)*
Volkov, D. M. (1937). Electron in the field of plane nonpolarized electromagnetic waves as viewed upon in light of the Dirac equation, *Zh. Eksp. Teor. Fiz.* **7**, pp. 1286–1289. *(p. 211)*
von Weizsäcker, C. F. (1935). Zur Theorie der Kernmassen, *Z. Phys.* **96**, pp.

431–458, doi:10.1007/BF01337700. *(p. 21)*

Voros, A. (1980). The zeta function of the quartic oscillator, *Nucl. Phys. B* **165**, 2, pp. 209–236, doi:http://dx.doi.org/10.1016/0550-3213(80)90085-1. *(p. 276)*

Wang, L., Bian, X., and Yang, H. (2002). Structural simulation of clusters in liquid Ni-Al alloy, *Phys. Lett. A* **302**, 5-6, pp. 318–324, doi:10.1016/S0375-9601(02)01074-5. *(p. 145)*

Washburn, S. and Webb, R. A. (1986). Aharonov-Bohm effect in normal metal quantum coherence and transport, *Adv. Phys.* **35**, 4, pp. 375–422, doi:10.1080/00018738600101921. *(p. 139)*

Watabe, M. and Hasegawa, M. (1973). in S. Takeuchi (ed.), *The properties of liquid metals. Proceedings of the Second International Conference held at Tokyo, Japan, September 1972.* (Taylor and Francis, London), p. 133. *(p. 131)*

Wehlitz, R., Juranić, P. N., Collins, K., Reilly, B., Makoutz, E., Hartman, T., Appathurai, N., and Whitfield, S. B. (2012). Photoemission of Cooper pairs from aromatic hydrocarbons, *Phys. Rev. Lett.* **109**, 19, p. 193001, doi:10.1103/PhysRevLett.109.193001. *(p. 42)*

Wiegmann, P. B., Ogievetskii, E. I., and Tsvelik, A. M. (1980). Exact solution of the degenerate Anderson model, *Sov. Phys. JETP Lett.* **37**, pp. 692–696. *(p. 248)*

Wiegmann, P. B. and Tsvelick, A. M. (1981). Exact solution of Kondo problem for alloys with rare-earth impurities, *Physica B+C* **107**, 1, pp. 379–380, doi:10.1016/0378-4363(81)90495-2. *(p. 248)*

Wierschem, K. and Sengupta, P. (2014). Quenching the Haldane gap in spin-1 Heisenberg antiferromagnets, *Phys. Rev. Lett.* **112**, 24, p. 247203, doi:10.1103/PhysRevLett.112.247203. *(p. 270)*

Wigner, E. (1934). On the interaction of electrons in metals, *Phys. Rev.* **46**, pp. 1002–1011, doi:10.1103/PhysRev.46.1002. *(p. 235)*

Wigner, E. and Seitz, F. (1933). On the constitution of metallic sodium, *Phys. Rev.* **43**, pp. 804–810, doi:10.1103/PhysRev.43.804. *(p. 64)*

Wilczek, F. (1982a). Magnetic flux, angular momentum, and statistics, *Phys. Rev. Lett.* **48**, pp. 1144–1146, doi:10.1103/PhysRevLett.48.1144. *(p. 46, 47, 49)*

Wilczek, F. (1982b). Quantum mechanics of fractional-spin particles, *Phys. Rev. Lett.* **49**, pp. 957–959, doi:10.1103/PhysRevLett.49.957. *(p. 46, 47, 49)*

Wilczek, F. (ed.) (1990). *Fractional Statistics and Anyon Superconductivity* (World Scientific, Singapore), ISBN 9789810200480. *(p. 47)*

Wilson, K. G. (1972). Feynman-graph expansion for critical exponents, *Phys. Rev. Lett.* **28**, 9, pp. 548–551, doi:10.1103/PhysRevLett.28.548. *(p. 178)*

Witten, E. (1998). Anti de Sitter space and holography, *Adv. Theor. Math. Phys.* **2**, pp. 253–291. *(p. 174, 219)*

Wu, F. (1971). Ising model with four-spin interactions, *Phys. Rev. B* **4**, 7, pp. 2312–2314, doi:10.1103/PhysRevB.4.2312. *(p. 184)*

Wu, T. T., McCoy, B. M., Tracy, C. A., and Barouch, E. (1976). Spin-spin correlation functions for the two-dimensional Ising model: Exact theory in the scaling region, *Phys. Rev. B* **13**, 1, pp. 316–374, doi:10.1103/PhysRevB.

13.316. *(p. 152)*

Wu, Y.-S. (1994). Statistical distribution for generalized ideal gas of fractional-statistics particles, *Phys. Rev. Lett.* **73**, pp. 922–925, doi:10.1103/PhysRevLett.73.922, [Erratum: *ibid.* **74**, 3906 (1995)]. *(p. 46, 55, 56, 57, 58, 59, 60, 61, 63, 67, 69, 71)*

Yang, C. (1967). Some exact results for the many-body problem in one dimension with repulsive delta-function interaction, *Phys. Rev. Lett.* **19**, 23, pp. 1312–1315, doi:10.1103/PhysRevLett.19.1312. *(p. 243)*

Yang, C. N. (1952). The spontaneous magnetization of a two-dimensional Ising model, *Phys. Rev.* **85**, 5, pp. 808–816, doi:10.1103/PhysRev.85.808. *(p. 148, 151, 152, 154)*

Yang, C. N. (1962). Concept of off-diagonal long-range order and the quantum phases of liquid He and of superconductors, *Rev. Mod. Phys.* **34**, 4, pp. 694–704, doi:10.1103/RevModPhys.34.694. *(p. 91)*

Yang, C. N. and Lee, T. D. (1952). Statistical theory of equations of state and phase transitions. I. Theory of condensation, *Phys. Rev.* **87**, pp. 404–409, doi:10.1103/PhysRev.87.404. *(p. 156, 157, 160, 169)*

Yang, H., Miao, H., Lee, D., Helou, B., and Chen, Y. (2013). Macroscopic quantum mechanics in a classical spacetime, *Phys. Rev. Lett.* **110**, 17, p. 170401, doi:10.1103/PhysRevLett.110.170401. *(p. 224)*

Yao, Z., da Costa, C., Karine P. Kiselev, M., and Prokof'ev, N. (2014). Critical exponents of the superfluid–Bose-glass transition in three dimensions, *Phys. Rev. Lett.* **112**, 22, p. 225301, doi:10.1103/PhysRevLett.112.225301. *(p. 97)*

Yessen, W. N. (2013). Properties of 1D classical and quantum Ising models: Rigorous results, *Annal. H. Poincaré* **15**, 4, pp. 793–828, doi:10.1007/s00023-013-0252-x. *(p. 171)*

Yoon, J., Li, C. C., Shahar, D., Tsui, D. C., and Shayegan, M. (1999). Wigner crystallization and metal-insulator transition of two-dimensional holes in GaAs at $B = 0$, *Phys. Rev. Lett.* **82**, pp. 1744–1747, doi:10.1103/PhysRevLett.82.1744. *(p. 237)*

Yosida, K. (1966). Bound state due to the $s - d$ exchange interaction, *Phys. Rev.* **147**, 1, pp. 223–227, doi:10.1103/PhysRev.147.223. *(p. 249)*

Yudson, V. I. (1985). Dynamics of integrable quantum systems, *Sov. Phys. JETP* **61**, pp. 1043–1050, [Zh. Eksp. Theor. Fyz. **88**, 1757 (1985)]. *(p. 111)*

Yudson, V. I. (1988). Dynamics of the integrable one-dimensional system "photons + two-level atoms", *Phys. Lett. A* **129**, 1, pp. 17–20, doi:10.1016/0375-9601(88)90465-3. *(p. 111)*

Yukalov, V. I. (2013). Self-consistent approach for Bose-condensed atoms in optical lattices, *Condens. Matter Phys.* **16**, p. 23002, doi:10.5488/CMP.16.23002. *(p. 103)*

Yukalov, V. I. and Yukalova, E. P. (2013). Order indices of density matrices for finite systems, *Comp. Theor. Chem.* **1003**, pp. 37–43, doi:10.1016/j.comptc.2012.08.002. *(p. 89, 100)*

Zaanen, J., Liu, Y., Sun, Y., and Schalm, K. (2015). *Holographic duality in condensed matter physics* (Cambridge University Press, Cambridge), ISBN 9781107080089. *(p. 147, 174)*

Zaccanti, M., Deissler, B., D'Errico, C., Fattori, M., Jona-Lasinio, M., Müller, S., Roati, G., Inguscio, M., and Modugno, G. (2009). Observation of an Efimov spectrum in an atomic system, *Nat. Phys.* **5**, 8, pp. 586–591, doi: 10.1038/nphys1334. *(p. 44)*

Zachar, O., Kivelson, S. A., and Emery, V. J. (1996). Exact results for a 1D Kondo lattice from bosonization, *Phys. Rev. Lett.* **77**, 7, pp. 1342–1345, doi:10.1103/PhysRevLett.77.1342. *(p. 251)*

Zener, C. (1951). Interaction between the *d* shells in the transition metals, *Phys. Rev.* **81**, 3, pp. 440–444, doi:10.1103/PhysRev.81.440. *(p. 247)*

Zhang, C.-H. and Joglekar, Y. N. (2007). Wigner crystal and bubble phases in graphene in the quantum Hall regime, *Phys. Rev. B* **75**, p. 245414, doi: 10.1103/PhysRevB.75.245414. *(p. 238)*

Zhang, Z.-D. (2007). Conjectures on the exact solution of three-dimensional (3D) simple orthorhombic Ising lattices, *Phil. Mag.* **87**, 34, pp. 5309–5419, doi: 10.1080/14786430701646325. *(p. 147, 178)*

Zhang, Z.-D. (2013). Mathematical structure of the three-dimensional (3D) Ising model, *Chinese Phys. B* **22**, 3, p. 030513, doi:10.1088/1674-1056/22/3/030513. *(p. 147)*

Zhang, Z. D. and March, N. H. (2013). Critical exponents proposed for marginal Fermi liquid, *Phys. Chem. Liq.* **51**, 2, pp. 261–264, doi:10.1080/00319104.2012.753545. *(p. 177, 179, 180, 191)*

Index

Printed in the United States
By Bookmasters